T0328422

CAMBRIDGE MONOGRAPHS
ON MECHANICS AND APPLIED MATHEMATICS

GENERAL EDITORS

G. K. BATCHELOR, PH.D., F.R.S.
Professor of Applied Mathematics at the University of Cambridge

J. W. MILES, PH.D.
Professor of Applied Mathematics, University of California, La Jolla

THE STRUCTURE OF
TURBULENT SHEAR FLOW

THE STRUCTURE OF
TURBULENT SHEAR FLOW

BY

A. A. TOWNSEND, F.R.S.

Reader in Experimental Fluid Mechanics
University of Cambridge

SECOND EDITION

CAMBRIDGE UNIVERSITY PRESS

CAMBRIDGE

LONDON   NEW YORK   NEW ROCHELLE
MELBOURNE   SYDNEY

CAMBRIDGE UNIVERSITY PRESS
Cambridge, New York, Melbourne, Madrid, Cape Town,
Singapore, São Paulo, Delhi, Tokyo, Mexico City

Cambridge University Press
The Edinburgh Building, Cambridge CB2 8RU, UK

Published in the United States of America by Cambridge University Press, New York

www.cambridge.org
Information on this title: www.cambridge.org/9780521298193

First published 1956
Second edition 1976
First paperback edition 1980

A catalogue record for this publication is available from the British Library

Library of Congress Catalogue Card Number: 74-14441

ISBN 978-0-521-20710-2 Hardback
ISBN 978-0-521-29819-3 Paperback

CONTENTS

Preface   *page* xi

I THE STATISTICAL DESCRIPTION OF TURBULENT FLOW

1.1	Introduction	1
1.2	The development of a theory for turbulent flow	3
1.3	The statistical description of turbulent flow	3
1.4	Notation for turbulent flows	5
1.5	Three-dimensional correlation and spectrum functions	6
1.6	One-dimensional correlation and spectrum functions	16
1.7	Correlations and spectra with time delay	23
1.8	Homogeneity and symmetry of turbulent flows	28

2 THE EQUATIONS OF MOTION FOR TURBULENT FLOW

2.1	Assumption of a continuous fluid	32
2.2	The equations of fluid motion	33
2.3	Approximate forms of the equations of motion	36
2.4	Mean value equations for momentum, energy and heat	38
2.5	Energy dissipation by viscosity	41
2.6	Conductive dissipation of temperature fluctuations	43
2.7	The relation between the pressure and velocity fields	43

3 HOMOGENEOUS TURBULENT FLOWS

3.1	Introduction	45
3.2	Eddy interactions in homogeneous turbulence	46
3.3	Experimental approximations to homogeneous turbulence	49
3.4	Isotropic turbulence: general	51
3.5	Reynolds number similarity in isotropic turbulence	53
3.6	Self-preserving development in isotropic turbulence	59
3.7	Space–time correlations in isotropic turbulence	62

3.8 The Taylor approximation of frozen flow 64
3.9 The tendency to isotropy of homogeneous turbulence 66
3.10 Uniform distortion of homogeneous turbulence 71
3.11 Irrotational distortion of grid turbulence 74
3.12 Unidirectional, plane shearing of homogeneous turbulence 80
3.13 Local isotropy and equilibrium of small eddies 88
3.14 Measurement of spectrum and structure functions 93
3.15 Energy transfer in the inertial subrange 99
3.16 The equilibrium spectrum in the viscous subrange 100
3.17 Local isotropy in non-Newtonian fluids 103

4 INHOMOGENEOUS SHEAR FLOW

4.1 Large eddies and the main turbulent motion 105
4.2 Structural similarity of the main turbulent motion 106
4.3 Nature of the main turbulent motion 118
4.4 Generation and maintenance of the main motion 120
4.5 Flow inhomogeneity and the large eddies 122
4.6 The dependence of Reynolds stress on mean velocity 124
4.7 Statistical distributions of velocity fluctuations 126

5 TURBULENT FLOW IN PIPES AND CHANNELS

5.1 Introduction 130
5.2 Equations of motion for unidirectional mean flow 131
5.3 Reynolds number similarity in pipe and channel flow 133
5.4 Wall similarity in the region of constant stress 135
5.5 Flow over rough walls 140
5.6 Mean flow in the central region 145
5.7 The turbulent motion in constant-stress equilibrium layers 150
5.8 Eddy structure in equilibrium layers 156
5.9 Motion in the viscous layer next the wall 163
5.10 Fluctuations of pressure and shear stress on a wall 165
5.11 The magnitude of the Kármán constant 168
5.12 Turbulent flow and flow constants 169

5.13 Similarity flows in channels and pipes of varying widths   172
5.14 Equilibrium layers with variable stress   176
5.15 Equilibrium layers with linear distributions of stress   180
5.16 Equilibrium layers with surface transpiration   184
5.17 Equilibrium layers with variable direction of flow   186

6 FREE TURBULENT SHEAR FLOWS

6.1 General properties of free turbulence   188
6.2 Equations of motion: the boundary-layer approximation   188
6.3 Integral constraints on free turbulent flows   193
6.4 Self-preserving development of free turbulent flows   195
6.5 The distributions of mean velocity and Reynolds stress   201
6.6 The balance of turbulent kinetic energy   205
6.7 The bounding surface of free turbulent flows   209
6.8 Distributions of turbulent intensity and Reynolds stress   214
6.9 Flow constants for self-preserving jets and wakes   220
6.10 The flow constants of plane mixing layers   227
6.11 The entrainment of ambient fluid   230
6.12 Basic entrainment processes   232
6.13 Entrainment eddies in plane wakes   241
6.14 Mechanism of the entrainment eddies   243
6.15 Control of the entrainment rate   247
6.16 Fluctuations outside the turbulent flow: sound radiation   248
6.17 Irrotational fluctuations in the near field   251
6.18 Development of nearly self-preserving flows   252
6.19 Development of a jet in a moving stream of constant   255
     velocity

7 BOUNDARY LAYERS AND WALL JETS

7.1 Wall layers in general   259
7.2 Self-preserving development of wall layers   262
7.3 General properties of self-preserving wall layers   263
7.4 Flow parameters of self-preserving wall layers   266
7.5 Development of self-preserving wall jets   268
7.6 Development of self-preserving boundary layers   272

7.7 Boundary-layer development with zero wall stress   276

7.8 Wall layers with convergent flow   280

7.9 Almost self-preserving development   283

7.10 Layers with nearly uniform velocity in the free stream   287

7.11 Turbulent flow in self-preserving boundary layers   289

7.12 Development of boundary layers in arbitrary external conditions   294

7.13 Boundary-layer development after a sudden change of external conditions   298

7.14 Development in a region of strong adverse pressure gradient   301

7.15 Layer development after a sudden change of roughness   307

7.16 Boundary layers with three-dimensional mean flow   312

7.17 Three-dimensional flow with negligible Reynolds stresses   316

7.18 Homogeneous three-dimensional flow – the Ekman layer   318

7.19 Secondary flow in a boundary layer with a free edge   323

7.20 Lateral variations of stress in boundary layers   328

7.21 Periodic structure of flow near the viscous layer   331

8 TURBULENT CONVECTION OF HEAT AND PASSIVE CONTAMINANTS

8.1 Governing equations and dimensional considerations   334

8.2 Diffusion by continuous movements: effect of molecular diffusive transport   336

8.3 Eulerian description of convective flows: mean value equations and correlation functions   338

8.4 Local forms of the Richardson number   341

8.5 Spectrum functions and local similarity   342

8.6 Scattering of light by density fluctuations in a turbulent flow   348

8.7 Self-preserving development of temperature fields in forced convection flows   350

8.8 Forced convection in wall flows   352

8.9 Rates of heat transfer in forced convection   356

8.10 Convection in a constant-stress layer after an abrupt change in wall flux or temperature   361

8.11 Longitudinal diffusion in pipe flow   364

8.12 Natural convection and energy transfer 366

8.13 Buoyant plumes and thermals 366

8.14 The effect of buoyancy forces on turbulent motion 372

8.15 Horizontal wall layers with heat transport 375

8.16 Nature of turbulence in strongly stable flows 378

8.17 Transient behaviour of boundary layers with heat transfer 379

8.18 Convective turbulence 380

8.19 Heat convection between horizontal, parallel planes 381

8.20 Heat transfer in Benard convection 384

8.21 Similarity and structure of Benard convection 386

8.22 Natural convection in wall layers 390

9 TURBULENT FLOW WITH CURVATURE OF THE MEAN VELOCITY STREAMLINES

9.1 Mean value equations for curved flow: the analogy between the effects of flow curvature and density stratification 393

9.2 Couette flow between rotating cylinders 398

9.3 Flow with the outer cylinder stationary 400

9.4 Turbulent motion with the outer cylinder stationary 404

9.5 Flow with the outer cylinder rotating 407

References 413

Index 425

PREFACE

Since my original monograph was published, so much new material has appeared that I am amazed to read my statement that (around 1950) 'experimental knowledge...was being accumulated very rapidly'. In the last ten years, papers have been appearing at such a rate that I must extend apologies to all those whose work I have either not read, ignored, or used without realising the origin. In spite of a considerable increase in size and the inclusion of sections on convection, the same approach is followed – to develop a consistent view of the nature of turbulence from observations of simple flows and then to use it to interpret and predict the behaviour of a variety of flows of more general interest. Although my current views have been developed from those I held in 1956, they have undergone considerable change. Perhaps I should thank especially Dr H. L. Grant who, as my research student, began the process by demolishing a complete chapter before the ink was wholly dry.

No doubt, I shall trouble some readers by my habit of omitting the density from most equations and by changing to the meteorological practice of using 'z' to denote displacement in the direction of shear. I offer them my sympathy but not my repentance.

My thanks are offered to the Cambridge University Press for their incredible patience, to many friends whose cries of 'When will it appear?' have flattered me into continued activity, and to Professor G. K. Batchelor for a line of tactful harassment. At the moment, I am more pleased than even my wife to have completed the writing.

CHAPTER I

THE STATISTICAL DESCRIPTION OF TURBULENT FLOW

1.1 Introduction

One of the few phenomena in the field of fluid motion that find their way into physics text-books is the existence of a critical velocity for the flow of a viscous fluid through a circular pipe. The critical velocity separates a regime of steady laminar flow from a regime of highly irregular turbulent flow in which the flow resistance is considerably greater than is indicated by the Poiseuille equation. The difference between the two kinds of flow can be seen if a filament of dye is injected near the centre-line of the pipe. In laminar flow, the filament remains straight and coherent but, with the onset of turbulent flow, it meanders, winds itself up into tight coils and is diffused rapidly over the whole section of the pipe. Although the transition from laminar to turbulent flow is not as simple as this and similar descriptions make it appear, the phenomenon illustrates very well the fundamental differences in character between laminar and turbulent flow, particularly the ability of a turbulent flow to transmit larger shear stresses and to diffuse heat and matter more rapidly than the corresponding laminar flow. It is well known that the differences arise from an intricate and eddying motion of the fluid which convects momentum, heat and matter from one part of the flow to another, the direction of net transport being in general down the gradient of the quantity concerned. Formally, the overall effect is equivalent to increasing greatly the effective coefficients of viscosity, heat conductivity and diffusion, and it is natural to draw an analogy between the turbulent motion and the molecular motion that is responsible for transport phenomena in gases. A similarity does exist but the analogy is imperfect in two important respects. First, at any moment the motion of a gas molecule is affecting the motion of at most one other molecule and mixing on the microscopic scale takes place freely. Turbulent diffusive movements of fluid particles are essentially part of the general motion of the fluid and the direct effect is to mingle rather than to mix parcels of fluid from different parts of the flow. Complete mixing depends on molecular diffusion which is intensified by increases of concentration gradient caused by mingling. Secondly, the turbulent motion is retarded by viscous stresses and requires a continuous supply of energy to maintain it, obtained from the working of the mean flow against the turbulent Reynolds stresses. The turbulent motion so depends for its kinetic energy on one of the quantities that it diffuses, the momentum of the mean flow, and the diffusing processes cannot be considered as small perturbations of an already existing motion as in the kinetic theory of gases. The necessary connection between the diffusion and the supply of energy to the turbulent motion is a fundamental characteristic of turbulent flow.

As a consequence of the irregularity and complexity of the motion, it is only practicable to consider mean values of functions of the instantaneous and local values of the fluid velocities and pressures, and all theoretical and experimental work uses mean values. For many years after Osborne Reynolds's formulation of the problem, measurements were confined to mean values of velocity and temperature and few measurements of the fluctuations were made. The theories developed at this time used speculative models of the fluid motion, sometimes derived by considering properties of the equations of fluid motion and sometimes by dimensional reasoning and analogies with the kinetic theory of gases. The most important were the various forms of the mixing-length theory developed by L. Prandtl and by G. I. Taylor, which served a purpose in providing a framework for current theoretical and experimental work, but they were admittedly incomplete and contained internal inconsistencies. About forty years ago, H. L. Dryden, A. M. Kuethe and others developed the hot-wire anemometer, making available for the first time a convenient means of studying the fluctuations of velocity, and its exploitation has provided detailed knowledge of the turbulent motion in a variety of flows. With so much information, it should now be possible to devise a physical theory of turbulence, based on dynamically consistent assumptions about the motion and capable of describing with fair accuracy the structure and properties of the simpler flows. The following account is an attempt in that direction.

1.2 The development of a theory for turbulent flow

Turbulent flow forms such a complicated mathematical problem that its solution must lean heavily on experimental data. One disadvantage is that experimental measurements commonly concern those quantities that are easy to measure rather than those that have an easily understood significance, and the sheer volume and detail of the data may be a bar to the understanding of the physical processes involved. The description of turbulent flow that is the theme of this book has grown in an irregular and haphazard way that may be appropriate to the subject but is a little difficult to follow in the course of its development. Its basis is a number of general principles and hypotheses about turbulent motion and it is convenient to introduce them as the results of study of a group of simple turbulent flows which lack one or more of the usual characteristics of 'complete' flows. The generalisations are of two kinds, those postulating the existence of various kinds of similarity and those dealing with the nature and mechanism of turbulent flow. Most of the first group are in no way new, having been at least implicit in most work for more than sixty years, but, without detailed assumptions about the nature of the motion, they lead to a number of important results which have been claimed to verify particular views about turbulent flow. Generalisations of the second kind depend on particular views of the structure and dynamics of turbulent motion, and their justification depends mostly on the success with which they may be used to predict the quantitative behaviour of turbulent flows. It is important to know which predictions about a particular flow may be derived solely from the similarity hypotheses before applying those assuming a particular kind of turbulent motion. It is also important not to apply the second kind of generalisation to the wrong kind of flow, a consideration that leads to a classification of turbulent flows by the restrictions placed on energy transfer from the mean flow to the fluctuations by the boundary conditions of the whole flow.

1.3 The statistical description of turbulent flow

The methods of statistical mechanics are used in the description of turbulent motion to the extent that use is made only of statistical mean values of the flow variables, i.e. of particle position, particle velocity, pressure and so on, but there are important differences. Unlike the molecular motion of gases, the motion at any point in a turbulent flow affects the motion at other distant points through the pressure field, and an adequate description cannot be obtained by considering only mean values associated with single fluid particles. This might be put by saying that turbulent motion is less random and more organised than molecular motion, and that to describe the organisation of the flow requires mean values of functions of the flow variables for two or more particles or at two or more positions. Even in the simplest (statistically) of turbulent flows – isotropic turbulence – the number of these functions necessary in the theory is large and, for normal turbulent flows whose asymmetry imposes still more organisation, an even larger number seems to be necessary. All this is true but, if the development of the theory is to be guided by experimental measurements, use must be made of practicable specifications that are very incomplete by the standards of the unaided theory.

A flow may be specified either in the Lagrangian way by $x(x_0, t_0; t)$, the position at time t of a particle which was at position x_0 at time t_0 or in the Eulerian way by $u(x, t)$, the velocity of the particle which is at position x at time t. No two realisations of a turbulent flow are identical and the complete statistical description is contained in the distribution function for the flow specification $x(x_0, t_0; t)$, which is the density in function-space of the points representing the realisations of the flow. In practice, measurements of Lagrangian flow quantities are extremely difficult and Eulerian ones are almost always used. If the distribution function for the Eulerian specification is $F[u(x, t)]$, the statistical or ensemble average of $M[u(x, t)]$, a function of the flow field in space and time, is

$$\langle M[u(x, t)] \rangle = \int F[u(x, t)]\, M[u(x, t)]\, \mathrm{d}V,$$

the integration being over function space. For some flows and for some measurements, the probability average may be the only possible average, but the majority of flows studied in the laboratory are statistically stationary with respect to time, i.e. in a co-ordinate system moving with a suitable uniform velocity, usually zero, the velocity components at a fixed point are stationary random functions of time. If this is so, the ergodic hypothesis asserts that the mean value with respect to time,

$$\overline{M} = \lim_{t_1 - t_2 \to \infty} \frac{1}{t_1 - t_2} \int_{t_2}^{t_1} M(t) \, dt, \qquad (1.3.1)$$

is identical with the ensemble average. If the flow possesses symmetry with respect to a plane or an axis, the flow variables may be stationary random functions of one or more space co-ordinates, and mean values over the appropriate directions are identical with time means. If the flow variables are stationary random functions of any space or time co-ordinate, mean values are independent of that co-ordinate and the flow is statistically homogeneous for the co-ordinate.

Mean values of velocity, pressure, temperature and concentration are comparatively easy to measure, and it is convenient to write their instantaneous values as sums of the mean value and the fluctuation from it. Necessarily, the mean of the fluctuation is zero. The physical significance of the distinction between mean values and fluctuations depends on the nature of turbulent transfer of energy and entropy from the mean flow to the fluctuations, which is normally a one-way process. Although the kinetic energy associated with the velocity fluctuations is, considered thermodynamically, free energy, it is most unusual for any substantial part of it to be changed back into kinetic energy of the mean flow and it may be regarded as a degraded form, intermediate between the organised and easily available energy of the mean flow and thermal energy.

1.4 Notation for turbulent flows

Two kinds of notation are in common use to describe turbulent flows, the compact suffix notation used for Cartesian tensors and the older notation used by Reynolds which differentiates more obviously between the various components of the vectors and tensors. If the motion is statistically isotropic, the choice of co-ordinate axes is unimportant and one component of a vector has no more and no less significance than either of the other two. In the inhomogeneous flows which are far more common, several directions are marked out by the symmetry and homogeneity of the flow and velocity components in these directions are physically distinct. It has become the custom to distinguish co-ordinates and components in these directions by using different letters as symbols rather than the less obvious suffixes, but, in the most general treatment of turbulent flow, directions are not specified and the suffix notation is more compact. For these reasons, the notation used below is mixed. Where it is convenient, the suffix notation will be used, i.e. a velocity vector **u** has components u_1, u_2, u_3 along the co-ordinate axes Ox_1, Ox_2 Ox_3, but if the axes are those appropriate to the particular flow the components become u, v, w parallel to Ox, Oy, Oz. The axes are chosen as a rule so that Ox is in the general direction of mean flow and Oz is in the direction of maximum gradient of mean velocity or mean temperature. For axisymmetric flows, cylindrical polar co-ordinates are used but the notation is similar.

1.5 Three-dimensional correlation and spectrum functions

The first problem is to obtain from experimental measurements a clear idea of the structure and motion of the turbulence. From now on, frequent references will be made to 'eddies' of the turbulent motion, a word intended to describe flow patterns with spatially limited distributions of vorticity and comparatively simple forms. Examples are the Hill spherical vortex and the simple vortex ring. It is supposed that real turbulent flows are the superposition of many such eddies of different kinds sizes and orientations.

Since the experimental data is always incomplete, the identification of eddy types must be by informed guesswork followed by measurements designed to confirm the guess, and then to fit the inferred structure into a coherent dynamical account of the motion. The first stage depends on familiarity with the meaning of the mean-value functions that are used and on an understanding of the relation between the form of the functions and the presence of particular forms of eddy. The mean-value function most used to examine the spatial structure of turbulence and its evolution in time is the co-variance between velocity components measured at two separated points in the flow, the *double-velocity correlation function* or, more briefly, the correlation function. It is defined as

$$R_{ij}(\mathbf{x}; \mathbf{r}, \tau) = \overline{u_i(\mathbf{x}, t)\, u_j(\mathbf{x}+\mathbf{r}, t+\tau)}, \qquad (1.5.1)$$

where $u_i(\mathbf{x}, t)$ is the instantaneous value of the ith component of the velocity fluctuation at the position $\mathbf{x}$ and time t. Restrictions on the form of R_{ij} are imposed by the condition of incompressibility, div $\mathbf{u} = 0$, and by the interchangeability of u_i and u_j in the definition. They are

$$\frac{\partial}{\partial r_j} R_{ij}(\mathbf{x}; \mathbf{r}, \tau) = 0 \qquad (1.5.2)$$

and

$$R_{ij}(\mathbf{x}; \mathbf{r}, \tau) = R_{ji}(\mathbf{x}+\mathbf{r}; -\mathbf{r}, -\tau). \qquad (1.5.3)$$

It is to be expected that the motion in one part of the fluid is statistically independent of the motion in a sufficiently distant part, and so R_{ij} should become negligibly small for $|\mathbf{r}|$ more than some value characteristic of the scale of the flow. For similar reasons, the correlation function becomes small for large values of the time interval.

The complete correlation function is a function of position in the flow and of four displacement variables, which places it beyond experimental measurement unless it exhibits an abnormal degree of symmetry and homogeneity. Most attention has been focused on the simultaneous correlation function with $\tau = 0$ and on the time–space correlation with spatial separation in the direction of mean flow. In this section we shall consider the relation of the simultaneous correlation function to the eddy patterns of the turbulence, using for illustration the contributions to the function of random distributions of eddies of similar forms. An eddy velocity distribution that can take many forms is defined by:

$$\left. \begin{aligned}
u_1 &= -\frac{\partial}{\partial x_2}[\mathrm{e}^{-\frac{1}{2}\alpha^2 x^2} \cos l_1 x_1 \cos l_2 x_2 \cos l_3 x_3], \\
u_2 &= \frac{\partial}{\partial x_1}[\mathrm{e}^{-\frac{1}{2}\alpha^2 x^2} \cos l_1 x_1 \cos l_2 x_2 \cos l_3 x_3], \\
u_3 &= 0,
\end{aligned} \right\} \qquad (1.5.4)$$

where

$$\alpha^2 x^2 = \alpha_1{}^2 x_1{}^2 + \alpha_2{}^2 x_2{}^2 + \alpha_3{}^2 x_3{}^2,$$

representing a finite, three-dimensional array of eddies. If a turbulent flow contains these eddies with their centres distributed randomly

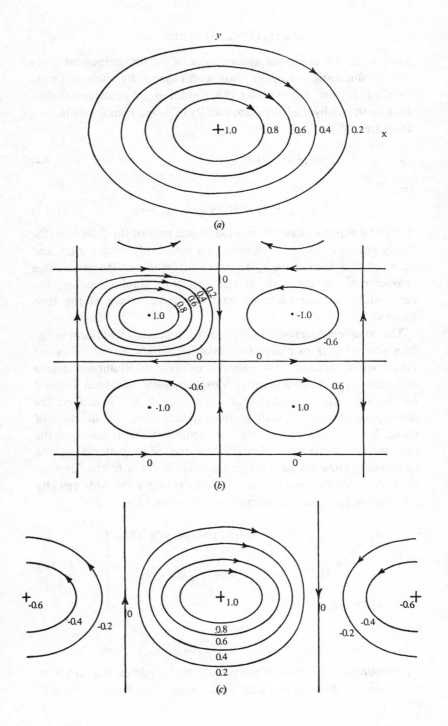

(a)

(b)

(c)

but statistically uniformly in space, it is easy to show that the contribution to the correlation function is a function of $\mathbf{r}$ with non-zero components,

$$R_{11}(\mathbf{r}) = -\frac{\partial^2}{\partial r_2{}^2} f(\mathbf{r}), \qquad R_{22}(\mathbf{r}) = -\frac{\partial^2}{\partial r_1{}^2} f(\mathbf{r}), \left.\vphantom{\frac{\partial^2}{\partial r_2{}^2}}\right\} \tag{1.5.5}$$
$$R_{12}(\mathbf{r}) = R_{21}(\mathbf{r}) = \frac{\partial^2}{\partial r_1 \, \partial r_2} f(\mathbf{r}),$$

where

$$f(\mathbf{r}) = A \, e^{-\frac{1}{4}\alpha^2 r^2}(\cos l_1 r_1 + e^{-l_1{}^2/\alpha_1{}^2})$$
$$\times (\cos l_2 r_2 + e^{-l_2{}^2/\alpha_2{}^2})(\cos l_3 r_3 + e^{-l_3{}^2/\alpha_3{}^2}),$$

A is a constant specifying the intensity of the eddy system, and

$$\alpha^2 r^2 = \alpha_1{}^2 r_1{}^2 + \alpha_2{}^2 r_2{}^2 + \alpha_3{}^2 r_3{}^2.$$

With particular values of the defining constants, the basic velocity pattern can take the form of several eddy structures of physical interest, including the isolated simple eddy, a periodic array of simple eddies and intermediate arrangements such as a finite row of eddies. Three interesting forms are sketched in fig. 1.1:

Type A: $l_1 = l_2 = l_3 = 0$. This represents a simple eddy with circulation at right angles to Ox_3. The non-zero components are

$$R_{11} = \tfrac{1}{2}A\alpha_2{}^2(1 - \tfrac{1}{2}\alpha_2{}^2 r_2{}^2) e^{-\frac{1}{4}\alpha^2 r^2}, \left.\vphantom{\tfrac{1}{2}}\right\}$$
$$R_{22} = \tfrac{1}{2}A\alpha_1{}^2(1 - \tfrac{1}{2}\alpha_1{}^2 r_1{}^2) e^{-\frac{1}{4}\alpha^2 r^2}, \left.\vphantom{\tfrac{1}{2}}\right\} \tag{1.5.6}$$
$$R_{12} = R_{21} = \tfrac{1}{4}A\alpha_1{}^2\alpha_2{}^2 r_1 r_2 \, e^{-\frac{1}{4}\alpha^2 r^2} \left.\vphantom{\tfrac{1}{2}}\right\}$$

and become very small for large values of $|\mathbf{r}|$ (greater than about $3\alpha^{-1}$).

Type B: $\alpha_1 = \alpha_2 = \alpha_3 = 0$. The motion is periodic in space and infinite in extent. The components are

$$R_{11} = Al_2{}^2 \cos l_1 r_1 \cos l_2 r_2 \cos l_3 r_3, \left.\vphantom{l_2{}^2}\right\}$$
$$R_{22} = Al_1{}^2 \cos l_1 r_1 \cos l_2 r_2 \cos l_3 r_3, \left.\vphantom{l_1{}^2}\right\} \tag{1.5.7}$$
$$R_{12} = R_{21} = Al_1 l_2 \sin l_1 r_1 \sin l_2 r_2 \cos l_3 r_3 \left.\vphantom{l_1{}^2}\right\}$$

Fig. 1.1. Simple eddy structures. (*a*) Isolated eddy (type A); section at $z = 0$, $\alpha_2 = \tfrac{3}{2}\alpha_1$. (*b*) Periodic array of eddies (type B); section at $z = 0$, $l_1 = \tfrac{2}{3}l_2$. (*c*) Finite row of simple eddies (type C); section at $z = 0$, $\alpha_1/l_1 = \pi^{-1}$. The numbers are the values of the stream function for two-dimensional flow, $l_3 = \alpha_3 = 0$.

and are likewise infinite in extent, oscillating with the same period as the velocity in the original disturbance.

Type C: $l_2 = l_3 = 0$. The motion is a finite array of simple eddies with centres along a line parallel to Ox_1. The components are:

$$
\left.
\begin{aligned}
R_{11} &= \tfrac{1}{2}A\alpha_2{}^2(1-\tfrac{1}{2}\alpha_2{}^2 r_2{}^2)(\cos l_1 r_1 + e^{-l_1{}^2/\alpha_1{}^2})\, e^{-\frac{1}{4}\alpha^2 r^2}, \\
R_{22} &= \tfrac{1}{2}A\alpha_1{}^2\left[(1-\tfrac{1}{2}\alpha_1{}^2 r_1{}^2)(\cos l_1 r_1 + e^{-l_1{}^2/\alpha_1{}^2})\right. \\
&\qquad\left. -2l_1 r_1 \sin l_1 r_1 - 2\frac{l_1{}^2}{\alpha_1{}^2}\cos l_1 r_1\right] e^{-\frac{1}{4}\alpha^2 r^2}, \\
R_{12} &= R_{21} = \tfrac{1}{2}A\alpha_2{}^2[\tfrac{1}{2}\alpha_1{}^2 r_1 r_2(\cos l_1 r_1 + e^{-l_1{}^2/\alpha_1{}^2}) \\
&\qquad + l_1 r_2 \sin l_1 r_1]\, e^{-\frac{1}{4}\alpha^2 r^2}
\end{aligned}
\right\} \quad (1.5.8)
$$

which oscillate in the Or_1 direction with nearly the period of the eddy system but decreasing amplitude. In the other directions, they fall off as $\exp(-\tfrac{1}{4}\alpha^2 r^2)$.

While the correlation function for turbulence composed of simple eddies must be a smooth and simple function of separation, a simple correlation function does not imply that the turbulence is composed of simple velocity patterns. For example, consider turbulence composed of quasi-periodic flow patterns of type C, each with different values for the characteristic wave number, l_1. If the values are normally distributed around a mean value l_0 with standard deviation β, the correlation component R_{11} is

$$
R_{11} = \tfrac{1}{2}A\alpha_2{}^2(1-\tfrac{1}{2}\alpha_2{}^2 r_2{}^2)\, e^{-\frac{1}{4}\alpha^2 r^2}
$$
$$
\times\left[e^{-\frac{1}{2}\beta^2 r_1{}^2}\cos l_0 r_1 + \frac{\alpha_1}{(\alpha_1{}^2+2\beta^2)^{\frac{1}{2}}}\, e^{-l_0{}^2/\alpha_1{}^2}\right] \quad (1.5.9)
$$

and the periodicity of the patterns is nearly undetectable in the correlation function if β/l_0 is more than about 0.4. Even if the ratio β/l_0 is smaller, the extent of the correlation function is considerably less than the length of the basic patterns. For the detection of periodic flow patterns with a wide range of periods, it is necessary to study higher-order multi-point correlations or their equivalents.

With simple eddies, the extent in **r**-space of appreciable values for the correlation function is comparable with the size of the largest eddies of the turbulence, and consideration of the correlation function for large separations may lead to valid conclusions about the form of these eddies. Typical turbulent flows contain eddies with a wide range of size, and the distribution of energy over the range of size is an important quantity for the discussion of the motion. Using the correlation function to assign energy to eddies of a particular size (or range of sizes) requires the construction of a function whose magnitude for argument r is clearly related to the energy of eddies with 'diameter' r. One method of adapting the correlation function to give a distribution function for eddy size is that of the *structure functions* used by A. N. Kolmogorov in his theory of local isotropy. A quadratic structure function may be defined as the mean product of velocity differences between points in the flow,

$$B_{ij}(\mathbf{x}; \mathbf{r}) = \overline{[u_i(\mathbf{x}) - u_i(\mathbf{x} + \mathbf{r})][u_j(\mathbf{x}) - u_j(\mathbf{x} + \mathbf{r})]}. \quad (1.5.10)$$

It may be expressed in terms of the correlation function as

$$B_{ij}(\mathbf{x}; \mathbf{r}) = R_{ij}(\mathbf{x}; 0) + R_{ij}(\mathbf{x} + \mathbf{r}; 0)$$
$$- R_{ij}(\mathbf{x}; \mathbf{r}) - R_{ij}(\mathbf{x} + \mathbf{r}; -\mathbf{r}). \quad (1.5.11)$$

It is argued that eddies of scale much larger than $|\mathbf{r}|$ contribute very little to the velocity difference, $u_i(\mathbf{x}) - u_i(\mathbf{x} + \mathbf{r})$, because their contributions to $u_i(\mathbf{x})$ and $u_i(\mathbf{x} + \mathbf{r})$ are almost identical. It is argued also that the contributions of eddies of scale much smaller than $|\mathbf{r}|$ is negligible but, although they may contribute less than an eddy 'matched' to the separation, the difference is not great and it is better to regard $B_{ij}(\mathbf{r})$ as determined by all eddies of size less than or comparable with $|\mathbf{r}|$. For homogeneous turbulence of the kind generated by the model eddies, we say that the contribution to the mean square velocity, $\overline{u_1^2} = R_{11}(0)$, from eddies of size r or less would be, say,

$$R_{11}(0) - R_{11}(r, 0, 0)$$

and that the contribution from eddies of size r in unit range of $\log r$ is

$$- r \frac{\partial}{\partial r} [R_{11}(0) - R_{11}(r, 0, 0)].$$

For simple model eddies (type A), the distribution function so obtained has a fairly sharp peak around $r = 2\alpha^{-1}$, and the function would give a reasonable description of the distribution of eddy size in turbulence containing similar eddies. A practical advantage of using structure functions is that velocity differences are comparatively insensitive to the slow and uncontrollable changes in the flow which always occur during measurements in the atmosphere or in the ocean.

Another way of analysing the distribution of eddy sizes is to use the three-dimensional Fourier transform of the correlation function, defined as

$$\Phi_{ij}(\mathbf{x}; \mathbf{k}) = (2\pi)^{-3} \int R_{ij}(\mathbf{x}; \mathbf{r}) \, e^{-i\mathbf{k}\cdot\mathbf{r}} \, dV(\mathbf{r}), \qquad (1.5.12)$$

the integration being over all r-space. The inverse relation,

$$R_{ij}(\mathbf{x}; \mathbf{r}) = \int \Phi_{ij}(\mathbf{x}; \mathbf{k}) \, e^{i\mathbf{k}\cdot\mathbf{r}} \, dV(\mathbf{k}) \qquad (1.5.13)$$

means that Φ_{ij} is the contribution to $\overline{u_i u_j} = R_{ij}(0; \mathbf{x})$ from Fourier components of the velocity field with wave numbers in unit volume of k-space. The assumption of incompressible flow leads to the restriction that

$$k_j \Phi_{ij}(\mathbf{x}; \mathbf{k}) = 0 \qquad (1.5.14)$$

equivalent to equation (1.5.2).

If the turbulence is inhomogeneous with scale L, the values of the spectrum function for values of $|\mathbf{k}|$ comparable with or smaller than L^{-1} are strongly dependent on the inhomogeneity and should not be used in a description of the turbulent motion. For separations considerably less than L, the variation of the correlation function with x is small compared with its variation with r and then the local spectrum function,

$$\Phi_{ij}{}^*(\mathbf{x}; \mathbf{k}) = (2\pi)^{-3} \iiint_{-B}^{B} R_{ij}(\mathbf{x}; \mathbf{r}) \, e^{-i\mathbf{k}\cdot\mathbf{r}} \, dV(\mathbf{r}) \qquad (1.5.15)$$

derived from correlation measurements within a cube of side $2B$, is related to the ordinary function by

$$\Phi_{ij}{}^*(\mathbf{x}; \mathbf{k}) = \left(\frac{B}{2\pi}\right)^3 \iiint_{-\infty}^{\infty} \left[\frac{\sin k_1{}'B}{k_1{}'B} \frac{\sin k_2{}'B}{k_2{}'B} \frac{\sin k_3{}'B}{k_3{}'B}\right]$$
$$\times \, \Phi_{ij}(\mathbf{k} + \mathbf{k}') \, dV(\mathbf{k}'). \qquad (1.5.16)$$

That is, the local spectrum function is the result of measuring Φ_{ij} with a resolution given by the bracketed factor, roughly an average over a volume $\pi^3 B^{-3}$ in k-space, and is nearly the same if $|\mathbf{k}|B$ is large. For values of the wave number considerably more than L^{-1}, the spectrum function has a local significance and is not directly dependent on the inhomogeneity.

For many purposes and particularly for the discussion of isotropic turbulence, the integrated spectrum function, defined as

$$E_{ij}(k) = \int_{|k|=k} \Phi_{ij}(\mathbf{k}) \, dS(\mathbf{k}), \qquad (1.5.17)$$

where the integration is over a spherical surface of radius k, is more useful. It represents the contributions to $\overline{u_i u_j}$ from wave numbers with magnitudes equal to k.

The spectrum function for the correlation function of equation (1.5.5) has non-zero components,

$$\left. \begin{array}{l} \Phi_{11} = k_2{}^2 g(\mathbf{k}), \qquad \Phi_{22} = k_1{}^2 g(\mathbf{k}), \\ \Phi_{12} = \Phi_{21} = -k_1 k_2 g(\mathbf{k}), \end{array} \right\} \qquad (1.5.18)$$

where

$$g(\mathbf{k}) = A \exp\left[-\frac{k_1{}^2 + l_1{}^2}{\alpha_1{}^2} - \frac{k_2{}^2 + l_2{}^2}{\alpha_2{}^2} - \frac{k_3{}^2 + l_3{}^2}{\alpha_3{}^2} \right]$$
$$\times \cosh^2\left(\frac{l_1 k_1}{\alpha_1{}^2} \right) \cosh^2\left(\frac{l_2 k_2}{\alpha_2{}^2} \right) \cosh^2\left(\frac{l_3 k_3}{\alpha_3{}^2} \right).$$

The components have maxima in the neighbourhood of the wave numbers $(\pm l_1, \pm l_2, \pm l_3)$, unless the α/l are large when maxima occur near the ellipsoid,

$$\frac{k_1{}^2}{\alpha_1{}^2} + \frac{k_2{}^2}{\alpha_2{}^2} + \frac{k_3{}^2}{\alpha_3{}^2} = 1.$$

The special forms of eddy pattern have the following spectrum functions:

Type A: $l_1 = l_2 = l_3 = 0$. The components of the spectrum function are

$$\left. \begin{array}{l} \Phi_{11} = A k_2{}^2 \exp -\left(\dfrac{k_1{}^2}{\alpha_1{}^2} + \dfrac{k_2{}^2}{\alpha_2{}^2} + \dfrac{k_3{}^2}{\alpha_3{}^2} \right), \\[3mm] \Phi_{22} = A k_1{}^2 \exp -\left(\dfrac{k_1{}^2}{\alpha_1{}^2} + \dfrac{k_2{}^2}{\alpha_2{}^2} + \dfrac{k_3{}^2}{\alpha_3{}^2} \right), \\[3mm] \Phi_{12} = \Phi_{21} = -A k_1 k_2 \exp -\left(\dfrac{k_1{}^2}{\alpha_1{}^2} + \dfrac{k_2{}^2}{\alpha_2{}^2} + \dfrac{k_3{}^2}{\alpha_3{}^2} \right). \end{array} \right\} \qquad (1.5.19)$$

If $\alpha_1 = \alpha_2 = \alpha_3$, which represents a 'spherical' eddy, the integrated spectrum function has non-zero components

$$E_{11} = E_{22} = A\pi k^4 \, e^{-k^2/\alpha^2} \qquad (1.5.20)$$

which have sharp maxima at $k = 2^{1/2}\alpha$. Since α^{-1} is a measure of the size of the eddies, it is seen that the dominant Fourier components for eddies of size L have wave numbers near L^{-1}.

Type B: $\alpha_1 = \alpha_2 = \alpha_3 = 0$. This infinite periodic array has a discrete spectrum function which is non-zero only for $\mathbf{k} = (\pm l_1, \pm l_2, \pm l_3)$.

Type C: $l_2 = l_3 = 0$, $\alpha_1 = \alpha_2 = \alpha_3$. Depending on the relative values of l_1 and α, the spectrum function has maxima near $\mathbf{k} = (\pm l, 0, 0)$, or, if $\alpha \gg l$, near $|\mathbf{k}| = \alpha^{-1}$. To show that turbulence composed of isolated eddies can be represented by a small range of wave numbers, the eddy velocity distributions for $\alpha/l_1 = 1/\pi$ and $\sqrt{(2)}/\pi$ are shown together with the corresponding integrated spectrum function in fig. 1.2.

Whether the basic eddies composing the turbulent flow resemble simple eddies or periodic arrays of eddies, the integrated spectrum function of a random distribution homogeneous in size is concentrated in wave numbers close to the inverse of the diameter or wavelength of the component eddies, and so the function can be used to express in quantitative form the relative intensities of physical eddies of different sizes. There are two exceptions to this conclusion:

(1) Any finite superposition of eddies with minimum size L leads to a form of $E_{ij}(k)$ such that

$$E_{ij}(k) = C_{ij}k^4 + \text{terms of order } k^6$$

for $kL \ll 1$, and no special meaning can be attached to the form of the spectrum at low wave numbers. Its magnitude depends mostly on the largest eddies of size around L.

(2) Viscous dissipation sets a lower limit to the size of physical eddies and the spectrum function for wave numbers larger than the reciprocal of the minimum size is determined by that size, say a, and varies nearly as $e^{-a^2k^2}$.

These two qualifications to the usual identification of eddies with Fourier components of the velocity field are due to the initial

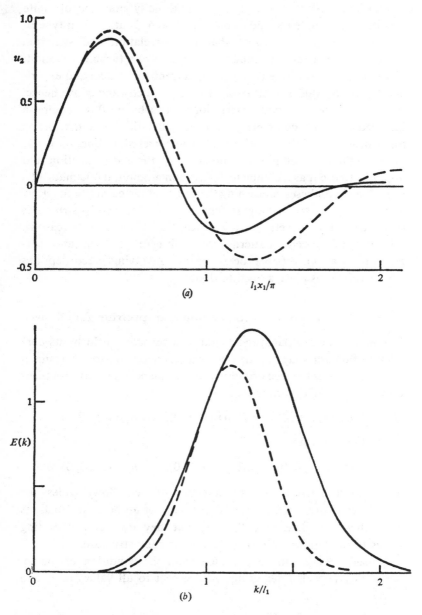

Fig. 1.2. Spectrum function of a simple isolated eddy. (a) Velocity distributions, (b) three-dimensional spectrum functions: for $---$, $\alpha = l_1/\pi$; ———, $\alpha = l_1\sqrt{2}/\pi$

assumption that a physically acceptable eddy must be of finite extent, i.e. resemble the eddies of types A and C rather than type B.

It is important to realise that the correlation and spectrum functions form a very incomplete description of a turbulent motion, and that the central role they play in current theoretical and experimental work is due to their comparative simplicity and convenience. In principle, the statistical description of a turbulent flow requires a knowledge of the complete joint-probability distribution function for realisations of the velocity field. The correlation function is one of the infinite set of integral moments of the basic function and attempts to use it as a complete description amount, mathematically, to making hypotheses concerning the nature of the complete function, and, physically, to making statements about the eddy structures that exist in fully developed turbulence. For this and other reasons, inference of velocity patterns from observed spectra and correlations is an uncertain process, usually involving preconceptions whose validity is always open to doubt.

1.6   One-dimensional correlation and spectrum functions

Incomplete though the specification of a turbulent field by its correlation function may be, the observations necessary to determine it are so numerous that the primary information nearly always concerns only the particular correlations

$$R_{11}(\mathbf{x}; r, 0, 0), \qquad R_{11}(\mathbf{x}; 0, r, 0), \qquad R_{11}(\mathbf{x}; 0, 0, r)$$

or the Fourier transforms of

$$R_{11}(\mathbf{x}; r, 0, 0), \qquad R_{22}(\mathbf{x}; r, 0, 0), \qquad R_{33}(\mathbf{x}; r, 0, 0)$$

with respect to r (Ox_1 is in the direction of mean flow). Unless the turbulence is isotropic, a knowledge of any or all of these functions is insufficient to determine $R_{ij}(\mathbf{x}; r)$, but they are useful for setting scales to the motion and for inferring the eddy structure.

A useful result can be obtained from the condition of incompressibility (1.5.2). Integrating with respect to all values of r_2 and r_3 gives

$$\frac{\partial}{\partial r_1} \iint_{-\infty}^{\infty} R_{i1}(\mathbf{x}; \mathbf{r}) \, dr_2 \, dr_3 = 0, \tag{1.6.1}$$

since $R_{ij}(x; r)$ is expected to be small for all large values of $|r|$. It follows that

$$\int_{-\infty}^{\infty} \int_{-\infty}^{\infty} R_{i1}(x; r_1, r_2, r_3) \, dr_2 \, dr_3 = \text{a constant} \qquad (1.6.2)$$

which must be zero. In practice, we are interested in $R_{11}(x; 0, r_2, r_3)$ for which

$$\int_{-\infty}^{\infty} \int_{-\infty}^{\infty} R_{11}(x; 0, r_2, r_3) \, dr_2 \, dr_3 = 0. \qquad (1.6.3)$$

The physical meaning is that the instantaneous flux across any closed surface is zero and that the compensating inflow across a plane $x_1 = $ constant in response to an outflow at a particular point takes place mostly at points displaced by values of r_2, r_3 such that $R_{11}(x; 0, r_2, r_3)$ is negative. In isotropic turbulence,

$$\int_0^{\infty} r R_{11}(0, r, 0) \, dr = 0 \qquad (1.6.4)$$

and the return flow takes place at distances that make $r R_{11}(0, r, 0)$ a minimum. In anisotropic turbulence, the return flow may be concentrated in a plane and then negative values of $R_{11}(0, r, 0)$ may be numerically much larger (or smaller) than those of $R_{11}(0, 0, r)$. Further, if the eddies are all much the same size, return flow takes place over a limited range of r, but a wide range of eddy sizes implies return flow over a wide range of r and consequently smaller negative values for the transverse correlations.

If eddies of the kinds considered in § 1.5 are taken to be typical of simple eddies, it is clear that the three-dimensional and one-dimensional correlation functions are smooth, i.e. the curvature is nowhere large if all the component eddies are of about the same size. The occurrence of locally high curvature implies the presence of a wide range of eddy sizes, which may take two forms. In the first form, a wide and continuous range of eddy sizes leads to large curvature near $r = 0$, a feature of nearly all correlation functions for turbulence of high Reynolds number. The other form occurs when there are two distinct ranges of eddy size present, and arises from the addition of two correlation functions of very different scales. Then the curvature is large near some positive value of r and the composite function has a characteristic two-component form. Examples occur in free turbulence and even more strikingly in correlation functions for atmospheric turbulence.

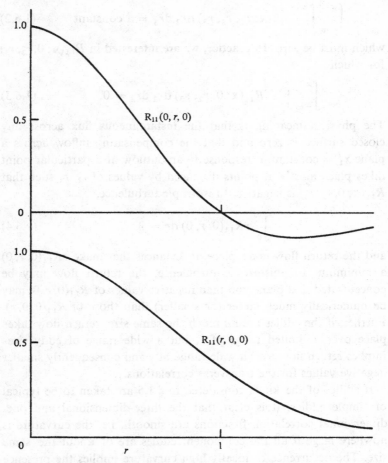

Fig. 1.3. Correlation functions for isotropic turbulence with eddies of uniform size.

The statements made above are illustrated in figs. 1.3–1.6, which show schematically correlation functions for the following types of turbulence:

(1) isotropic turbulence of uniformly sized eddies,

(2) isotropic turbulence of a wide range of eddy sizes,

(3) isotropic turbulence with two distinct ranges of eddy sizes,

(4) 'two-dimensional' turbulence in which $u_3 = 0$,

(5) 'two-dimensional' turbulence for which $u_1 = u_2$ everywhere.

If the velocity fluctuations are small compared with the mean velocity of flow, the changes in the velocity pattern as it sweeps past

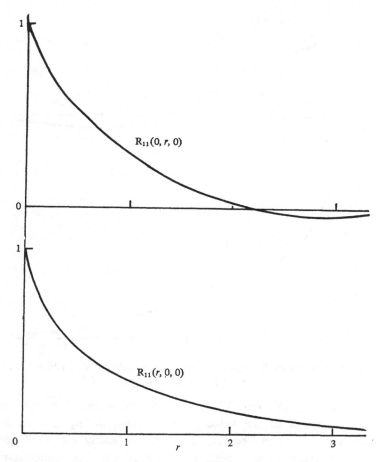

Fig. 1.4. Typical transverse and longitudinal correlation functions for isotropic turbulence with a wide spectrum of eddy sizes.

a fixed point are negligible and a time displacement of τ is equivalent to a displacement in the flow direction of $-U_1\tau$, where U_1 is the mean velocity. Then the Fourier transforms of $R_{11}(\mathbf{x}; r, 0, 0)$,

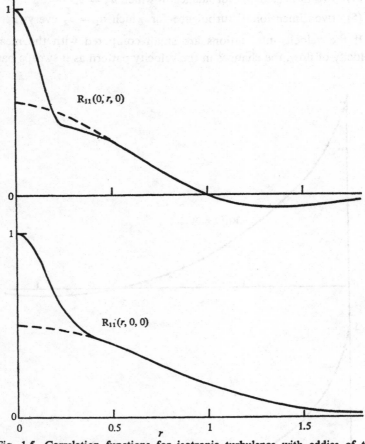

Fig. 1.5. Correlation functions for isotropic turbulence with eddies of two distinct sizes. (Equal intensities, size ratio 5:1.)

$R_{22}(\mathbf{x}; r, 0, 0)$, $R_{33}(\mathbf{x}; r, 0, 0)$ are nearly proportional to the frequency spectra of the velocity components, u_1, u_2 and u_3, for frequency kU_1. The frequency spectrum is readily measured with electrical spectrum analysers. In terms of the three-dimensional function, the one-dimensional spectrum function is

$$\phi_{ij}(k_1) = \frac{1}{\pi} \int_{-\infty}^{\infty} R_{ij}(r, 0, 0)\, e^{-ik_1 r}\, dr \qquad (1.6.5)$$

$$= 2 \int_{-\infty}^{\infty} \int_{-\infty}^{\infty} \Phi_{ij}(\mathbf{k})\, dk_2\, dk_3,$$

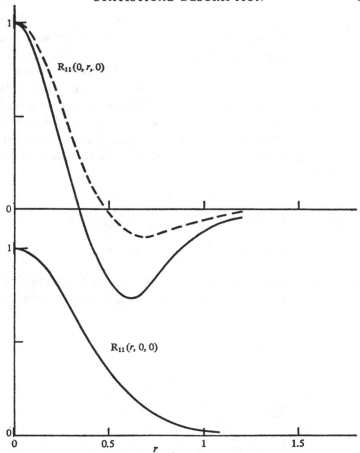

Fig. 1.6. Correlation functions for turbulence with motion parallel to a fixed plane. ———, $u_3 = 0$ (motion parallel to $x_1 O x_2$); – – –, $u_1 = u_2$.

showing that a single value of $\phi_{ij}(k_1)$ represents the sum of values of Φ_{ij} for a wide range of $|\mathbf{k}|$, from k_1 to infinite wave number. So while a simple eddy structure leads to a three-dimensional spectrum function which is large only over a limited range of $|\mathbf{k}|$, it leads to a one-dimensional spectrum function hardly more confined in extent than the corresponding correlation function. For isotropic turbulence, the integrated spectrum function (1.5.14) is related to $\phi_{11}(k_1)$ by

$$E_{11}(k) = \frac{1}{3}\left[k^2 \frac{\mathrm{d}^2\phi_{11}(k)}{\mathrm{d}k^2} - k \frac{\mathrm{d}\phi_{11}(k)}{\mathrm{d}k} \right] \tag{1.6.6}$$

which shows the size distribution to be determined more by the curvature of the spectrum function than its magnitude at any particular wave number. For example, $\phi_{ij}(0)$ is almost always not zero and its value is determined exclusively by values of Φ_{ij} at non-zero values of $\mathbf{k}$. It does not indicate a finite intensity of very large eddies.

The one-dimensional spectrum function may be used in a similar way to the correlation function to reveal anisotropy of the component eddies. If the motion of the turbulence is almost confined to the $x_1 O x_2$ plane, backflow will take place in the $O x_1$ direction for a u_2 outflow and

$$\int_{-\infty}^{\infty} R_{22}(r, 0, 0)\, \mathrm{d}r$$

should be nearly zero. The corresponding spectrum function

$$\phi_{22}(k) = \frac{1}{\pi} \int_{-\infty}^{\infty} R_{22}(r, 0, 0)\, \mathrm{e}^{-ikr}\, \mathrm{d}r$$

will also be nearly zero at $k_1 = 0$. Roughly, spectra of the shape I in fig. 1.7 imply motion in the $x_1 O x_2$ plane, of shape II, motion in

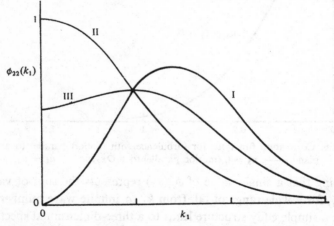

Fig. 1.7. One-dimensional spectrum functions for isotropic and two-dimensional turbulence. Curve I, $u_3 = 0$; curve II, $u_1 = 0$; curve III, isotropic.

the $x_2 O x_3$ plane, and of shape III, motion either equally distributed or at 45° to $O x_1$. Isotropic turbulence gives spectra of shape III.

If two distinct ranges of eddy size exist, the spectrum functions take characteristic two-component forms similar to those of the correlation function. It is worth noting that a low-intensity group of large eddies in a background of smaller eddies is much easier to detect from the appearance of the spectrum function than from the appearance of the correlation function.

1.7 Correlations and spectra with time delay

The simultaneous correlation and spectrum functions provide information about the instantaneous flow patterns, but a description of their growth and development requires comparison of flow patterns at different times, conveniently by appropriate use of the complete space–time correlation function. Changes of flow pattern that affect the correlation can arise either by displacement of individual eddy patterns or by change of the patterns themselves. In turbulent flow, localised eddies move with respect to the surrounding fluid with a self-induced velocity that depends on their structure, a familiar example being the motion of a vortex ring. So the 'centre' moves with a velocity compounded of the local mean velocity, the instantaneous local velocity of the larger eddies in which it is imbedded, and the self-induced velocity. If it were possible to follow an eddy centre, the evolution of a single eddy could be studied but Eulerian measurements allow only a determination of the average displacement. In general, it is not possible to say whether the apparent change of eddy pattern after allowing for the average displacement is a real change or merely the effect of a deviation of the eddy centre from its mean position.

Addition of a time variable adds to the complexity of experimental measurements, and most of the available measurements refer to the space–time correlation function for a single velocity component and spatial separation in the direction of mean flow. The general appearance of the function, $R_{11}(\mathbf{x}; r, 0, 0, \tau)$, is shown in fig. 1.8 as variations with r for several fixed time delays. With increasing time delay,

(1) the heights of the maxima with respect to variation of r become less;

(2) r_m, the position of the maximum, increases nearly proportionally to time delay, the ratio r_m/τ being defined as the *convection velocity*;

(3) the radius of curvature at the maximum becomes greater;

(4) the variation about the maximum may become asymmetric.

The first two effects may be interpreted as the consequence of a velocity pattern that moves with the convection velocity but also changes and loses identity in a way described by the variation of maximum correlation with time delay. Since u_1 is a characteristic of the whole motion, the convection velocity and loss of correlation

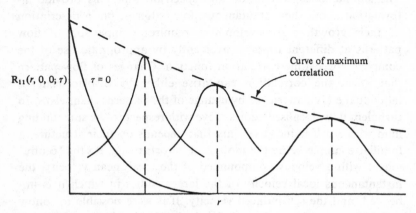

Fig. 1.8. Time delay correlations, as functions of streamwise separation for various time intervals.

are averages over eddy components of all sizes but the last two effects are indicative of different behaviour of smaller-scale components and suggest that more information can be gained from a closer study. The object is to recast the correlation function into a form that discriminates between eddies of different sizes.

In § 1.5, it was argued that the velocity difference, $u_i(\mathbf{x}, t) - u_i(\mathbf{x} + \mathbf{r}, t)$, is determined mostly by eddies of size $|\mathbf{r}|$ or less. As a rule, the smaller eddies of turbulent flow hold less energy than the larger ones and the major contribution to the difference comes always from eddies of size comparable with the separation. Hence the velocity difference may be interpreted as a measure of the influence of eddies of size r, on the flow velocity at the point $\mathbf{x} + \frac{1}{2}\mathbf{r}$ and the space–time structure function,

$$
\begin{aligned}
B_{11}(\mathbf{x}, \mathbf{r}; \mathbf{s}, \tau) &= \overline{[u_1(\mathbf{x}, t) - u_1(\mathbf{x} + \mathbf{r}, t)]} \\
&\quad \times \overline{[u_1(\mathbf{x} + \mathbf{s}, t + \tau) - u_1(\mathbf{x} + \mathbf{s} + \mathbf{r}, t + \tau)]} \\
&= R_{11}(\mathbf{x}; \mathbf{s}, \tau) + R_{11}(\mathbf{x} + \mathbf{r}; \mathbf{s}, \tau) \\
&\quad - R_{11}(\mathbf{x}; \mathbf{s} + \mathbf{r}, \tau) - R_{11}(\mathbf{x} + \mathbf{r}; \mathbf{s} - \mathbf{r}, \tau), \quad (1.7.1)
\end{aligned}
$$

i.e. the covariance between velocity differences distant apart s in space and τ in time, measures the changes in eddies of size r. The relation between the changes in the eddies and the changes of the structure function can be discussed by supposing that the velocity pattern for eddies of size r is known so that a single eddy with centre at x_0 has the velocity differences,

$$u_1(\mathbf{x}) - u_1(\mathbf{x} + \mathbf{r}) = af(\mathbf{x} - \mathbf{x}_0, \mathbf{r}),$$

where f is the same for all eddies of size r and a is the velocity amplitude of the eddy. (Variations of orientation and type can be introduced if it is thought necessary.) For a particular eddy, the amplitude and centre position are functions of time and the contribution of one eddy to the structure function is

$$\{a(t)a(t + \tau)\} \times \{f[\mathbf{x} - \mathbf{x}_0(t)]f[\mathbf{x} + \mathbf{s} - \mathbf{x}_0(t + \tau)]\}.$$

The first factor depends on a real change in eddy amplitude and the second depends on translation of the eddy. To make up the whole structure function, we need the probability distribution function,

$$P[a(t), a(t + \tau), x_0(t), x_0(t + \tau)],$$

defining the probability of observing particular values of the amplitudes and centre positions in one realisation of the flow. For a steady flow,

$$B_{11}(\mathbf{x}, \mathbf{r}; \mathbf{s}, \tau) = \int P[a(0), a(\tau) - a(0), \mathbf{x}_0(0), \mathbf{x}_0(\tau) - \mathbf{x}_0(0)]$$
$$\times a(0)a(\tau)f(\mathbf{x} - \mathbf{x}_0(0))f(\mathbf{x} - \mathbf{x}_0(\tau) + \mathbf{s}) \, dV, \quad (1.7.2)$$

the integration being over all values of $a(0)$, $a(\tau)$, $\mathbf{x}_0(0)$ and $\mathbf{x}_0(\tau)$.

Consider now a homogeneous distribution of eddies of size r, with centres distributed randomly in space with uniform number density N and with amplitudes independent of the centre positions. The distribution function factorises to

$$NP_1[a(0), a(\tau) - a(0)]P_2[\mathbf{x}_0(\tau) - \mathbf{x}_0(0)]$$

and

$$B_{11}(\mathbf{s}, \tau) = \iint P_1[a(0), a(\tau) - a(0)]a(0)a(\tau) \, da(0) \, da(\tau)$$
$$\times N \iint P_2[\mathbf{x}_0(\tau) - \mathbf{x}_0(0)] f[\mathbf{x} - \mathbf{x}_0(0)]$$
$$\times f[\mathbf{x} - \mathbf{x}_0(\tau) + \mathbf{s}] \, d[\mathbf{x}_0(\tau) - \mathbf{x}_0(0)]$$
$$= \frac{\overline{a(t)a(t + \tau)}}{\overline{(a(t))^2}} \int P_2[\mathbf{x}_0(\tau) - \mathbf{x}_0(0)]B_{11}[\mathbf{s} - \mathbf{x}_0(\tau) + \mathbf{x}_0(0), 0]$$
$$\times d[\mathbf{x}_0(\tau) - \mathbf{x}_0(0)]. \quad (1.7.3)$$

The first factor is simply the autocorrelation coefficient for individual eddy amplitudes with time delay τ and describes the intrinsic changes in the eddies. The second factor depends on the movements of the eddy centres in the time interval, and the effect can be described as the combination of a translation in s-space of $\overline{\mathbf{x}_0(\tau)} - \mathbf{x}_0(0)$ and a diffusive spread in s-space by random movements of the eddy centres. The diffusive spread is a problem in Lagrangian diffusion, but experimental studies suggest that a fair approximation to P_2 is

$$P_2[\mathbf{x}_0(\tau) - \mathbf{x}_0(0)]$$
$$= (2\pi)^{-3/2}\sigma^{-3}\exp[-\tfrac{1}{2}(\mathbf{x}_0(\tau) - \mathbf{x}_0(0) - \overline{(\mathbf{x}_0(\tau) - \mathbf{x}_0(0))})^2/\sigma^2] \quad (1.7.4)$$

where σ is the standard deviation of the centre displacement about its mean value. Then the maximum value of $B_{11}(\mathbf{s}, \tau)$ occurs where

$$\mathbf{S}_m \equiv \mathbf{x}_0(\tau) - \mathbf{x}_0(0) = \mathbf{s} \quad (1.7.5)$$

defining a convection velocity, $\mathbf{U}_c = \mathbf{S}_m/\tau$, and the magnitude of σ could be found by comparing the shapes of the structure functions for $\tau = 0$ and for time delay τ. With a knowledge of σ, the reduction in height of the maximum by diffusion can be calculated and used to find the autocorrelation coefficient for the amplitude of individual eddies.

The smaller eddies are carried around by the larger ones and, for them, the convection velocity of their eddy centres is nearly the local velocity. If u_0 and L_0 are the scales of velocity and length for the main turbulent motion, particle velocities are expected to remain nearly unchanged for times short compared with L_0/u_0 and then

$$\mathbf{x}_0(\tau) - \mathbf{x}_0(0) = \mathbf{U}\tau + \mathbf{u}(0)\tau. \quad (1.7.6)$$

It follows that

$$\sigma = (\overline{u_1^2})^{1/2}\tau \quad (1.7.7)$$

and we see that the effects of diffusion are small if $(\overline{u_1^2})^{1/2}\tau \ll r$, and that they reduce the structure function to a negligible value if $(\overline{u_1^2})^{1/2}\tau \gg r$. Unless the autocorrelation coefficient changes appreciably in a time interval of $r/(\overline{u_1^2})^{1/2}$, it is difficult to distinguish the effect of its variation on the delayed structure function from the effect of diffusion.

In inhomogeneous turbulence with spatial variation of mean velocity, the interpretation is less simple. For example, eddies of a particular size may tend to aggregate near a plane parallel to the direction of mean flow and the convection velocity of the centres will be nearly the mean velocity at the plane. If the convection velocity is determined from structure functions for **x** in another part of the flow, it will differ from the local mean velocity, being in general intermediate between the local velocity and the convection velocity of the centres. Further, the maximum value of the structure function taken over a plane through **x** and parallel to the preferred plane can be increased if the point of delayed measurement at **x** + **s** is moved towards the preferred plane where it receives a larger contribution from each eddy. The apparent convection velocity defined by the condition of maximum structure function is then directed towards the preferred plane although the eddies move parallel to it. Usually it is only the larger eddies that show a preference for a particular location and show convection velocities different from the local mean velocity either in magnitude or direction.

For some purposes, it is better to represent the contribution to the flow from eddies of a particular size by a group of Fourier components which may be isolated from an electrical signal by a band-pass filter. Several forms have been used, in particular the space–time spectrum function,

$$\phi_{11}(k, \omega) = \frac{1}{(2\pi)^2} \iint_{-\infty}^{\infty} R_{11}(r, 0, 0, \tau)\, e^{-i(kr + \omega\tau)}\, dr\, d\tau, \quad (1.7.8)$$

where the wave number k refers to displacement in the direction of mean flow. The identification of eddies of size k^{-1} with Fourier components of wave number k is justified by the observation that eddies of size much larger than k^{-1} make negligible contributions while the contributions from smaller eddies are widely dispersed in wave number and are less in total energy. Then the evolution in time of an eddy is described by the spectrum function for a range of k covering perhaps an octave of wave number. The usual appearance of the spectrum function is indicated in fig. 1.9 by lines of constant spectral intensity on the $k\omega$ plane. Typically, large intensities are concentrated near a line defined by the condition,

$$\frac{\partial\phi_{11}(k, \omega)}{\partial\omega} = 0, \quad (1.7.9)$$

and the slope of the line, $-d\omega_m/dk$, is the convection velocity for eddies of size k^{-1}. The variance of spectral intensity about the central line is caused both by the variability of convection velocity and by real decay of individual eddies, but the separation of the two effects is not as clear as with the structure function. The value of the spectral representation lies in the simple and direct presentation of the magnitudes of convection velocity for eddies of the whole range of eddy size.

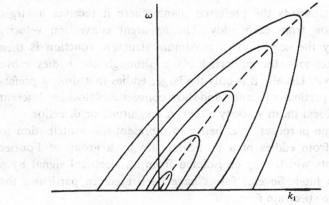

Fig. 1.9. Contours of equal spectrum intensity in (k_1, ω) plane.

Other representations of the information contained in the space–time correlation function are possible and have been used. Some of these are discussed by Wills (1964).

1.8　Homogeneity and symmetry of turbulent flows

Study of turbulent flows is much easier if they possess properties of symmetry and homogeneity that reduce the complexity of their statistical description. The symmetry and homogeneity may be properties of the flow boundaries and the forces that drive the flow, or it may arise from the tendency of turbulent flows to 'forget' details of their initiation and to assume as homogeneous and symmetrical conformation as is possible. For example, a jet from a round nozzle is hardly distinguishable from one of similar momentum flux issuing from a nozzle of irregular shape except close to the plane of exit. Although the boundary conditions of practical

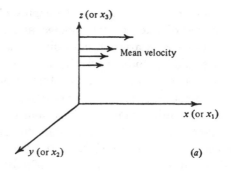

(a)

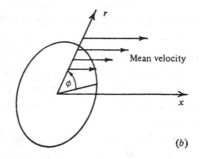

(b)

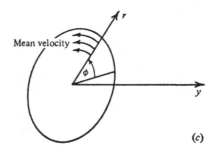

(c)

Fig. 1.10. Co-ordinate systems for (a) Two-dimensional mean flow. (b) Axisymmetric mean flow. (c) Flow with circular or helical streamlines (between rotating cylinders.)

flows may have no particular symmetry, the majority of them resemble ideal flows which are homogeneous and symmetrical in some respects. The ideal flows are best described in co-ordinate systems so chosen that the boundary conditions are easily stated and any symmetry or homogeneity can be expressed as invariance with respect to co-ordinates or interchange and reversal of co-ordinates. The system of Cartesian co-ordinates used to describe flows with plane symmetry and the cylindrical polar co-ordinates for axisymmetric flows is shown in fig. 1.10. Most turbulent flows are strongly inhomogeneous with respect to variation of one co-ordinate, z in plane flow and r in axisymmetric flow, and the velocity component in this direction is w. Many are unidirectional (or nearly so over the significant part of the flow) flows and the co-ordinate axis Ox is in the direction of flow and is the axis of axisymmetric flows. u is the corresponding velocity component. The remaining co-ordinate, y for plane flow and ϕ for axisymmetric flow, defines a direction in which the flow is homogeneous and the corresponding velocity component is v. Flows with strongly curved streamlines, e.g. the flow between concentric rotating cylinders without axial pressure gradient, have their velocity in the direction of variation of ϕ, and it is logical to use u for the velocity component in this direction and to use v for the axial component of velocity and y for the axial co-ordinate.

The more important flows may be classified by their homogeneity and symmetry, and the table shows the principle classes with the appropriate co-ordinate systems and the more important members.

Classification of turbulent flows by homogeneity and symmetry

Description	Members of class	System of co-ordinates	Homogeneous in	Mean flow
Homogeneous	Isotropic axisymmetric turbulence	Cartesian	x, y, z	x direction
Grid turbulence	Flow behind a uniform grid in a uniform stream	Cartesian	y, z	x direction
Unidirectional flow	Flow between parallel planes, plane Couette, pressure flows	Cartesian	x, y	x direction
	Flow in a circular pipe or between concentric cylinders	Cylindrical polar (b)	x, ϕ	x direction
Rotating flow	Flow between rotating cylinders without pressure gradient	Cylindrical polar (c)	ϕ, y	ϕ direction
Skewed flow	Combined pressure and Couette flow	Cartesian	x, y	in xy plane
Two-dimensional or plane flows	Plane jets and wakes, boundary layers, mixing layers	Cartesian	y	in xy plane but nearly in x-direction
Skewed flow	Boundary layer on yawed wing	Cartesian	y	nearly in xy plane
Axisymmetric developing flows	Circular jets and wakes, flow in conical diffusers	Cylindrical polar (b)	ϕ	In axial plane but nearly in x direction
Swirling flow	Swirling jets and wakes	Cylindrical polar (b)	ϕ	–
Secondary flow	Secondary flows non-circular pipes	Cartesian	x	Nearly in x direction

CHAPTER 2

THE EQUATIONS OF MOTION FOR TURBULENT FLOW

2.1 Assumption of a continuous fluid

It is usual to assume that the turbulent motion of gases and liquids can be described by the equations of motion for a continuous fluid without a molecular structure. The assumption has been queried but no inconsistencies have been found so far, and, in gases, whose structure is well understood, it may be shown that departures can occur only for scales of motion and fluid velocities outside normal experience. The essence of the continuum approximation is that the flow velocities and other continuum properties can be defined as averages over regions of space and intervals of time that are large compared with the scales of the molecular motion and small compared with the scales of the continuum flow. Separate molecular realisations of a particular flow deviate from the development predicted by the continuum equations in a way that may be represented by 'random noise' terms describing molecular fluctuations of the continuum averages. If a continuum flow is essentially uniform over lengths less than l_0 and times less than t_0, the averages may be taken over cubes of side l_0 and time intervals of t_0. Then the density at a 'point' will have a standard deviation from its expected value ρ of order

$$\rho' = \rho(nl_0{}^3)^{-1/2}(c_1{}^2 t_0/v)^{-1/2}, \qquad (2.1.1)$$

where n is the number-density of the molecules, c_1 is the root-mean-square of one component of the molecular velocity, and v is the kinematic viscosity. Similarly, a component of fluid momentum has a standard deviation from its expected value of about

$$(\rho u)' = \rho c_1 (nl_0{}^3)^{-1/2}(c_1{}^2 t_0/v)^{-1/2} \qquad (2.1.2)$$

Finally, the thermodynamic equation of state can be used to relate density, temperature and pressure only if the number of molecules in the volume $l_0{}^3$ is sufficient to define a molecular distribution function and if little change takes place in a time comparable with the relaxation time v/c_1^2. The two conditions are

$$nl_0^3 \gg 1, \qquad c_1^2 t_0/v \gg 1. \tag{2.1.3}$$

If the largest eddies of a turbulent flow have a characteristic size L and a characteristic velocity V, it is known that the smallest scales of motion are of size $(v^3 L/V^3)^{1/4} = l_0$, of duration $(vL/V^3)^{1/2} = t_0$, and with characteristic velocity $(vV^3/L)^{1/4}$. Substitution in (2.1.1, 2.1.2, 2.1.3) shows that deviations from the continuum values due to molecular fluctuations are small if both

$$\left.\begin{array}{l} nl_0^3 \approx nL^3 \left(\dfrac{VL}{v}\right)^{-9/4} \gg 1 \\[3mm] \text{and} \qquad \dfrac{c_1^2 t_0}{v} \approx \dfrac{c_1^2}{V^2}\left(\dfrac{VL}{v}\right)^{1/2} \gg 1. \end{array}\right\} \tag{2.1.4}$$

In the flow of air along a pipe of 1 cm radius with a central velocity of 10^4 cm s^{-1}, the velocity and size of the smallest eddies near the wall are nearly 200 cm s^{-1} and 5×10^{-3} cm. In standard conditions $nl_0^3 \approx 3 \times 10^{12}$ and $c_1^2 t_0/v \approx 2 \times 10^5$, both large numbers. Only when the velocities involved exceed greatly the molecular velocities will the continuum approximation fail to describe turbulent motion.

2.2 The equations of fluid motion

The equations describing the motion of a continuous fluid are derived from the conservation laws and the fluid properties, in the form of the equation of state and the relations between stress and rate of strain and between conducted heat-flux and temperature gradient. The vector flux of mass at any point in the fluid is ρu_i, and conservation of mass is expressed by equating the divergence of the mass flux to the rate of decrease of local density, i.e. by

$$\left.\begin{array}{l} \dfrac{\partial}{\partial x_i}(\rho u_i) + \dfrac{\partial \rho}{\partial t} = 0, \\[3mm] \text{or by} \qquad \dfrac{\partial u_i}{\partial x_i} + \dfrac{1}{\rho}\dfrac{D\rho}{Dt} = 0, \end{array}\right\} \tag{2.2.1\dagger}$$

† The usual summation convention is used, that a repeated free suffix implies summation over the three possible values, e.g.

$$\frac{\partial u_i}{\partial x_i} = \frac{\partial u_1}{\partial x_1} + \frac{\partial u_2}{\partial x_2} + \frac{\partial u_3}{\partial x_3}.$$

where

$$\frac{D}{Dt} \equiv \frac{\partial}{\partial t} + u_l \frac{\partial}{\partial x_l}$$

measures the rate of change following a fluid particle, i.e. a point that moves always with the continuum velocity at its current position.

The tensor $\rho u_i u_j$ is the flux of fluid momentum in the Ox_i direction across a surface element normal to Ox_j, and conservation of momentum is expressed by equating the difference between the rate of increase of momentum and the divergence of the flux to the rate of gain from external forces and by molecular forces and migration across the boundaries of the control volume. For external forces described by a potential ϕ, the condition for conservation of momentum is

$$\frac{\partial(\rho u_i)}{\partial t} + \frac{\partial}{\partial x_l}(\rho u_i u_l) = -\frac{\partial p}{\partial x_i} + \frac{\partial p_{il}}{\partial x_l} - \rho \frac{\partial \phi}{\partial x_i}, \qquad (2.2.2)$$

where p is the thermodynamic pressure, and p_{ij} is the stress tensor arising from departure from thermodynamic equilibrium consequent on straining of fluid elements. Using the condition for conservation of mass (2.2.1),

$$\rho \frac{Du_i}{Dt} \equiv \rho\left(\frac{\partial u_i}{\partial t} + u_l \frac{\partial u_i}{\partial x_l}\right) = -\frac{\partial p}{\partial x_i} + \frac{\partial p_{il}}{\partial x_l} - \rho \frac{\partial \phi}{\partial x_i}, \qquad (2.2.3)$$

also obtainable directly by considering the acceleration of a fluid element. From (2.2.3), an equation for the kinetic energy of the streaming motion can be obtained,

$$\rho \frac{D}{Dt}(\tfrac{1}{2}u_i^2) = -u_i\frac{\partial p}{\partial x_i} + u_i\frac{\partial p_{il}}{\partial x_l} - \rho u_i \frac{\partial \phi}{\partial x_i}. \qquad (2.2.4)$$

The remaining conservation law is that of energy. The energy per unit mass of the fluid is $E + \tfrac{1}{2}u_i^2 + \phi$, the sum of the internal energy E, the kinetic energy of streaming and the potential energy. The total energy flux is the sum of the convected flux of energy, $\rho u_i(E + \tfrac{1}{2}u_i^2 + \phi)$, the flux by working of the stresses on the fluid velocities $u_j(\delta_{ij}p - p_{ij})$, and the conducted flux of thermal energy, $-k\, \partial T/\partial x_i$, where T is the local temperature and k is the thermal conductivity. Conservation of energy is expressed by

$$\frac{\partial}{\partial t}(\rho(E+\tfrac{1}{2}u_i{}^2+\phi))+\frac{\partial}{\partial x_i}[\rho u_i(E+\tfrac{1}{2}u_i{}^2+\phi)]$$
$$=-\frac{\partial(pu_i)}{\partial x_i}+\frac{\partial(p_{il}u_l)}{\partial x_i}+\frac{\partial}{\partial x_i}\left(k\frac{\partial T}{\partial x_i}\right). \tag{2.2.5}$$

By using equations (2.2.1) and (2.2.4), equations for the internal energy E may be obtained,

$$\rho\frac{DE}{Dt}=-p\frac{\partial u_l}{\partial x_l}+p_{il}\frac{\partial u_i}{\partial x_l}+\frac{\partial}{\partial x_i}\left(k\frac{\partial T}{\partial x_l}\right) \tag{2.2.6}$$

and for the total heat (enthalpy), $H = E + p/\rho$,

$$\rho\frac{DH}{Dt}=\frac{Dp}{Dt}+p_{il}\frac{\partial u_i}{\partial x_l}+\frac{\partial}{\partial x_i}\left(k\frac{\partial T}{\partial x_l}\right). \tag{2.2.7}$$

The interpretation is simply that changes of internal energy are composed of work done by expansion against the pressure, heat generated by work against the fluid stresses p_{ij}, and the net gain of heat by conduction. The heat generated by working against fluid stresses is denoted by

$$\rho\varepsilon = p_{ij}\,\partial u_i/\partial x_j. \tag{2.2.8}$$

For the present purposes, the equation of state of the fluid is assumed to be the ideal gas equation,

$$p = RT\rho, \tag{2.2.9}$$

where, if the fluid is not a perfect gas, the zeros of pressure and temperature are chosen so that the average value of the pressure is the isothermal bulk modulus and the average 'absolute temperature' is the reciprocal of the coefficient of thermal expansion at constant pressure. Nearly all the following refers to fluids with Newtonian viscosity, i.e. the fluid stresses are linearly related to the rate of distortion by

$$p_{ij}=\mu\left(\frac{\partial u_i}{\partial x_j}+\frac{\partial u_j}{\partial x_i}\right)-\lambda\delta_{ij}\frac{\partial u_k}{\partial x_k}, \tag{2.2.10}$$

where μ and λ are coefficients of viscosity. The second law of thermodynamics requires that the rate of conversion of mechanical energy to heat by viscous action should always be positive, and so, since

$$\rho\varepsilon=p_{ij}\frac{\partial u_i}{\partial x_j}=\tfrac{1}{2}\mu\left(\frac{\partial u_i}{\partial x_j}+\frac{\partial u_j}{\partial x_i}-\tfrac{2}{3}\delta_{ij}\frac{\partial u_k}{\partial x_k}\right)^2+(\tfrac{2}{3}\mu-\lambda)\left(\frac{\partial u_k}{\partial x_k}\right)^2, \tag{2.2.11}$$

μ must be positive and λ must be less than $\tfrac{2}{3}\mu$. For a perfect gas, kinetic theory predicts that $\lambda = \frac{2}{3}\mu$, indicating that uniform dilatation produces no viscous stress. If the viscosity is Newtonian, the viscous term in the equation of motion (2.2.2, 2.2.3) is

$$\frac{\partial p_{ij}}{\partial x_j} = \mu \frac{\partial^2 u_i}{\partial x_j \partial x_j} - (\mu - \lambda)\frac{\partial}{\partial x_i}\left(\frac{\partial u_k}{\partial x_k}\right). \tag{2.2.12}$$

2.3 Approximate forms of the equations of motion

In most turbulent flows, the variations of velocity are small compared with the speed of sound and variations of density are small compared with the average density. Then the variations of density may be ignored in so far as they affect the inertia of the fluid or its heat capacity, and the equations of motion take a simpler form. The mass density in the equations for the fluid momentum and internal energy is replaced by ρ_a, an average density over the horizontal plane of constant gravitational potential ϕ through the point considered. Similar average values for the pressure and temperature are defined by the hydrostatic equation,

$$\frac{\partial p_a}{\partial x_i} = -\rho_a \frac{\partial \phi}{\partial x_i} \tag{2.3.1}$$

and by the equation of state,

$$p_a = R\rho_a T_a \tag{2.3.2}$$

if the variations of pressure and density are small compared with the average values at the particular level. These average values are independent of time in ordinary circumstances. Using the average values for pressure and density related by (2.3.1), the momentum equation (2.2.3) becomes

$$\frac{\partial u_i}{\partial t} + u_l \frac{\partial u_i}{\partial x_l} = -\frac{1}{\rho}\frac{\partial(p - p_a)}{\partial x_i} - \frac{\rho - \rho_a}{\rho}\frac{\partial \phi}{\partial x_i} + \frac{1}{\rho}\frac{\partial p_{il}}{\partial x_l} \tag{2.3.3}$$

without approximation. The buoyancy term,

$$-\frac{\rho - \rho_a}{\rho}\frac{\partial \phi}{\partial x_i},$$

involves the variations of density but, by considering the orders of magnitude of the terms, it can be shown that either the density variations are too small to cause appreciable buoyancy forces or the fractional variations of pressure are small compared with those of density and temperature. Then,

$$\rho T = \rho_a T_a, \quad \text{i.e.} \quad \frac{\rho - \rho_a}{\rho} = -\frac{T - T_a}{T_a} \tag{2.3.4}$$

and, substituting ρ_a for ρ in the other terms, the momentum equation becomes

$$\frac{\partial u_i}{\partial t} + u_l \frac{\partial u_i}{\partial x_l} = -\frac{\partial}{\partial x_i}\left(\frac{p - p_a}{\rho_a}\right) + \frac{T - T_a}{T_a}\frac{\partial \phi}{\partial x_i} + \frac{\partial}{\partial x_i}\left(\frac{p_{il}}{\rho_a}\right) \tag{2.3.5}$$

if the vertical variation of ρ_a is small over the characteristic scale-length of the flow.

Over small ranges of temperature, total heat is related to temperature by

$$H = c_p T + \text{constant},$$

and, to the approximation of constant density, equation (2.2.7) becomes

$$\frac{DT}{Dt} = \frac{1}{\rho_a c_p}\left[\frac{D(p - p_a)}{Dt} + \frac{Dp_a}{Dt}\right] + \frac{p_{il}}{\rho_a c_p}\frac{\partial u_i}{\partial x_l} + \frac{k}{\rho c_p}\frac{\partial^2 T}{\partial x_c^2}. \tag{2.3.6}$$

Since p_a and ϕ are independent of time,

$$\frac{Dp_a}{Dt} = u_l \frac{\partial p_a}{\partial x_l} = -u_l \rho_a \frac{\partial \phi}{\partial x_l} = -\rho_a \frac{D\phi}{Dt}$$

and the equation for the total heat turns into an equation for the potential temperature, $T + \phi/c_p$,

$$\frac{D}{Dt}(T + \phi/c_p) = k \frac{\partial^2}{\partial x_l^2}(T + \phi/c_p) \tag{2.3.7}$$

omitting terms

$$\frac{1}{\rho_a c_p}\frac{D(p - p_a)}{Dt} \quad \text{and} \quad \frac{p_{il}}{\rho_a c_p}\frac{\partial u_i}{\partial x_l} = \frac{\varepsilon}{c_p}.$$

Here $\kappa = k/(\rho_a c_p)$ is the thermometric conductivity.

To the approximation of constant density, mass conservation is expressed by the continuity equation,

$$\frac{\partial u_l}{\partial x_l} = 0 \tag{2.3.8}$$

omitting terms $(D/Dt)(\rho - \rho_a)$ and Dp_a/Dt. For a Newtonian fluid, the equation of motion becomes

$$\frac{\partial u_i}{\partial t} + u_l \frac{\partial u_i}{\partial x_l} = -\frac{\partial p}{\partial x_l} + v \frac{\partial^2 u_i}{\partial x_l^2} + \frac{T - T_a}{T_a}\frac{\partial \phi}{\partial x_i} \tag{2.3.9}$$

with a change of notation so that p now denotes $(p - p_a)/\rho_a$, the 'kinematic' pressure variation. In all that follows, kinematic pressures and stresses are used and mechanical values can be recovered by multiplying by the fluid density.

The three equations (2.3.7–9) describe most kinds of turbulent motion. If the flow has characteristic scales for variation of velocity, position and temperature, u_0, L and θ_0, the ratio of the terms omitted in equation (2.3.6) to those retained is of order u_0^2/a^2 (where a is the speed of sound) or θ_0/T_a. The terms omitted from the continuity equation are of order u_0^2/a^2, θ_0/T_a or L/L_s, where $L_s = RT/g$ is the scale-height of the fluid. The terms omitted from the temperature equation (2.3.7) are smaller in ratios of order u_0^2/a^2, θ_0/T_a, L/L_s or $u_0^2/(c_p\theta_0)$. The conditions that the approximate equations should describe the *motion* accurately are:

(1) that the square of the flow Mach number, u_0^2/a^2, is small,

(2) that the temperature loading, θ_0/T_a, is small,

(3) that the scale of the flow is small compared with the scale-height of the fluid.

The temperature field described by the equations refers only to variations induced by external heat sources or by interaction with a non-uniform distribution of the ambient potential temperature, $T_a + \phi c_p = T_p$. The equations do not describe the temperature fluctuations induced by pressure changes or by viscous dissipation of mechanical energy. The ratio of these fluctuations to the ambient temperature is of order u_0^2/a^2 for turbulent flows.

2.4   Mean value equations for momentum, energy and heat

From now on, the flow variables are expressed as the sum of a mean value and the fluctuation from the mean value, i.e. the velocity is $U_i + u_i$ where U_i is the mean velocity. By definition the mean value of the fluctuation is zero. Then taking the mean value of the continuity equation (2.3.8),

$$\frac{\overline{\partial(U_l + u_l)}}{\partial x_l} = \frac{\partial U_l}{\partial x_l} + \frac{\overline{\partial u_l}}{\partial x_l} = \frac{\partial U_l}{\partial x_l} = 0 \qquad (2.4.1)$$

and the velocity fluctuations satisfy a continuity equation,

$$\frac{\partial u_l}{\partial x_l} = 0. \tag{2.4.2}$$

Taking the mean value of the momentum equation (2.3.9) and using the continuity condition for the fluctuations, we obtain the equation for the mean velocity in the standard form,

$$\frac{\partial U_i}{\partial t} + U_l \frac{\partial U_i}{\partial x_l} + \frac{\overline{\partial u_i u_l}}{\partial x_l} = -\frac{\partial P}{\partial x_i} + \frac{g_i(T - T_a)}{T_a} + v \frac{\partial^2 U_i}{\partial x_l^2} \tag{2.4.3}$$

where $P + p$ is the pressure difference from the ambient pressure, $T + \theta$ is the potential temperature, and T_a the ambient potential temperature. The equation for the mean velocity may be put into the form,

$$\frac{\partial U_i}{\partial t} + U_l \frac{\partial U_i}{\partial x_l} = \frac{g_i(T - T_a)}{T_a} + \frac{\partial}{\partial x_l}\left[-\delta_{il}P + v\left(\frac{\partial u_i}{\partial x_l} + \frac{\partial u_l}{\partial x_i}\right) - \overline{u_i u_l} \right] \tag{2.4.4}$$

showing that the mean flow is accelerated by forces arising from the mean buoyancy, the gradient of the mean pressure, the viscous stresses developed by the mean flow alone, and by a virtual force which is the gradient of the Reynolds stress, $-\overline{u_i u_j}$. The additional force describes the effect of the turbulent fluctuations on the mean flow and is simply interpreted as a consequence of the mean rates of transport of momentum by the turbulent movements of the fluid. If it could be determined, the mean flow would be known and the first problem of turbulent flow, the nature of the mean motion, could be solved.

The kinetic energy of the velocity fluctuations is a quantity of clear physical significance, and an equation for it can be obtained by multiplying the equation for the total velocity,

$$\frac{\partial(U_i + u_i)}{\partial t} + (U_l + u_l)\frac{\partial(U_i + u_i)}{\partial x_l}$$

$$= -\frac{\partial(P + p)}{\partial x_i} + \frac{g_i(T - T_a + \theta)}{T_a} + v\frac{\partial^2(U_i + u_i)}{\partial x_l^2}$$

by the velocity fluctuation and taking the mean value. It is

$$\frac{\partial}{\partial t}(\tfrac{1}{2}\overline{q^2}) + U_l \frac{\partial(\tfrac{1}{2}\overline{q^2})}{\partial x_l} + \frac{\partial}{\partial x_l}(\overline{pu_l} + \tfrac{1}{2}\overline{q^2 u_l}) + \overline{u_i u_l}\frac{\partial U_i}{\partial x_l}$$

$$= \frac{g_i}{T_a}\overline{\theta u_i} + v\overline{u_i \frac{\partial^2 u_i}{\partial x_l^2}} \tag{2.4.5}$$

where $q^2 = u_i u_i$. The equation amounts to a statement of energy conservation for the velocity fluctuations and the terms have simple interpretations. They are:

(1) rate of increase of fluctuation energy,
(2) gain through advection of energy by the mean flow,
(3) production of turbulent energy by working of the mean flow on the turbulent Reynolds stresses,
(4) transport of turbulent energy by turbulent pressure gradients and by turbulent convection,
(5) gain of energy through working of the buoyancy forces, and
(6) transformation of fluctuation energy to heat plus a smaller amount of energy diffusion by the working of viscous stress fluctuations.

An equation for the total kinetic energy of the flow is also useful. It is

$$\frac{\partial}{\partial t}[\tfrac{1}{2}(\overline{q^2}+U_i{}^2)] + U_i\frac{\partial}{\partial x_l}[\tfrac{1}{2}(\overline{q^2}+U_i{}^2)]$$
$$+ \frac{\partial}{\partial x_l}[\tfrac{1}{2}\overline{q^2 u_l}+\overline{u_i u_l}U_i+PU_l+\overline{pu_l}]$$
$$= \nu\left[U_i\frac{\partial^2 U_i}{\partial x_l{}^2}+\overline{u_i\frac{\partial^2 u_i}{\partial x_l{}^2}}\right]+\frac{g_i}{T_a}[(T-T_a)U_i+\overline{\theta u_i}] \qquad (2.4.6)$$

and equates the rate of change to the divergence of an energy flux, with contributions from the buoyancy forces and the viscous dissipation of mechanical energy.

The equation for the potential temperature, $T + \theta$, leads to the equation for the mean temperature,

$$\frac{\partial T}{\partial t}+U_l\frac{\partial T}{\partial x_l}+\frac{\partial \overline{u_l\theta}}{\partial x_l} = k\frac{\partial^2 T}{\partial x_l{}^2}. \qquad (2.4.7)$$

Compared with the equation for a flow without fluctuations, the only difference is the presence of the term, $(\partial/\partial x_l)(\overline{u_l\theta})$, representing the effect of turbulent transport of heat and analogous to the Reynolds stress term in the equation for the mean velocity. Again a knowledge of the term would permit calculation of the distribution of mean temperature.

For the temperature field, the mean square of the temperature fluctuation plays a part similar to that of the fluctuation energy. Physically, it is closely related to the contribution of the fluctuations to the mean entropy. To the approximation in use, the difference of the local entropy from the entropy of fluid at pressure p_a and temperature T_a is

$$S - S_a = c_p \log_e (T + \theta)/T_a), \tag{2.4.8}$$

and the mean entropy is

$$S = S_a + c_p \log_e T/T_a - \tfrac{1}{2} c_p \overline{\theta^2}/T^2. \tag{2.4.9}$$

The quantity $\tfrac{1}{2}\overline{\theta^2}$ is proportional to the (negative) contribution to the mean entropy arising from the presence of temperature fluctuations, and the destruction of fluctuation entropy by molecular conduction of heat is analogous to the dissipation of fluctuation kinetic energy by viscous stresses. An equation for $\tfrac{1}{2}\overline{\theta^2}$ is easily obtained by multiplying the equation for the temperature by the temperature fluctuation and taking the mean value. It is

$$\frac{\partial}{\partial t}(\tfrac{1}{2}\overline{\theta^2}) + U_l \frac{\partial}{\partial x_l}(\tfrac{1}{2}\overline{\theta^2}) + \overline{u_l \theta} \frac{\partial T}{\partial x_l} + \frac{\partial}{\partial x_l}(\tfrac{1}{2}\overline{\theta^2 u_l}) = k\overline{\theta \nabla^2 \theta} \tag{2.4.10}$$

and the interpretation of the terms is similar to that of the terms in the energy equation, i.e. the term $\overline{u_l \theta}\, \partial T/\partial x_l$ represents production of 'entropy' by turbulent flux of heat along the gradient of mean temperature, the term $\tfrac{1}{2}\overline{\theta^2 u_l}$ turbulent convection of fluctuation entropy, and the term $k\overline{\theta \nabla^2 \theta}$ mostly a destruction of fluctuation entropy by heat conduction down temperature gradients.

The equation for $\tfrac{1}{2}[(T - T_a)^2 + \overline{\theta^2}]$ is sometimes useful. It is

$$\frac{\partial}{\partial t}[\tfrac{1}{2}(T - T_a)^2 + \tfrac{1}{2}\overline{\theta^2}] + U_l \frac{\partial}{\partial x_l}[\tfrac{1}{2}(T - T_a)^2 + \tfrac{1}{2}\overline{\theta^2}] + \frac{\partial}{\partial x_l}(\overline{\theta u_l}(T - T_a))$$

$$+ \frac{\partial}{\partial x_l}(\tfrac{1}{2}\overline{\theta^2 u_l}) + \frac{\partial T_a}{\partial x_l}[(T - T_a)U_l + \overline{\theta u_l}]$$

$$= k[(T - T_a)\nabla^2(T - T_a) + \overline{\theta \nabla^2 \theta}]. \tag{2.4.11}$$

2.5 Energy dissipation by viscosity

The equation for the kinetic energy of the fluctuations (2.4.5) contains the term $\overline{vu_i(\partial^2 u_i/\partial x_l^2)}$ representing the effect of viscous forces on the energy. It is related closely to the local rate of conversion of mechanical energy to heat, in kinematic form,

$$E + \varepsilon = \tfrac{1}{2}\nu\left[\left(\frac{\partial U_i}{\partial x_j} + \frac{\partial U_j}{\partial x_i}\right)^2 + \overline{\left(\frac{\partial u_i}{\partial x_j} + \frac{\partial u_j}{\partial x_i}\right)^2}\right], \qquad (2.5.1)$$

where

$$E = \tfrac{1}{2}\nu\left(\frac{\partial U_i}{\partial x_j} + \frac{\partial U_j}{\partial x_i}\right)^2 \qquad (2.5.2)$$

is the part of the dissipation of mechanical energy due to the mean velocity gradients, and

$$\varepsilon = \tfrac{1}{2}\nu\overline{\left(\frac{\partial u_i}{\partial x_j} + \frac{\partial u_j}{\partial x_i}\right)^2} \qquad (2.5.3)$$

is the mean turbulent energy dissipation. The viscous term in the energy equation can be expressed as

$$\left.\begin{aligned}
\overline{\nu u_i \frac{\partial^2 u_i}{\partial x_j^2}} &= \nu \frac{\partial^2}{\partial x_j^2}(\tfrac{1}{2}\overline{u_i^2}) - \nu\overline{\left(\frac{\partial u_i}{\partial x_j}\right)^2} \\
&= \nu\left[\frac{\partial^2}{\partial x_j^2}(\tfrac{1}{2}\overline{q^2}) + \frac{\partial^2 \overline{u_i u_j}}{\partial x_i \partial x_j}\right] - \varepsilon,
\end{aligned}\right\} \qquad (2.5.4)$$

after using the continuity equation to show that

$$\overline{\frac{\partial u_i}{\partial x_j}\frac{\partial u_j}{\partial x_i}} = \frac{\partial^2 \overline{u_i u_j}}{\partial x_i \partial x_j}. \qquad (2.5.5)$$

The first term of (2.5.4) involves only second derivatives of mean values of velocity products, and has the nature of a generalised diffusion term. If the flow field is bounded or homogeneous, the volume integral of the term is zero and the term contributes nothing to actual dissipation of energy. Further, if u_0 is a typical velocity fluctuation of the flow and the width of the flow is L, the magnitude of the first term is $\nu u_0^2/L^2$ while the actual energy dissipation is of order u_0^3/L. Provided that $u_0 L/\nu$, the Reynolds number of the turbulent flow is large, the turbulent energy dissipation is

$$\varepsilon = -\overline{\nu u_i \frac{\partial^2 u_i}{\partial x_j^2}} = \tfrac{1}{2}\nu\overline{\left(\frac{\partial u_i}{\partial x_j} + \frac{\partial u_j}{\partial x_i}\right)^2}$$

$$= \nu\overline{\left(\frac{\partial u_i}{\partial x_j}\right)^2} = \nu\overline{\omega_i^2}. \qquad (2.5.6)$$

2.6 Conductive dissipation of temperature fluctuations

The equation for the intensity of the temperature fluctuations contains a conductive term which is almost proportional to the rate of entropy production by heat conduction down gradients of the temperature fluctuation. It may be written in the form,

$$k\overline{\theta\nabla^2\theta} = k\nabla^2(\tfrac{1}{2}\overline{\theta^2}) - k\overline{\left(\frac{\partial\theta}{\partial x_i}\right)^2}, \qquad (2.6.1)$$

where the first part describes a viscous diffusion of fluctuation intensity by conduction, and the second measures the rate of entropy production. The quantity,

$$\varepsilon_\theta = k\overline{\left(\frac{\partial\theta}{\partial x_i}\right)^2} \qquad (2.6.2)$$

is the dissipation rate for temperature fluctuations and, in the discussion of the fluctuations, plays the same role as ε does for velocity fluctuations. The 'diffusion' part of $k\overline{\theta\nabla^2\theta}$ is negligible if the Péclét number of the turbulence is large, and then

$$\varepsilon_\theta = k\overline{(\partial\theta/\partial x_i)^2}. \qquad (2.6.3)$$

2.7 The relation between the pressure and velocity fields

Taking the divergence of the equation of motion (2.4.3) leads to

$$\frac{\partial(U_i+u_i)}{\partial x_j}\frac{\partial(U_j+u_j)}{\partial x_i} = -\frac{\partial^2(P+p)}{\partial x_i\,\partial x_i} + \frac{g_i}{T}\frac{\partial(T+\theta)}{\partial x_i} \qquad (2.7.1)$$

after using the continuity condition (2.3.8). The pressure is so determined by the velocity and temperature fields, and a formal solution of (2.7.1) is

$$P+p = \frac{1}{4\pi}\int\left[\frac{\partial^2}{\partial x_i\,\partial x_j}\{(U_i+u_i)(U_j+u_j)\}\right.$$
$$\left. -\frac{g_i}{T}\frac{\partial(T+\theta)}{\partial x_i}\right]\frac{dV(\mathbf{x})}{|\mathbf{x}'-\mathbf{x}|}, \qquad (2.7.2)$$

where $\mathbf{x}'$ is the position at which the pressure is to be found. It should be noticed that pressure is not a local quantity but depends on an integral over the entire field of velocity and temperature. The pressure fluctuation, p, is given by

$$p(\mathbf{x}') = \frac{1}{4\pi} \int \left[2 \frac{\partial U_i}{\partial x_j} \frac{\partial u_j}{\partial x_i} + \frac{\partial^2}{\partial x_i \partial x_j}(u_i u_j - \overline{u_i u_j}) - \frac{g_i}{T} \frac{\partial \theta}{\partial x_i} \right] \frac{dV(\mathbf{x})}{|\mathbf{x}' - \mathbf{x}|} \quad (2.7.3)$$

as the sum of contributions from three effects:

(1) the term in the bracket, $2(\partial U_i/\partial x_j)(\partial u_j/\partial x_i)$, represents an interaction between the mean velocity gradients and turbulent velocity gradients,

(2) the term, $(\partial^2/\partial x_i \partial x_j)(u_i u_j - \overline{u_i u_j})$, represents the effect of fluctuations of Reynolds stress about the mean value and, in shear flows, it is usually much smaller than the first term,

(3) the term, $-(g_i/T)(\partial \theta/\partial x_i)$, represents the induction of pressures in response to the buoyancy forces.

CHAPTER 3

HOMOGENEOUS TURBULENT FLOWS

3.1 Introduction

Making a distinction between the mean values and the fluctuations of velocity and temperature carries with it an implication that there is a physical difference between the parts of the kinetic energy and entropy densities that are associated respectively with the mean fields and with the fluctuation fields, essentially because transfer from the mean value forms to the fluctuation forms is normally irreversible and is the first stage in a cascade process ending in transformation or destruction by molecular transport processes. Both the broad features and the details of the transfer processes are of vital importance for the full understanding of turbulent flow, and the necessary information is obtained most easily from study of homogeneous turbulent flows. Most turbulent flows are inhomogeneous but the more important features of the energy transfer process are the same whether the flow is homogeneous or not, and the comparative simplicity of the statistical description of homogeneous flows makes possible experimental and theoretical studies in a detail that is not feasible for inhomogeneous flow. We shall examine some flows which are homogeneous, or nearly so, in the turbulent fluctuations and in the gradients of mean velocity, that is to say, the mean values of functions of the fluctuations and of the mean velocity gradients are independent of position in the flow. In order of increasing complexity, they are:

(1) nearly homogeneous and isotropic turbulence with uniform mean velocity;

(2) nearly homogeneous but strongly anisotropic turbulence with uniform mean velocity;

(3) nearly homogeneous turbulence with uniform gradient of mean velocity.

3.2 Eddy interactions in homogeneous turbulence

For homogeneous turbulence with uniform gradients of mean velocity, equation (2.4.5) for the turbulent kinetic energy, $\frac{1}{2}\overline{q^2}$, becomes

$$\frac{\partial}{\partial t}(\tfrac{1}{2}\overline{q^2}) = -\overline{u_i u_j}\frac{\partial U_i}{\partial x_j} - v\overline{\left(\frac{\partial u_i}{\partial x_j}\right)^2}, \qquad (3.2.1)$$

showing that energy is generated by working of the mean flow against the Reynolds stresses and is dissipated as heat by working of the turbulent velocity gradients against the viscous stresses. Since the Reynolds stress tensor specifies the turbulent energy, the eddies containing most of the energy also contribute most to the Reynolds stresses and presumably receive most of the energy that is transferred from the mean flow. On the other hand, the rate of energy dissipation is proportional to the mean square of the velocity gradients which is determined by eddies much smaller than those containing most of the energy. So a simple consideration of the energy budget raises two problems – the nature of the energy transfer from the mean flow to the turbulent eddies and the nature of the transfer from the large energy-containing eddies to the much smaller dissipating eddies.

Some understanding of the problems, though unfortunately not their solution, comes from examining the equations for the rates of change of Fourier components of the velocity field. In chapter 1, it was shown that an eddy of limited spatial extent can be described by a group of Fourier components with wave numbers of comparable magnitudes and that, with some caution, conclusions about Fourier components of wave number k can be applied to eddies of 'size' k^{-1}. The velocity fluctuations within a large volume V can be expressed as the sum of Fourier components,

$$u_i(\mathbf{x}) = \sum a_i(\mathbf{k}) \exp(i\,\mathbf{k}\,.\,\mathbf{x}), \qquad (3.2.2)$$

where the allowed values of $\mathbf{k}$ satisfy cyclic boundary conditions and are distributed uniformly in wave number space with number density $(2\pi)^{-3}V$. The condition of incompressibility, $\partial u_i/\partial x_i = 0$, requires that

$$k_i a_i(\mathbf{k}) = 0 \qquad (3.2.3)$$

and, since $u_i(\mathbf{x})$ is real,

$$a_i(\mathbf{k}) = a_i^*(-\mathbf{k}). \qquad (3.2.4)$$

In homogeneous turbulence, the equation for the velocity fluctuation is

$$\frac{\partial u_i}{\partial t} + U_l \frac{\partial u_i}{\partial x_l} + u_l \frac{\partial U_i}{\partial x_l} + u_l \frac{\partial u_i}{\partial x_l} = -\frac{\partial p}{\partial x_i} + \nu \frac{\partial^2 u_i}{\partial x_l^2}, \qquad (3.2.5)$$

where

$$-\frac{\partial^2 p}{\partial x_l^2} = 2 \frac{\partial U_l}{\partial x_m} \frac{\partial u_m}{\partial x_l} + \frac{\partial u_l}{\partial x_m} \frac{\partial u_m}{\partial x_l} \qquad (3.2.6)$$

(see 2.3.8, 2.7.1). It follows that the variation with time of a single Fourier component of the fluctuation field is described by an equation for its amplitude,

$$\frac{da_i(\mathbf{k})}{dt} = -\nu k^2 a_i(\mathbf{k}) - \frac{\partial U_i}{\partial x_l} a_l(\mathbf{k}) + 2\frac{k_i k_l}{k^2} \frac{\partial U_l}{\partial x_m} a_m(\mathbf{k})$$

$$+ i \sum_{\mathbf{k}' + \mathbf{k}'' = \mathbf{k}} \left(k_i \frac{k_l k_m}{k^2} - \delta_{im} k_l \right) a_l(\mathbf{k}') a_m(\mathbf{k}'') \qquad (3.2.7)\dagger$$

and by an equation for the rate of change of the wave number of the component,

$$\frac{dk_i}{dt} = -\frac{\partial U_l}{\partial x_i} k_l. \qquad (3.2.8)$$

The second equation describes the rotation and distortion of the velocity pattern by the mean velocity gradients and, since the divergence of the 'velocity' dk_i/dt in wave-number space is

$$\frac{\partial}{\partial k_i}\left(\frac{dk_i}{dt}\right) = -\frac{\partial U_l}{\partial x_l} = 0, \qquad (3.2.9)$$

the motion of the allowed wave numbers in wave-number space is solenoidal and the originally constant number-density is preserved.

The first three terms in (3.2.7) are linear in the component amplitude and describe changes in amplitude due to viscous stresses and to interaction between the turbulent motion and the mean flow. The remaining terms are quadratic in the amplitude and describe changes due to interactions between components of different wave numbers, here expressed as a sum over pairs of components whose wave numbers satisfy

$$\mathbf{k}' + \mathbf{k}'' = \mathbf{k}.$$

† δ_{im} is the substitution tensor, $= 1$ if $i = m$ and 0 if $i \neq m$.

Since each Fourier component is a plane wave pattern with motion at right angles to the wave normal, the scalar amplitude $(a_i a_i^*)^{1/2}$ and the direction and ellipticity of the polarisation determine the amplitude. Using (3.2.7) to form an equation for the square of the scalar amplitude, we find

$$\frac{d}{dt}(a_i a_i^*) = -2\nu k^2 a_i a_i^* - \frac{\partial U_i}{\partial x_j}(a_i a_j^* + a_i^* a_j)$$

$$-i\, k_l \sum [a_i(\mathbf{k}')a_i(\mathbf{k}'')a_i^*(\mathbf{k}) - a_i^*(\mathbf{k}')a_i^*(\mathbf{k}'')a_i(\mathbf{k})], \quad (3.2.10)$$

noticing that the net contributions of the terms

$$2\frac{k_i k_l}{k^2}\frac{\partial U_l}{\partial x_m} a_m(\mathbf{k}) + i \sum k_l \frac{k_l k_m}{k^2}a_i(\mathbf{k}')a_m(\mathbf{k}'')$$

in (3.2.7) vanish since $k_i a_i(\mathbf{k}) = k_i a_i^*(\mathbf{k}) = 0$. The implication is that these terms, which describe the changes in a_i caused by fluid acceleration by pressure gradients, alter the polarisation of the component without affecting its scalar amplitude. It does *not* follow that the interactions causing changes in scalar amplitude, i.e. in kinetic energy, are physically distinct and separable from those causing changes in polarisation. For example, the second term on the right of (3.2.10) may be written in the form

$$-\frac{\partial U_i}{\partial x_j}\frac{a_i a_j^* + a_i^* a_j}{a_i a_i^*}a_i a_i^*.$$

where $(a_i a_j^* + a_i^* a_j)/(a_i a_i^*)$ is a dimensionless quantity specifying the polarisation, and we see that the rate of variation of $a_i a_i^*$ depends on the scalar product of the polarisation tensor and the mean velocity gradient tensor.

The quantity, $\frac{1}{2}(a_i a_j^* + a_i^* a_j)$, averaged over all realisations of the flow, is the mean contribution of the component with instantaneous wave number $\mathbf{k}$ to the Reynolds stress $\overline{u_i u_j}$, and, since the distortion in wave-number space described by (3.2.8) preserves the original density of allowed wave numbers, the three-dimensional spectrum function is

$$\Phi_{ij}(\mathbf{k}) = \frac{V}{16\pi^3}\langle a_i a_j^* + a_i^* a_j\rangle, \quad (3.2.11)$$

where the angle brackets signify an ensemble average over all realisations. It is not difficult to derive equations for the variations of Φ_{ij} using (3.2.7), but the most interesting one is the equation for $\Phi_{ii}(\mathbf{k})$, the spectrum function for the mean square velocity fluctuation, satisfying

$$\int \Phi_{ii}(\mathbf{k})\, d\mathbf{k} = q^2.$$

It is

$$\frac{\partial}{\partial t} \Phi_{ii}(\mathbf{k}) = -2\nu k^2 \Phi_{ii}(\mathbf{k}) - 2\frac{\partial U_i}{\partial x_j} \Phi_{ij}(\mathbf{k})$$

$$+ k_l \frac{\partial U_l}{\partial x_m} \frac{\partial \Phi_{ii}(\mathbf{k})}{\partial k_m} - \frac{\partial}{\partial k_l} S_i(\mathbf{k}) \quad (3.2.12)$$

and the terms on the right represent (1) loss by direct viscous dissipation, (2) energy transfer from the mean flow to components of wave number $\mathbf{k}$, (3) redistribution of energy in wave-number space consequent on distortion of the velocity pattern, and (4) redistribution of energy by non-linear interactions between different components. To emphasise the essentially conservative nature of the energy transfer between components, the last term has been written as the divergence in wave-number space of $S_i(\mathbf{k})$, the flux vector of the total intensity $\overline{q^2}$. One interpretation of the spectrum equation (3.2.12) is that turbulent energy flows in wave-number space from the region of comparatively small wave numbers, where most of the energy is produced and resides, towards much larger wave numbers where the rate of viscous dissipation is sufficient to convert the flow to heat.

A fundamental problem of turbulent motion is the relation between the flow of energy in wave-number space and the energy distribution in that space. No satisfactory solution for the region containing most of the turbulent energy has been found, but conditions are simpler for the larger wave numbers which contain only a small part of the total energy but which include nearly all the components involved in the process of viscous conversion of energy to heat.

3.3 Experimental approximations to homogeneous turbulence

In truly homogeneous flow, all the mean values in the full equation for the turbulent energy (2.4.5) are independent of position in the flow and the equation reduces to the simple form (3.2.1). Unless production of turbulent energy by the Reynolds stresses happens to be exactly equal to the rate of dissipation, the energy varies with time and the flow is not stationary. Even if the practical problems of initiating spatially homogeneous turbulent motion could be overcome, the transience of the flow would make measurements very tedious and all experimental studies of 'homogeneous turbulence' have used flows which are stationary with respect to time but necessarily inhomogeneous in one direction. These flows are produced by placing grids of suitable forms across the entrance to ducts of constant or varying section, and the turbulent motion near any cross-section of the flow is assumed to be similar to that in a truly homogeneous flow which has existed for a time equal to the 'time-of-flight' from the grid to the particular section,

$$t_d = \int_0^{x_1} \frac{dx_1'}{U_1(x_1')}, \tag{3.3.1}$$

where U_1 is the local mean velocity. In flows with gradients of mean velocity, a clear definition of t_d is possible only if the lateral variations of U_1 are small.

The equivalence of the stationary grid flow and the theoretical homogeneous flow depends on the possibility of defining a volume that moves with the local mean velocity and that has dimensions small compared with the scale of the streamwise inhomogeneity but large compared with the size of the turbulent eddies. Then the motion within the volume interacts with gradients of mean velocity and with a turbulent motion outside that are much the same as they would be in a homogeneous, non-stationary flow after the decay time t_d determined by its present position. The condition for equivalence is that the scale of inhomogeneity, say

$$L_h = \overline{q^2} \left/ \left| \frac{\partial \overline{q^2}}{\partial x_1} \right| \right. \tag{3.3.2}$$

should be large compared with the size of the energy-containing eddies, specified by an *integral scale*

$$L_0 = (\overline{u^2})^{-1} \int_0^\infty R_{11}(r, 0, 0)\, dr. \tag{3.3.3}$$

As we shall see, the rate of turbulent energy dissipation,

$$\varepsilon = \nu \overline{\left(\frac{\partial u_i}{\partial x_j}\right)^2},$$

is approximately

$$\varepsilon = \tfrac{1}{3}(\overline{q^2})^{3/2}/L_0 \qquad (3.3.4)$$

and substitution in the energy equation for stationary grid turbulence,

$$U_1 \frac{\partial}{\partial x_1}(\tfrac{1}{2}\overline{q^2}) + \overline{u_i u_j}\frac{\partial U_i}{\partial x_j} + \frac{\partial}{\partial x_1}(\tfrac{1}{2}\overline{q^2 u_1} + \overline{pu_1})$$

$$= \nu \frac{\partial^2}{\partial x_1^2}(\tfrac{1}{2}\overline{q^2}) - \nu \overline{\left(\frac{\partial u_i}{\partial x_j}\right)^2}, \qquad (3.3.5)$$

leads to the relation

$$\frac{L_0}{L_h} = -2\frac{\overline{u_i u_j}}{\overline{q^2}}\frac{L_0}{U_1}\frac{\partial U_i}{\partial x_j} + \frac{2}{3}\frac{(\overline{q^2})^{1/2}}{U_1}$$

$$+ \text{terms of order } (\overline{q^2})^{1/2}/U_1 \text{ or less.} \qquad (3.3.6)$$

For equivalence, it is necessary that

(1) $(\overline{q^2})^{1/2} \ll U_1$; and

(2) $\dfrac{L_0}{U_1}\left|\dfrac{\partial U_i}{\partial x_j}\right| \ll 1$

i.e. both the turbulent velocity fluctuations and the variation of mean velocity over an eddy 'diameter' should be small compared with the local mean velocity.

3.4 Isotropic turbulence: general

The study of isotropic turbulence began in 1935 when G. I. Taylor defined it by the condition that all mean values of functions of the flow variables should be independent of translation, rotation and reflexion of the axes of reference. It is the simplest form of turbulence that is relevant to 'complete' turbulent flows such as jets and boundary layers, but it still presents unsolved problems in spite of intensive theoretical and experimental study. The importance of the theory of isotropic turbulence lay in the demonstration that it is possible to derive from the equations of motion and continuity relations between mean values connected with the intensity and scale of the turbulence, mean values whose measurement had become possible through the previous development of the hot-wire anemometer by H. L. Dryden and A. M. Kuethe. The presently accepted view that a satisfactory theory of turbulent flow must be based on an adequate and realistic account of the turbulent motion may be attributed equally to the development of the theory and of the hot-wire techniques. A comprehensive account of the present knowledge of isotropic turbulence would occupy considerable space and we will consider mostly features that are common to all turbulent motion. For more detail and, indeed, for an exact statement of much of the following, reference should be made to the literature (in particular, Batchelor 1953).

The usual experimental approximation to isotropic turbulence is produced by placing a uniform grid across the entrance to the working section of a wind tunnel. The grids most commonly used are of the 'biplane' type, with two layers of uniformly spaced, circular cylinders with axes in the two layers at right angles. The grid is specified by the mesh length M, the separation between the axes of adjacent cylinders, and by d the diameter of the cylinders. The ratio M/d is usually in the range 4–6. Within a few mesh lengths of the grid, the individual wakes of the cylinders merge and the flow becomes statistically homogeneous over planes at right angles to the direction of mean flow along the Ox_1 axis.†

Grid turbulence produced in this way is noticeably anisotropic, the ratio of the turbulent intensity in the stream direction to that normal to the stream direction being approximately 1.25. Comte-Bellot & Corrsin (1966) have shown that the intensities may be made equal by passing the grid turbulence through a 1.27:1 contraction, but most of the experimental work has been analysed assuming complete isotropy. The consequent errors are unlikely to affect the conclusions drawn here.

The special feature of isotropic turbulence is the simplicity of its specification. By definition, the correlation tensor, $R_{ij}(\mathbf{x}; \mathbf{r})$, has a

† If the flow resistance of the grid is too large (too small values of M/d), flow through it is not uniform and persistent lateral variations of turbulent intensity are found (Grant & Nisbet 1957, Bradshaw 1965).

form independent of the position and orientation of the axes of reference, i.e.

$$R_{ij} = r_i r_j A(r) + \delta_{ij} B(r), \tag{3.4.1}$$

where $A(r)$ and $B(r)$ are functions of the scalar separation r. The condition of continuity for R_{ij}, that

$$\frac{\partial}{\partial r_j} R_{ij}(\mathbf{x}; \mathbf{r}) = 0,$$

leads to a differential equation connecting the two functions, and the whole correlation tensor may be expressed in terms of a single scalar function of r. The usual choice of function is the longitudinal correlation function,

$$f(r) = R_{11}(r, 0, 0)/R_{11}(0, 0, 0) \tag{3.4.2}$$

and then

$$R_{ij}(\mathbf{r}) = \overline{u_1^2}\left[-\tfrac{1}{2}\frac{r_i r_j}{r}f' + \delta_{ij}(f + \tfrac{1}{2}rf') \right]. \tag{3.4.3}$$

In a similar way, the three-dimensional spectrum function can be expressed in terms of the (scalar) integrated spectrum function defined in § 1.5,

$$\Phi_{ij}(\mathbf{k}) = \frac{3}{8\pi k^4}(\delta_{ij}k^2 - k_i k_j)E_{11}(k). \tag{3.4.4}$$

The equations for the rate of change of the correlation and spectrum functions include third-order tensors describing the non-linear interactions between components of the motion. The triple velocity correlation tensor is

$$T_{ijk}(\mathbf{r}) = \overline{u_i(\mathbf{x})u_j(\mathbf{x})u_k(\mathbf{x} + \mathbf{r})}$$

and may also be expressed in terms of a single scalar function of the separation, conventionally

$$h(r) = \overline{u_2^2(x_1, x_2, x_3)u_1(x_1 + r, x_2, x_3)}/(\overline{u_1^2})^{3/2}. \tag{3.4.5}$$

3.5 Reynolds number similarity in isotropic turbulence

Perhaps the most significant fact about turbulent flows is that, while geometrically similar flows are expected to be dynamically and structurally similar if their Reynolds numbers are the same, their structures are also very nearly similar for all Reynolds numbers which are large enough to allow turbulent flow. The nature of the phenomenon of Reynolds number similarity can be made clear by considering the integrated spectrum function, $E(k) = E_{11}(k)$ defined in § 1.5, in decaying isotropic turbulence. From (3.2.12), the rate of change of the spectrum function is

$$\frac{\partial E(k, t)}{\partial t} = -2\nu k^2 E(k, t) - \frac{\partial}{\partial k} S(k, t), \qquad (3.5.1)$$

where $\frac{1}{2}S(k, t)$ is the flux of kinetic energy from components of wave numbers less than k to components of larger wave numbers. Fig. 3.1, which is based on measurements of power spectra in grid turbulence, shows on a logarithmic scale of k the distributions of kinetic energy, viscous energy dissipation and energy transfer for two Reynolds numbers of flow, made non-dimensional by using the mesh length and the stream velocity as scales. As could be foreseen from the overall energy equation (3.2.1), the Fourier components of small wave number (eddies of large size) that contribute most to the kinetic energy,

$$\tfrac{1}{2}\overline{q^2} = \tfrac{3}{2} \int_0^\infty E(k)\, \mathrm{d}k = \tfrac{3}{2} \int_0^\infty kE(k)\, \mathrm{d}(\log k),$$

contribute comparatively little to the viscous dissipation,

$$\varepsilon = \nu \overline{\left(\frac{\partial u_i}{\partial x_j}\right)^2} = 3\nu \int_0^\infty k^2 E(k)\, \mathrm{d}k = 3\nu \int_0^\infty k^3 E(k)\, \mathrm{d}(\log k)$$

i.e. the sizes of the energy-containing eddies are different by orders of magnitude from the sizes of the dissipating eddies. In this non-dimensional presentation, the only effect of a change in Reynolds number is to extend the large-wave-number end of the spectral distributions so that the total dissipation remains unchanged in spite of the change in (non-dimensional) viscosity. For very large Reynolds numbers the ratio of the sizes of the dissipating eddies to those of the energy-containing eddies becomes very small, but there are no significant changes in the eddies not contributing to the dissipation.

In 'isotropic' grid turbulence, measurements have been made of turbulent intensity, of correlation functions, of one-dimensional spectrum functions and of the triple correlation functions related to the transfer term, $S(k)$. In ordinary conditions, the flow around the grid is hardly affected by the fluid viscosity and is determined by the stream velocity and by the mesh length. Ignoring the contribution of the viscous eddies, which is appreciable only for small values of r, the correlation function should be independent of Reynolds number and have the form

$$R_{ij}(\mathbf{x};\mathbf{r}) = U_1{}^2 R_{ij}{}^*(\mathbf{x}/M;\mathbf{r}/M). \qquad (3.5.2)$$

Figs. 3.2 and 3.3 show the variations of the turbulent intensity, $\overline{u_1{}^2} = R_{11}(x_1;0)$, with distance from the grid and the variation of

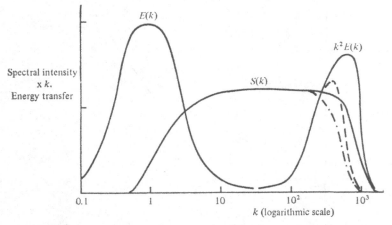

Fig. 3.1. Spectral distributions of turbulent intensity, energy transfer and viscous dissipation. (The broken curves indicate the effect of an increase of fluid viscosity.)

the defining scalar for the correlation function at fixed x_1/M, in each case over a range of Reynolds numbers. The only systematic variation with Reynolds number arises from the expected deficiency of small eddies at the lowest Reynolds numbers and has been allowed for in the correlation measurements. Fig. 3.4 shows some one-dimensional spectra. The invariance of form for wave numbers below the dissipating range and the extension of the spectrum as the Reynolds number increases are clear. Also shown in fig. 3.3 is the triple correlation function, $h(r)$, which has a similar behaviour.

To explain the existence of Reynolds number similarity, it is supposed that turbulent eddies of different sizes affect one another only if their sizes are comparable. Then interaction between the large, energy-containing eddies and the much smaller viscous eddies takes place between a considerable number of intermediary stages,

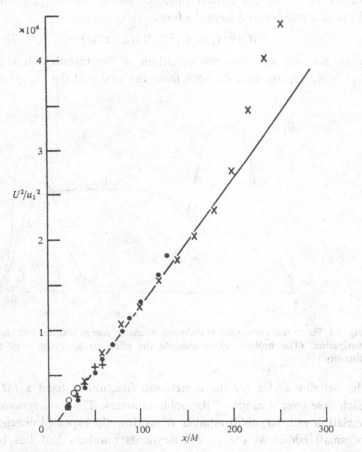

Fig. 3.2. Similarity of energy decay at different Reynolds numbers (from Batchelor & Townsend 1948). $\times$, $M = 0.635$ cm; $\bullet$, $M = 1.27$ cm; $+$, $M = 2.54$ cm; $\bigcirc$, $M = 5.08$ cm: $U = 1286$ cm s^{-1}.

and they are completely independent except that the viscous eddies dissipate the energy lost by the large eddies. It follows that the motion on the large scale is essentially inviscid and so independent of the Reynolds number.

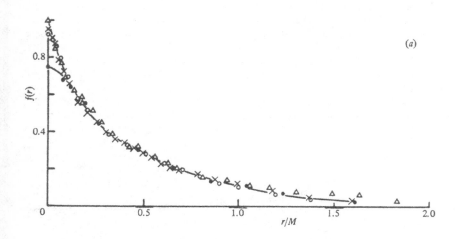

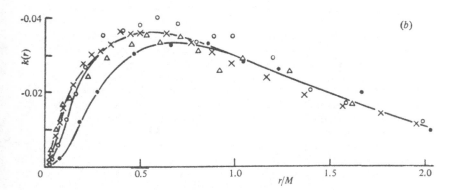

Fig. 3.3. Similarity of (a) double and (b) triple correlation functions at different Reynolds numbers (from Stewart & Townsend 1951). The values have been adjusted to allow for the deficiency of small eddies at the lower Reynolds numbers. $x/M = 30$. R_M: ●, 5250; ○, 21,000; ×, 42,000; △, 115,000.

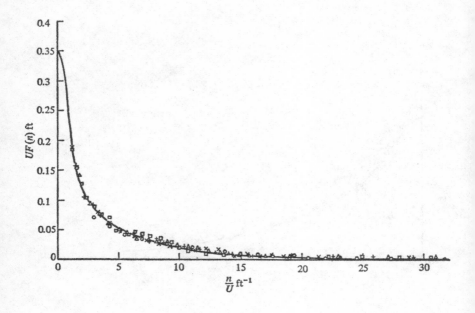

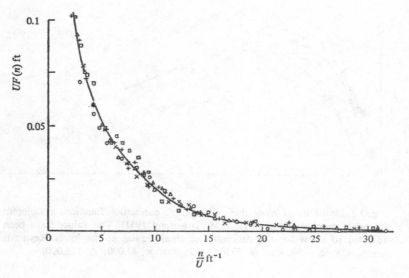

Fig. 3.4. (part 1).

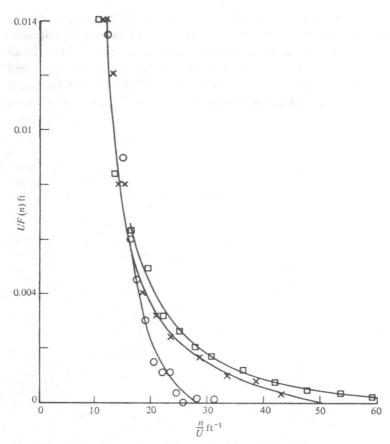

Fig. 3.4. Similarity of one-dimensional spectra at different Reynolds numbers from Taylor 1938). U: $\bigcirc$, $15\,\mathrm{ft\,s^{-1}}$; $\times$, $20\,\mathrm{ft\,s^{-1}}$; $\triangle$, $25\,\mathrm{ft\,s^{-1}}$; $+$, $30\,\mathrm{ft\,s^{-1}}$; $\square$, $35\,\mathrm{ft\,s^{-1}}$.

3.6 Self-preserving development in isotropic turbulence

Measurements of the correlation function for different values of the decay-time show that its shape does not change greatly, particularly if the contribution of the viscous eddies is ignored. For example, Stewart (1951) has shown that the defining scalar $f(x; r)$ is very nearly of the form,

$$f(x; r) = f^*(r/L_0), \qquad (3.6.1)$$

where $L_0(x) = \displaystyle\int_0^\infty f(r)\,\mathrm{d}r$ is the integral scale and the function f^* is

independent both of Reynolds number and x/M. It appears that
the decaying turbulence reaches some sort of moving equilibrium,
one set of eddy structures giving way to another set similar in all
respects except for changes in their common scales of velocity and
length. A moving equilibrium, once attained, should be independent
of its origin and turbulent flows produced by grids of different

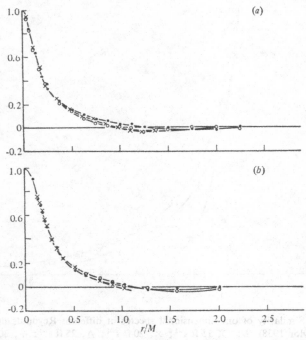

Fig. 3.5. Effect of grid geometry on the double-velocity correlation function
(from Stewart & Townsend 1951). ●, $R_{11}(0, r, 0)$; (a) parallel cylinder, (b) slats:
○, $R_{11}(0, 0, r)$; (a) parallel cylinder, (b) slats: ×, $R_{11}(0, r, 0) = R_{11}(0, 0, r)$,
square mesh.

geometry should approach similar moving equilibria and have the
same shapes for their correlation functions. Measurements behind
grids of different shapes (fig. 3.5) do show a considerable degree of
similarity if the shapes are not too complex.

The hypothesis of self-preserving development of a turbulent flow
assumes that all aspects of the motion except those directly influenced by viscosity have similar forms at all stages, the differences

being described wholly by changes of velocity and length scales which are functions of time (in decaying turbulence) or of position in the flow direction. For decaying turbulence, the spectrum function will have the form,

$$E(k) = \overline{u_1^2} L_0 E^*(kL_0) \qquad (3.6.2)$$

and the transfer function the form,

$$S(k) = (\overline{u_1^2})^{3/2} L_0^{-1} S^*(kL_0) \qquad (3.6.3)$$

where the time-dependent scales are $(\overline{u_1^2})^{1/2}$ and L_0. Substitution in the spectrum equation (3.5.1) leads to

$$\frac{d}{dt}(\overline{u_1^2} L_0) E^* + \overline{u_1^2} \frac{dL_0}{dt} E^{*\prime} = -(\overline{u_1^2})^{3/2} S^{*\prime} \qquad (3.6.4)$$

(a dash indicates differentiation with respect to kL_0), for wave numbers not contributing appreciably to the viscous dissipation. Since the non-dimensional functions are independent of time, the equation may be satisfied only if

$$\frac{d\overline{u_1^2}}{dt} \propto \frac{(\overline{u_1^2})^{3/2}}{L_0} \quad \text{and} \quad \frac{dL_0}{dt} \propto (\overline{u_1^2})^{1/2}. \qquad (3.6.5)$$

Batchelor (1953) has analysed the rather scattered measurements of integral scale to show that

$$\varepsilon = -\frac{3}{2} \frac{d\overline{u_1^2}}{dt} = A(\overline{u_1^2})^{3/2}/L_0 \quad (A \approx 0.8), \qquad (3.6.6)$$

for a wide range of decay-time and Reynolds number, in agreement with the first condition of (3.6.5). More recently, Comte-Bellot & Corrsin (1966) have shown that the decay of isotropic turbulence is described accurately by

$$\overline{u_1^2} \propto (t - t_0)^{-1.28} \qquad (3.6.7)$$

over a very wide range of decay-times. The conditions (3.6.5) imply a power-law (or exponential) variation of intensity with time.

The concept of self-preserving development has wide applications in the theory of turbulent flow and conditions similar to those of (3.6.5) will occur again. The expression (3.6.6) for the dissipation is often used in the form,

$$\varepsilon = (\overline{q^2})^{3/2}/L_\varepsilon \qquad (3.6.8)$$

where L_ε is the *dissipation length-scale*. For many free turbulent flows, the dissipation length-scale is nearly three times the integral scale.

3.7   Space–time correlations in isotropic turbulence

The observed changes in the correlation function during the decay of grid turbulence could arise either from the large energy-containing eddies slowing down without losing their identity or by their breaking-up and replacement by similar but new eddies of less intensity and larger scale. Rates of rotation and rates of shear of the large eddies are of order $(\overline{u_1^2})^{1/2} L_0^{-1}$, and so it is unlikely that they break up in times less than $L_0/(\overline{u_1^2})^{1/2}$. A substantial decrease in turbulent energy (see equation (3.6.6)) occurs in this time and it seems likely that the lifetime of individual eddies is comparable with the duration of the whole turbulent motion.

Measurements of the space–time correlation function, $R_{11}(r, 0, 0; \tau)$, can provide a more quantitative estimate of the average life-time of individual eddies. In § 1.7, it was explained that the differences between the correlation function with and without time delay arise both from real changes in individual eddies and from convection of the eddies by the mean velocity and by the velocity fields of neighbouring eddies. Taking the total velocity fluctuation to be a characteristic velocity of the energy-containing eddies, the arguments used to derive (1.7.3) show that

$$R_{11}(r; \tau) = \frac{\overline{a(t)a(t+\tau)}}{\overline{(a(t))^2}} \int P_2(\Delta x_0) R_{11}(r - \Delta x_0; 0)\, d\Delta x_0, \quad (3.7.1)$$

where $P_2(\Delta x_0)$ is the probability distribution function for $\Delta x_0(\tau)$ $= x_0(\tau) - x_0(0)$, the displacement of an eddy centre in time τ, and the coefficient outside the integral is the autocorrelation function for the 'amplitude' of an individual eddy. The effect of centre displacement is to disperse the correlation function while the effect of amplitude changes is to reduce it in overall magnitude. In terms of the measured correlation function, centre displacement causes the ratio of the integral of the delayed correlation to its maximum value,

$$\int_{-\infty}^{\infty} R_{11}(r, 0, 0; \tau)\, dr / R_{11}(0, 0, 0; \tau),$$

to change by an amount dependent on the magnitude of the displacements. Assuming a normal distribution of the displacements of an eddy centre, with standard deviation d,

$$P_2(\Delta \mathbf{x}_0) = (2\pi d^2)^{-3/2} \exp(-\tfrac{1}{2}\Delta \mathbf{x}_0^2/d^2) \tag{3.7.2}$$

and an initial, simultaneous correlation of 'exponential' form,

$$R_{11}(r; 0) = \overline{u_1^2}\, e^{-r/L_0}\,[1 - \tfrac{1}{2}(r - r_1)/L_0] \tag{3.7.3}$$

it is found that

$$R_{11}(0; \tau) = \overline{u_1^2}\,\frac{\overline{a(t)a(t+\tau)}}{(\overline{a(t)})^2}\left[\left(1 + 2\frac{d^2}{L_0^2} + \frac{1}{3}\frac{d^4}{L_0^4}\right)\exp[\tfrac{1}{2}(d^2/L_0^2)]\right.$$
$$\left. \times\left(1 - \mathrm{erf}\,\frac{d}{\sqrt{2}L_0}\right) - \sqrt{\left(\frac{2}{\pi}\right)}\left(\frac{5}{3}\frac{d}{L_0} + \frac{1}{3}\frac{d^3}{L_0^3}\right)\right] \tag{3.7.4}$$

$$\int_{-\infty}^{\infty} R_{11}(r_1, 0, 0; \tau)\,\mathrm{d}r_1 = L_0\overline{u_1^2}\,\frac{\overline{a(t)a(t+\tau)}}{(\overline{a(t)})^2}$$
$$\times\left\{1 + \frac{1}{2}\frac{d^2}{L_0^2} - \frac{d^2}{L_0^2}\left(1 + \frac{1}{4}\frac{d^2}{L_0^2}\right)\right.$$
$$\left. \times \exp[\tfrac{1}{2}(d^2/L_0^2)]E_i\left(-\frac{1}{2}\frac{d^2}{L_0^2}\right)\right\}.$$

Favre, Gaviglio & Dumas (1962) have measured space–time correlations behind square-mesh, biplane grids, and one set of their results is shown in fig. 3.6. The ratio,

$$\frac{\displaystyle\int_{-\infty}^{\infty} R_{11}(r, 0, 0; \tau)\,\mathrm{d}r/R_{11}(0; \tau)}{\displaystyle\int_{-\infty}^{\infty} R_{11}(r, 0, 0; 0)\,\mathrm{d}r/R_{11}(0; 0)}$$

found from the measurements for a time delay, $U_1\tau/M = 7.57$, has been compared with values calculated from (3.7.4). In this way, the effective value of d/L_0 is found to be 0.36, and the estimated value of the autocorrelation coefficient is 0.85, compared with the measured maximum correlation coefficient at this time delay of 0.41. Although an exponential form for the simultaneous correlation function is not a perfect representation a different form is unlikely to do more than change the value of d/L_0 without affecting the conclusion that most of the reduction in value of $R_{11}(0; \tau)$ is caused by convection of the eddy centres.

The inferred value of 0.85 for the eddy autocorrelation should be compared with an energy decay by a factor of 0.80 over the same time interval. Part at least of the change of the autocorrelation is caused by increase of eddy size (the integral scale increases by 10% over the time interval), and it is apparent that the energy of the turbulent motion will have fallen to a very low level before no trace can be found of an original eddy. In other words, the energy-containing eddies of isotropic turbulence persist through most of the life of the motion. A considerable degree of permanence of the larger eddies is a universal characteristic of turbulent flow.

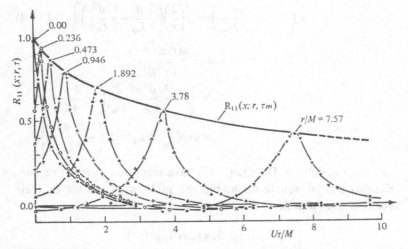

Fig. 3.6. Space–time correlations with fixed spatial separations and variable time interval (from Favre, Gaviglio & Dumas 1962). $R_M = 21,500$, $x/M = 40$.

3.8 The Taylor approximation of frozen flow

If the time-delay correlation function is converted to a structure or spectrum function to give information about eddies of various sizes, the decrease of maximum autocorrelation with time is caused mostly by the random movements of eddy centres, and, if the random displacements are small compared with the eddy diameters, the change in the structure function (or other specification of eddy intensity) is substantially that produced by simple translation by a convection velocity. In homogeneous turbulence, the convection velocity is the same for all sizes of eddy and equal to the mean flow velocity. The Taylor approximation (Taylor 1938) asserts that

$$u_i(\mathbf{x}, t) = u_i(\mathbf{x} - \mathbf{U}\tau, t + \tau) \tag{3.8.1}$$

for not too large values of τ. The approximation in this form is a good one if the random displacements of eddy centres in τ are small compared with the diameter of the smallest eddies, say l_s. Since the random convective velocities are of order $(\overline{u_1{}^2})^{1/2}$, the condition is that

$$(\overline{u_1{}^2})^{1/2}\tau/l_s \ll 1. \tag{3.8.2}$$

If (3.8.1) is used to obtain statistical mean values rather than to convert spatial to temporal variations, the condition is unnecessarily restrictive. Velocity variations due to eddies of size l are completely uncorrelated for separations large compared with l/U_1, and inaccuracy from the use of (3.8.1) to calculate spatial correlation functions from frequency spectra will be appreciable only if the condition is not satisfied for $\tau \leqslant l/U_1$. No serious error is expected if

$$(\overline{u_1{}^2})^{1/2} \ll U_1. \tag{3.8.3}$$

The approximation is widely used for calculating one-dimensional spectrum functions from measured frequency spectra of signals from a hot-wire anemometer. If a velocity pattern in the neighbourhood of the wire is moving with velocity $\mathbf{U} + \mathbf{u}$, Fourier components of frequency ω are derived from components whose wave numbers satisfy

$$\mathbf{k}\cdot(\mathbf{U} + \mathbf{u}) = k_1 U_1 + \mathbf{k}\cdot\mathbf{u} = \omega \tag{3.8.4}$$

or, to the approximation, those with a component of wave number in the flow direction of ω/U_1. For even mean-value functions such as the spectrum function, the error from using the Taylor approximation is of the second order in $(\overline{u_1{}^2})^{1/2}/U_1$. If $P(\mathbf{u})$ is the probability distribution function for the random convective velocity $\mathbf{u}$,

$$\psi_{ij}(\omega) = \frac{1}{U_1}\int_{-\infty}^{\infty}\iint \Phi_{ij}(\omega/U_1 - \mathbf{u}\cdot\mathbf{k}/U_1, k_2, k_3)P(\mathbf{u})\,\mathrm{d}k_2\,\mathrm{d}k_3\,\mathrm{d}\mathbf{u} \tag{3.8.5}$$

or, after expanding $\Phi_{ij}(\mathbf{k})$ as a power series in $k_1 - \omega/U_1$,

$$\psi_{ij}(\omega) = \frac{1}{U_1}\Phi_{ij}(\omega_1/U_1) + \frac{1}{2U_1}\int_{-\infty}^{\infty}\iint \frac{\partial^2\Phi_{ij}}{\partial k_1{}^2}\overline{\left(\frac{\mathbf{u}\cdot\mathbf{k}}{U_1}\right)^2}\,\mathrm{d}k_2\,\mathrm{d}k_3$$
$$+ \text{higher order terms.} \tag{3.8.6}$$

Another application is to the determination of mean values of functions of turbulent velocity gradients from single anemometers, using electrical circuits to perform time differentiation. To the approximation,

$$\frac{\partial}{\partial t} = -U_1 \frac{\partial}{\partial x_1} \tag{3.8.7}$$

and, if u is the velocity component detected by the anemometer,

$$\left.\begin{array}{l}\overline{\left(\frac{\partial u_1}{\partial x_1}\right)^2} = U_1^{-2}\overline{\left(\frac{\partial u}{\partial t}\right)^2}, \qquad \overline{\left(\frac{\partial u}{\partial x_1}\right)^3} = -U_1^{-3}\overline{\left(\frac{\partial u}{\partial t}\right)^3} \\[3mm] \overline{\left(\frac{\partial^2 u}{\partial x_1^2}\right)^2} = U_1^{-4}\overline{\left(\frac{\partial^2 u}{\partial t^2}\right)^2}, \qquad \text{etc.} \end{array}\right\} \tag{3.8.8}$$

Use of the frozen-flow approximation is not restricted to homogeneous turbulence, and it applies to all flows whose variations of mean velocity and fluctuating velocity are both small compared with the average velocity over the whole flow. The condition is not satisfied in turbulent jets and boundary layers, and there the convection velocities of the larger-scale patterns of velocity may be considerably different from the local mean velocity. Even then, the small eddies that determine the velocity gradients are convected with the local velocity of the fluid and the approximations of (3.8.8) are substantially accurate if $\overline{u^2}/U_1^2$ is small.

3.9 The tendency to isotropy of homogeneous turbulence

The evidence that the energy-containing eddies of isotropic turbulence retain their identities over periods of time sufficient for substantial decay of energy shows that interaction between the larger eddies is almost limited to displacement by each other's velocity fields and that interactions with smaller eddies remove energy without substantially affecting their structure. If the turbulent motion were initially anisotropic as a consequence of conditions at its formation, it would be expected that the anisotropic, energy-containing eddies would persist with nearly unchanged structure during decay and that the anisotropy would be slow to disappear. On the other hand, the smaller eddies draw energy from the larger ones and pass it to still smaller ones. They are not stable, persistent

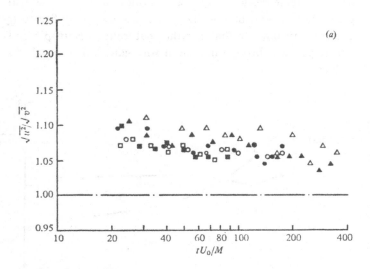

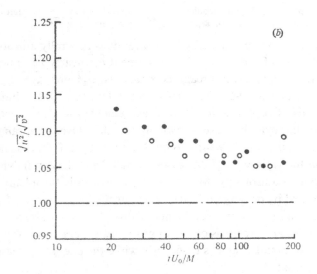

Fig. 3.7. Approach to isotropy of decaying turbulence generated by square-mesh grids (from Comte-Bellot & Corrsin 1966). (*a*) biplane, square-rod grids. (*b*) biplane, round-rod grids. $M = 2.56$ cm: ▲, $U_0 = 10$ ms^{-1}; △, $U_0 = 20$ ms^{-1}. $M = 5.08$ cm: ●, $U_0 = 10$ ms^{-1}; ○, $U_0 = 20$ ms^{-1}. $M = 10.16$ cm: ■, $U_0 = 10$ ms^{-1}; □, $U_0 = 20$ ms^{-1}.

structures and their motion may resemble more closely the statistically probable state of isotropy. The expectations are confirmed by studies of grid turbulence that is either naturally anisotropic or is made so by passage through fine-mesh wire gauzes or through distorting ducts.

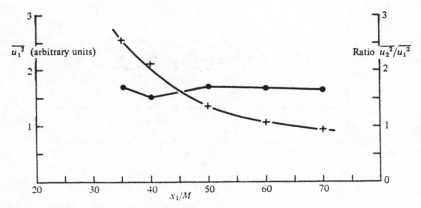

Fig. 3.8. Approach of grid turbulence to isotropy after passage through a wire gauze at $x/M = 30$. ●, ratio $\overline{u_2^2}/\overline{u_1^2}$; +, intensity $\overline{u_1^2}$.

Turbulent flow produced by biplane grids is naturally anisotropic with intensity of the streamwise component $\overline{u_1^2}$ nearly 1.2 times the intensities of the cross-stream components, $\overline{u_2^2}$ and $\overline{u_3^2}$. Comte-Bellot & Corrsin (1966) have measured intensity ratios behind biplane grids of several kinds, and measurements for grids of square and circular cylinders are shown in fig. 3.7. Over the range of measurement, the turbulent energy decreases by a factor of nearly 25 while the ratio $\overline{u_1^2}/\overline{u_2^2}$ decreases from 1.25 to 1.12, a halving of the degree of anisotropy. More strongly anisotropic turbulence may be generated by passing grid turbulence through wire gauzes of small mesh size, and some measurements of intensity ratios are shown in fig. 3.8. The anisotropy is of the opposite sign, i.e. $\overline{u_1^2}/\overline{u_2^2} < 1$, and the ratio does not change significantly while the energy decreases by a factor of 2.5.

The behaviour of turbulence made anisotropic by passage through distorting ducts is more complex. During passage through the duct, larger eddies are distorted by the mean velocity gradients induced by the duct walls but eddies one size smaller are distorted by velocity gradients of the larger eddies as well as by those of the mean velocity. They too become anisotropic but less so than the larger ones. After emergence from the duct, the smaller eddies continue to be distorted by and to receive energy from the larger ones but the anisotropy that was induced by the distortion disappears as their energy is replaced. Since the energy of the smaller eddies is a considerable fraction of the total, we may expect the ratio of intensities to approach one initially as the smaller eddies adjust themselves to the

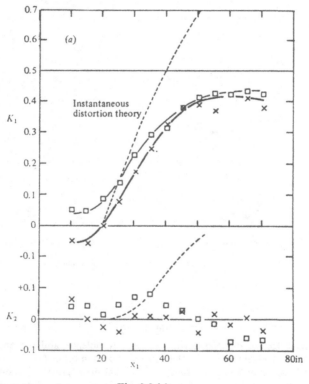

Fig. 3.9 (a).

change, but that, after the adjustment, it will tend to a nearly constant value. This kind of behaviour is shown by the measurements of Tucker & Reynolds (1968) for turbulence made anisotropic by irrotational, plane distortion (fig. 3.9). On the other hand, measurements by Uberoi (1956) of turbulence after emergence from an axisymmetric contraction show a slow but steady approach to the isotropic value of one. Reasons for the difference are not known.

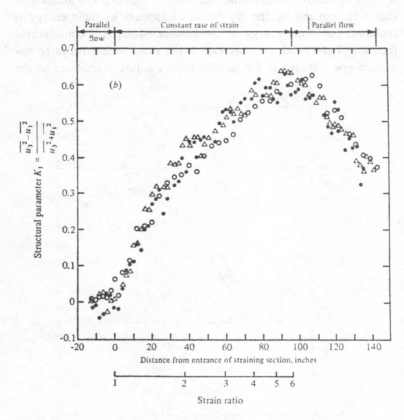

Fig. 3.9 (b).

Fig. 3.9. Effect of plane distortion on grid turbulence, and the approach to isotropy after emergence from the distorting section (after (a) Townsend 1954 and (b) Tucker & Reynolds 1968). (a) ×, half-inch grid; □, one-inch grid. (b) ○, square perforated metal grid, 20 ft s^{-1}; ●, diamond expanded metal grid, 20 ft s^{-1}; △, diamond expanded metal grid, 40 ft s^{-1}.

The behaviour of the smallest eddies contributing to the viscous energy dissipation is studied by measuring intensities of gradients of the velocity fluctuations, conveniently gradients in the downstream direction. In isotropic turbulence,

$$\overline{\left(\frac{\partial u_1}{\partial x_1}\right)^2} = \frac{1}{2}\overline{\left(\frac{\partial u_2}{\partial x_1}\right)^2} = \frac{1}{2}\overline{\left(\frac{\partial u_3}{\partial x_1}\right)^2} \qquad (3.9.1)$$

and any lack of isotropy in the dissipating eddies may be measured by the difference of the ratio,

$$\overline{\left(\frac{\partial u_2}{\partial x_1}\right)^2} \Big/ \overline{\left(\frac{\partial u_1}{\partial x_1}\right)^2},$$

from two. In the turbulence produced by grids, before and after passage through gauzes, the ratio is sufficiently close to the isotropic value to conclude that the dissipating eddies may be very nearly isotropic even though the energy-containing eddies are not. The experiments on distortion of grid turbulence were made with large rates of distortion and the flow emerges from the distorting duct with the dissipating eddies in a state of moderate anisotropy. After emergence, the gradient ratio approaches the isotropic value much more rapidly than does the intensity ratio (fig. 3.10), confirming that the passage of energy from larger to smaller ones involves a redistribution of energy among the three components and a relatively rapid approach towards isotropy (Townsend 1954, Uberoi 1956).

3.10 Uniform distortion of homogeneous turbulence

In ordinary shear flows, neither the turbulent motion nor the gradients of mean velocity are spatially uniform, but the scale of the turbulent motion appears to be somewhat less than the lateral scale of the inhomogeneity and it is possible to argue that the turbulence is approximately homogeneous and exists in an environment of approximately uniform strain. Then it would be informative to determine the rheological properties of 'turbulent fluid' by subjecting it to known distortions and measuring the changes in its structure and, in particular, the changes in the Reynolds stress tensor. Measurements of uniform distortion fall into two groups:

(1) irrotational distortions, produced by passage through ducts of changing section, and

(2) unidirectional, plane shearing flow for which

$$U_1 = \alpha x_3, \qquad U_2 = U_3 = 0.$$

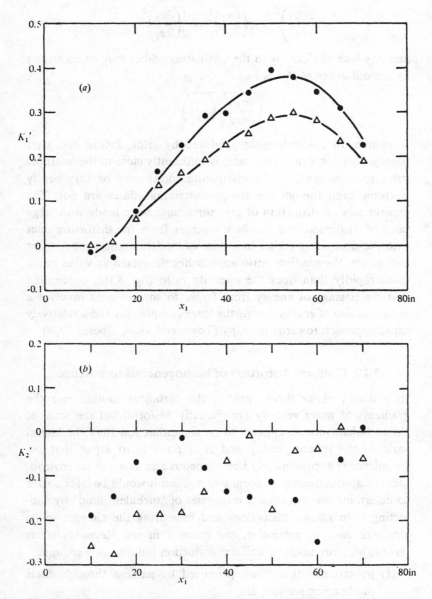

Fig. 3.10. Effect of plane distortion on the dissipating eddies of grid turbulence, and their approach to isotropy after emergence from the distorting section. ●, half-inch grid; △, one-inch grid.

The first group are technically simpler to realise, but the second kind of distortion is a closer approximation to that found in ordinary shear flows and has the greater interest.

Equations (3.2.7) and (3.2.8) describe the development of turbulent motion in a uniform gradient of mean velocity but the non-linear terms representing interactions between Fourier components of the velocity field prevent any simple solution. Observations of shear-free turbulence have shown that the greater part of the turbulent energy resides in a group of larger eddies which are stable, persisting structures and which transfer energy to smaller eddies without significant change of structure. In terms of the Fourier components, it means that interactions between pairs of Fourier components are negligible, if both contribute to the larger eddies, and that inter-action between these components and components of larger wave number (smaller eddies) has the general effect of a viscous damping. So the development equations may be approximated by

$$\frac{\mathrm{d}a_i(\mathbf{k})}{\mathrm{d}t} = -v_t k^2 a_i(\mathbf{k}) - \frac{\partial U_i}{\partial x_l} a_l(\mathbf{k}) + 2\frac{k_i k_l}{k^2}\frac{\partial U_l}{\partial x_m} a_m(\mathbf{k}) \qquad (3.10.1)$$

and

$$\frac{\mathrm{d}k_i}{\mathrm{d}t} = -\frac{\partial U_l}{\partial x_i} k_l, \qquad (3.10.2)$$

where v_t is an effective viscosity which represents the effects of the smaller eddies. Unless the effective viscosity is large, its principal effect is to change the intensity of the motion without affecting the structure and its omission leads to the 'rapid-strain' equations,

$$\frac{\mathrm{d}a_i(\mathbf{k})}{\mathrm{d}t} = -\frac{\partial U_i}{\partial x_l} a_l(\mathbf{k}) + 2\frac{k_i k_l}{k^2}\frac{\partial U_l}{\partial x_m} a_m(\mathbf{k}) \qquad (3.10.3)$$

and (3.10.2). The equations are linear in the component amplitude and its wave number, and they can be solved for any uniform velocity gradient. It is apparent that the changes of amplitude depend on the total strain but not otherwise on the history of strain, and the effects of rapid distortion on turbulence are completely reversible. In other words, the response of turbulent fluid to rapid distortion is *elastic*, though possibly anisotropic and not obeying Hooke's law.

For approximate validity of predictions of the behaviour of homogeneous turbulence under uniform strain, the rapid-strain equations would be satisfactory if:

(1) changes in absolute intensity are not considered,

(2) larger eddies remain stable and identifiable during distortion in spite of their changes of shape, and

(3) the contributions of the smaller eddies to the observed quantities can be ignored.

To obtain definite predictions, it is necessary to assume the initial state of the turbulence before distortion. All the quoted results refer to initially isotropic turbulence with a spectrum function of the proper form, i.e.

$$\Phi_{ij}(\mathbf{k}) = \frac{3}{8\pi k^4}[\delta_{ij}k^2 - k_i k_j]E(k), \qquad (3.10.4)$$

where $E(k)$ is the integrated spectrum function defined in § 1.5. For initially isotropic turbulence, the changes in components of the Reynolds stress tensor are independent of the form of the spectrum function. One general result of some interest is that for small total strains,

$$(-\overline{u_i u_j})_t - (-\overline{u_i u_j})_{t=0} = \tfrac{2}{5}\overline{u_1^2} \int_0^t \left(\frac{\partial U_i}{\partial x_j} + \frac{\partial U_j}{\partial x_i}\right) dt', \qquad (3.10.5)$$

showing that isotropically turbulent fluid behaves as an isotropic elastic material with shear modulus $0.4\overline{u_1^2}$.

3.11 Irrotational distribution of grid turbulence

If the mean flow causing the distortion is irrotational, principal axes of strain exist and it is convenient to use them as axes of reference so that the velocity field is

$$U_1 = \alpha_1 x_1, \qquad U_2 = \alpha_2 x_2, \qquad U_3 = \alpha_3 x_3. \qquad (3.11.1)$$

By the condition of incompressibility, $\alpha_1 + \alpha_2 + \alpha_3 = 0$. The effect of rapid distortion is most simply expressed in terms of the vorticity. Vortex lines move with the fluid, and so the ratio of the final to the initial vorticity at a fluid particle equals the extension ratio of a vortex-line element at the particle, i.e.

$$\omega_1(e_1 x_1, e_2 x_2, e_3 x_3; t) = e_1 \omega_1(x_1, x_2, x_3; 0), \qquad (3.11.2)$$

with similar relations for ω_2 and ω_3, where

$$e_i(t) = \exp \int_0^t \alpha_i(t') \, dt' \qquad (3.11.3)$$

are the *total strain ratios* along the axes. The changes in amplitude of Fourier components of the vorticity can then be written down, and it is not difficult to show that changes in amplitude of velocity components are given by

$$k^2 a_1(t) = \left(k_{20}^2 \frac{e_3}{e_2} + k_{30}^2 \frac{e_2}{e_3} \right) a_1(0)$$
$$- k_{10} k_{20} \frac{e_3}{e_2} a_2(0) - k_{10} k_{30} \frac{e_2}{e_3} a_3(0) \qquad (3.11.4)$$

and the two similar expressions obtained by cyclic interchange of the suffices. Here,

$\mathbf{k}_0 = (k_{10}, k_{20}, k_{30})$ is the wave number of the component at time $t = 0$, and

$\mathbf{k} = (e_1^{-1} k_{10}, e_2^{-1} k_{20}, e_3^{-1} k_{30})$ is its wave number at time t. Starting with isotropic turbulence, the spectrum function for the u_1 component is

$$\Phi_{11}(\mathbf{k}) = \frac{3E(k_0)}{8\pi k^4 k_0^2} \left[\left(\frac{e_3}{e_2} \right)^2 k_{20}^2 \, (k_{10}^2 + k_{20}^2) \right.$$
$$\left. + \left(\frac{e_2}{e_3} \right)^2 k_{30}^2 \, (k_{10}^2 + k_{30}^2) + 2 k_{20}^2 k_{30}^2 \right] \qquad (3.11.5)$$

with similar expressions for $\Phi_{22}(\mathbf{k})$ and $\Phi_{33}(\mathbf{k})$. Changes of intensity are found by integrating over all wave numbers, and the ratio of final to initial intensity does not depend on the form of $E(k)$. For an arbitrary strain, the results involve complicated integrals but the more important results can be expressed in terms of elementary functions (Batchelor & Proudman 1954).

The changes in the integral scales of the motion also do not depend on the form of the initial spectrum. For any irrotational strain of isotropic turbulence, it may be shown from equation (3.10.5) that

$$\left. \begin{aligned} \overline{u_1^2} L_{11} &\equiv \int_0^\infty R_{11}(r, 0, 0) \, dr = e_1^{-1} [\overline{u_1^2} L_{11}]_{t=0}, \\ \overline{u_1^2} L_{12} &\equiv \int_0^\infty R_{11}(0, r, 0) \, dr = e_2^2 [\overline{u_1^2} L_{12}]_{t=0}, \end{aligned} \right\} \qquad (3.11.6)$$

where L_{11} and L_{12} are the longitudinal and transverse integral scales of the u_1 component in the Ox_1 and Ox_2 direction. Similar relations exist for the other integral scales. The changes may occur either by a change in the extent of the correlation, i.e. the range of r for which it is appreciable, or by a change in form such as the development or suppression of negative values.

Two kinds of irrotational distortion have been studied experimentally, *axisymmetric extension* with

$$e_1 = \beta, \qquad e_2 = e_3 = \beta^{-1/2},$$

produced by flow in an axisymmetric contracting section, and *plane straining* with

$$e_1 = 1, \qquad e_2 = e_3^{-1} = \beta,$$

produced by flow in a duct of constant area but changing cross-sectional shape. Uberoi (1956) has measured turbulent intensities and spectra in grid turbulence undergoing axisymmetric extension, and he finds that the changes of intensity are nearly those predicted by the rapid distortion theory, i.e.

$$
\left.
\begin{aligned}
\overline{u_1^2} &= \tfrac{3}{4}\beta^{-2}\left[\frac{2-\beta^{-3}}{(1-\beta^{-3/2})^{3/2}} \right. \\
&\qquad \left. \times \tanh^{-1}(1-\beta^{-3})^{1/2} - \frac{1}{1-\beta^{-3}}\right](\overline{u_1^2})_{t=0}, \\
\overline{u_2^2} = \overline{u_3^2} &= \tfrac{3}{8}\beta\left[\frac{2-\beta^{-3}}{1-\beta^{-3}} - \frac{\beta^{-6}}{(1-\beta^{-3})^{3/2}}\right. \\
&\qquad \left. \times \tanh^{-1}(1-\beta^{-3})^{1/2}\right](\overline{u_2^2})_{t=0}
\end{aligned}
\right\} \quad (3.11.7)
$$

if the contraction ratio, β, is less than four. The changes in the (one-dimensional) spectrum functions of $\overline{u_1^2}$ and $\overline{u_2^2}$ agree moderately well with those calculated from the theory. The form of the changes could be inferred from the changes in scale consequent on the results in equations (3.11.6) and (3.11.7). Together, they show that

$$L_{11} = [\tfrac{4}{3}\beta/(\log 4\beta^3 - 1)](L_{11})_{t=0}, \quad L_{21} = \tfrac{4}{3}\beta(L_{21})_{t=0} \quad (3.11.8)$$

for large values of β. Since the extent of correlation is unlikely to be changed by distortion, the ratio of a scale to the relevant extension ratio can change only by appearance or disappearance of negative correlations. The increase of L_{21} and the decrease of L_{11} arise in this way, and the changes in the spectrum functions are those expected.

Distortion by plane straining is of more interest because the mean velocity gradients in most shear flows are nearly everywhere a plane strain with a superimposed rotation. The effects of the rotation on the motion are not negligible but the similarity of the distortion has led to several experimental studies of grid turbulence subjected to plane straining. The rapid-distortion theory shows that initially isotropic turbulence would change its intensity in the way specified by

$$
\left.
\begin{aligned}
\overline{u_1{}^2} &= [1+\tfrac{8}{35}(\beta-\beta^{-1})^2](\overline{u_1{}^2})_{t=0}, \\
\overline{u_2{}^2} &= \left[1-\frac{4}{5}\left(\frac{\beta-\beta^{-1}}{\beta+\beta^{-1}}\right)+\tfrac{3}{35}(\beta-\beta^{-1})^2\right](\overline{u_2{}^2})_{t=0}, \\
\overline{u_3{}^2} &= \left[1+\frac{4}{5}\left(\frac{\beta-\beta^{-1}}{\beta+\beta^{-1}}\right)+\tfrac{3}{35}(\beta-\beta^{-1})^2\right](\overline{u_3{}^2})_{t=0}, \\
\overline{q^2} &= [1+\tfrac{2}{15}(\beta-\beta^{-1})^2](\overline{q^2})_{t=0},
\end{aligned}
\right\}
\tag{3.11.9}
$$

if the strain ratio, β, is small, and by

$$
\left.
\begin{aligned}
\overline{u_2{}^2} &= \tfrac{3}{4}\beta^{-1}(\log 4\beta-1)(\overline{u_2{}^2})_{t=0}, \\
\overline{u_1{}^2} &= [\tfrac{3}{4}\beta-\tfrac{3}{8}\beta^{-1}(\log 4\beta-\tfrac{3}{2})](\overline{u_1{}^2})_{t=0}, \\
\overline{u_3{}^2} &= [\tfrac{3}{4}\beta+\tfrac{3}{8}\beta^{-1}(\log 4\beta-\tfrac{1}{2})](\overline{u_3{}^2})_{t=0},
\end{aligned}
\right\}
\tag{3.11.10}
$$

if it is large (in practice, more than three). The most extensive measurements of intensities are those by Tucker & Reynolds (1968) who find that the ratios of the intensities behave in much the predicted way for strain ratios less than two, but that the intensity ratios are closer to one than predicted if the strain ratio is larger. The intensity ratio of most interest is the ratio $\overline{u_2{}^2}/\overline{u_3{}^2}$, conveniently expressed by the structural parameter,

$$
K_1 = (\overline{u_3{}^2} - \overline{u_2{}^2})/(\overline{u_3{}^2} + \overline{u_2{}^2}).
\tag{3.11.11}
$$

Typical measurements of K_1 are shown in fig. 3.9. The important feature, which, with some allowance for differences in flow arrangements and in experimental techniques, is confirmed by Maréchal (1967), is that K_1 increases steadily with increasing strain and shows no sign of approaching an asymptotic value. Turbulent fluid subjected to prolonged plane straining (with a strain ratio of as much as fourteen in the experiments of Maréchal) does not approach a steady condition with unvarying Reynolds stresses as it would if its behaviour were viscous.

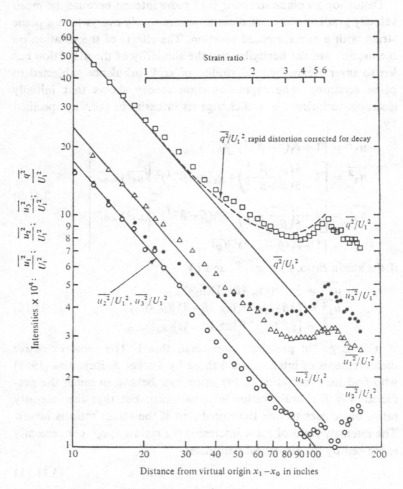

Fig. 3.11. Variation of intensities of the velocity components during and after plane distortion (after Tucker & Reynolds 1968).

Although the rapid-distortion theory has only limited success in predicting the intensity ratios, it gives a good general description of the form of the correlation and spectrum functions. Grant (1958) has measured the two transverse correlation functions, $R_{22}(0, 0, r)$

and $R_{33}(0, r, 0)$, for a strain ratio of just less than four, and fig. 3.12 compares his measured curves with correlations calculated from the rapid-distortion theory assuming initial correlations of the exponential form,

$$R_{ij}(\mathbf{r}) = \overline{u_1}^2\left[\delta_{ij}\left(1 - \frac{1}{2}\frac{r}{L_0}\right) + \frac{1}{2}\frac{r}{L_0}\frac{r_i r_j}{r^2}\right]e^{-r/L_0}. \qquad (3.11.12)$$

Townsend (1954) has measured the spectrum function $\phi_{11}(k_1)$ and finds good agreement with the calculated form. Considerable changes of $R_{22}(0, 0, r)$ and of $\phi_{11}(k_1)$ occur during straining and it is likely that a considerable proportion of the eddies are distorted in the way described by the rapid-distortion equations.

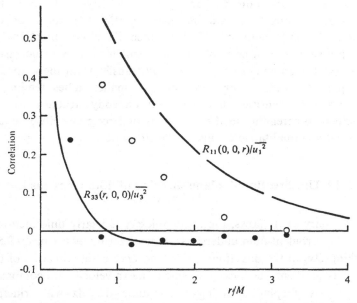

Fig. 3.12. Comparison of calculated correlations for irrotational plane shear with measurements by Grant (1958). (Ox_1 is the direction of expansion, Ox_3 is the direction of compression, and the calculations are for a total strain ratio of four.)

The experimental study of prolonged, irrotational distortion of grid turbulence shows that the rapid-distortion theory provides a good description only of the changes in the spectrum and correlation functions. If it is assumed that the rate of viscous dissipation of turbulent energy continues unchanged during the distortion, the changes in $\overline{q^2} = \overline{u_1^2} + \overline{u_2^2} + \overline{u_3^2}$, the total turbulent intensity, can be described fairly well by the theory, but the ratios of the intensities are those predicted by the theory only for small total strains. For large strains, the anisotropy is considerably less than that predicted by the theory but there is no indication that a balance is attained between a tendency to isotropy of the turbulence and the reverse effect of the distortion.

It is likely that the reasons for the differences and agreements lie in the changed nature of energy transfer from the larger eddies after distortion. During distortion, individual eddies are extended in directions of positive rate of strain and compressed in directions of negative rate of strain. In consequence, velocity gradients in the compression directions are increased, particularly those of the amplified components of velocity which lie along directions of compression, and transfer of energy to smaller eddies is increased. The result is an enhancement of energy transfer from the amplified components to the rather more isotropic smaller eddies, tending to equalise the intensities, but the effect on eddy structure will be confined to spreading in the directions of strong velocity gradient without any qualitative change in the structure.

3.12 Unidirectional, plane shearing of homogeneous turbulence

In ordinary shear flows, the mean velocity is nearly unidirectional and the turbulent fluid undergoes simple shearing by the mean flow. Taking Ox_1 in the direction of flow and Ox_3 in the direction of the velocity gradient, the local gradients of mean velocity are composed very nearly of a plane straining of magnitude $\frac{1}{2}\mathrm{d}U_1/\mathrm{d}x_3$ with principal axes in the $(1/\sqrt{2}, 0, 1/\sqrt{2})$, $(-1/\sqrt{2}, 0, 1/\sqrt{2})$ and $(0, 1, 0)$ directions, and a rotation about the Ox_2 axis with angular velocity $\frac{1}{2}\mathrm{d}U_1/\mathrm{d}x_3$ (fig. 3.13). Unlike irrotational plane strain, the rotation continuously turns vortex lines away from the direction of positive rate of strain, which is the optimum direction for energy gain, and configurations with vorticity almost entirely aligned along this direction are not possible. Consequently, the generation of Reynolds stress and the transfer of energy to the turbulent motion are less efficient processes than in irrotational straining, and there are significant and important differences in the nature of the motion after prolonged strain.

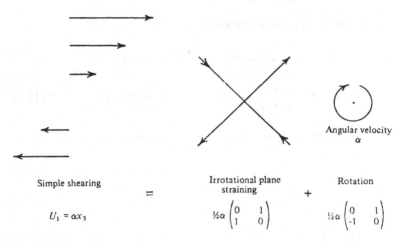

Fig. 3.13. Decomposition of simple shearing into rotation and an irrotational plane straining.

It is useful to compare experimental observations with calculations of the development of initially isotropic turbulence using the 'weak turbulence' approximation† of (3.10.1), which assumes that interactions between the larger and smaller components of the motion is equivalent to the action of an effective viscosity. For uniform plane shearing defined by

$$U_1 = \alpha x_3, \qquad U_2 = U_3 = 0 \tag{3.12.1}$$

the equations for the amplitude of one Fourier component of the velocity fluctuation become

$$\left.\begin{array}{l} \dfrac{da_1(\mathbf{k})}{dt} + v_t k^2 a_1(\mathbf{k}) = \alpha(2k_1^2/k^2 - 1)a_3(\mathbf{k}), \\[2mm] \dfrac{da_2(\mathbf{k})}{dt} + v_t k^2 a_2(\mathbf{k}) = (2\alpha k_1 k_2/k^2)a_3(\mathbf{k}), \\[2mm] \dfrac{da_3(\mathbf{k})}{dt} + v_t k^2 a_3(\mathbf{k}) = (2\alpha k_1 k_3/k^2)a_3(\mathbf{k}), \end{array}\right\} \tag{3.12.2}$$

† Pearson (1959), Deissler (1958) and others use the approximation to calculate the behaviour of weak turbulence under shearing motion. Here weak turbulence means turbulence with a small Reynolds number.

with

$$\frac{dk_1}{dt} = \frac{dk_2}{dt} = 0, \qquad \frac{dk_3}{dt} = -\alpha k_1.$$

Integration from zero time leads to

$$\left.\begin{aligned}
a_1(\mathbf{k}, t) &= \frac{k_0{}^2}{k_1{}^2 + k_2{}^2} D Q_1 a_3(\mathbf{k}_0, 0) + D a_1(\mathbf{k}_0, 0), \\
a_2(\mathbf{k}, t) &= \frac{k_1 k_2}{k_1{}^2 + k_2{}^2} D Q_2 a_3(\mathbf{k}_0, 0) + D a_2(\mathbf{k}_0, 0), \\
a_3(\mathbf{k}, t) &= D \frac{k_0{}^2}{k_0{}^2 - 2\beta k_1 k_{30} + \beta^2 k_1{}^2} a_3(\mathbf{k}, 0),
\end{aligned}\right\} \qquad (3.12.3)$$

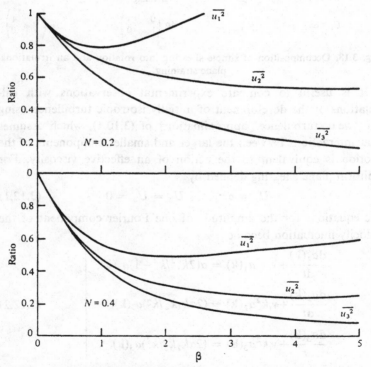

Fig. 3.14. Calculated dependence of the intensities of velocity components on total shear, using the rapid-distortion approximation for two values of N (from Townsend 1970).

where $\mathbf{k}_0 = (k_{10}, k_{20}, k_{30})$ is the initial wave number of the component,

$\mathbf{k} = (k_{10}, k_{20}, k_{30} - \beta k_{10})$ is the final wave number,

$$Q_1 = -\frac{k_2{}^2}{k_1(k_1{}^2+k_2{}^2)^{1/2}} \arctan\left(\frac{\beta k_1(k_1{}^2+k_2{}^2)^{\frac{1}{2}}}{k_0{}^2 - \beta k_1 k_{30}}\right)$$
$$+ \frac{\beta k_1{}^2(k_0{}^2 - 2k_{30}{}^2 + \beta k_1 k_{30})}{k_0{}^2(k_0{}^2 - 2\beta k_1 k_{30} + \beta^2 k_1{}^2)},$$

$$Q_2 = \frac{k_0{}^2}{k_1(k_1{}^2+k_2{}^2)^{1/2}} \arctan\left(\frac{\beta k_1(k_1{}^2+k_2{}^2)^{\frac{1}{2}}}{k_0{}^2 - \beta k_1 k_{30}}\right)$$
$$+ \frac{\beta(k_0{}^2 - 2k_{30}{}^2 + \beta k_1 k_{30})}{k_0{}^2 - 2\beta k_1 k_{30} + \beta^2 k_1{}^2},$$

$$D = \exp[-v_t t(k_0{}^2 - \beta k_1 k_{30} + \tfrac{1}{3}\beta^2 k_1{}^2)],$$
$$\beta = \alpha t.$$

Using the relation (3.2.11), expressions for components of the spectrum function can now be written down in terms of the spectrum function at zero time. If the initial motion has the isotropic spectrum function (3.10.4), the spectrum function after distortion is given by

$$\Phi_{11}(\mathbf{k}) = \frac{3E(k_0)}{8\pi k_0{}^2}\left[\frac{k_0{}^2}{k_1{}^2+k_2{}^2}Q_1\left(Q_1 - \frac{2k_1 k_{20}}{k_0{}^2}\right)\right.$$
$$\left. + \frac{k_2{}^2+k_{30}{}^2}{k_0{}^2}\right]D^2,$$

$$\Phi_{22}(\mathbf{k}) = \frac{3E(k_0)}{8\pi k_0{}^2}\left[\frac{k_1{}^2 k_2{}^2}{k_1{}^2+k_2{}^2}Q_2{}^2\right.$$
$$\left. - \frac{2k_1 k_2{}^2 k_{30}}{k_0{}^2(k_1{}^2+k_2{}^2)}Q_1 + \frac{k_1{}^2+k_{30}{}^2}{k_0{}^2}\right]D^2,$$

$$\Phi_{33}(\mathbf{k}) = \frac{3E(k_0)}{8\pi k_0{}^2}\frac{k_0{}^2(k_1{}^2+k_2{}^2)}{(k_0{}^2 - 2\beta k_1 k_{30} + \beta^2 k_1{}^2)^2}D^2,$$

$$\Phi_{13}(\mathbf{k}) = \frac{k_0{}^2 Q_1 - k_1 k_{30}}{k_0{}^2 - 2\beta k_1 k_{30} + \beta^2 k_1{}^2}\frac{3E(k_0)}{8\pi}D^2.$$

(3.12.4)

A first use of these equations is to calculate the turbulent intensities and Reynolds stress by integrating the spectrum functions over all wave numbers. If the effective viscosity v_t is zero, the intensities and Reynolds stress, expressed as functions of the initial intensity are

4

independent of the form of $E(k)$. In particular, for small values of the total strain,

$$\left.\begin{array}{rcl}\overline{u_1}^2 &=& [1+\tfrac{2}{7}\beta^2+O(\beta^4)](\overline{u_1}^2)_{t=0}, \\ \overline{u_2}^2 &=& [1+\tfrac{8}{35}\beta^2+O(\beta^4)](\overline{u_2}^2)_{t=0}, \\ \overline{u_3}^2 &=& [1-\tfrac{4}{35}\beta^2+O(\beta^4)](\overline{u_3}^2)_{t=0}, \\ -\overline{u_1u_3} &=& [\tfrac{2}{5}\beta+O(\beta^5)](\overline{u_1}^2)_{t=0}.\end{array}\right\} \quad (3.12.5)$$

For larger strains, values have been calculated and are shown in figs. 3.14 and 3.15.

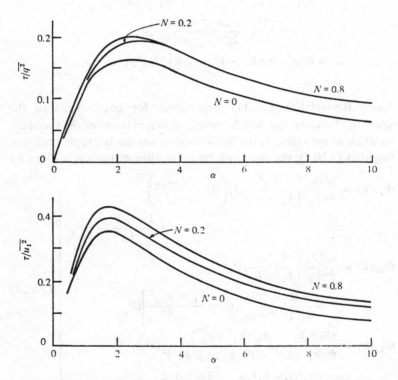

Fig. 3.15. Calculated dependence of the ratio of Reynolds stress to total intensity, using the rapid-distortion approximation for three values of N (from Townsend 1970).

If the effective viscosity is not zero, the ratios do depend on the form of $E(k)$, but most of the variation is dependent on the parameter $v_t/(\alpha L_{11}^2)$, where L_{11} is the longitudinal integral scale of the initial motion. To approximate the turbulent transfer from the larger eddies by a coefficient of viscosity implies a neglect of the smaller eddies, and so it is appropriate to use a form for $E(k)$ that decreases very rapidly with increasing wave number. The form used here is

$$E(k) = \left(\frac{2}{9\pi}\right)^{1/2} k^4 L^5 \, e^{-\frac{1}{4}k^2L^2} \, \overline{(u_1{}^2)}_{t=0}, \qquad (3.12.6)$$

where $L = (\pi/2)^{-1/2} L_{11}$, and figs. 3.14 and 3.15 show the results of calculations for several values of the parameter

$$N = 4\frac{v_t}{\alpha L^2},$$

chosen to cover the likely range in shear flows. For strain ratios less than ten, the general effect is to reduce or to reverse the gain of energy by the u_1 and u_2 components without considerable change in the structure of the eddies. In particular, the ratio of the Reynolds stress to the total turbulent intensity is almost unaffected by the magnitude of the effective viscosity (fig. 3.15).

A flow which approximates to the uniform shearing of homogeneous turbulence may be obtained behind a grid of parallel cylinders with non-uniform spacing, placed across the entrance to a wind-tunnel working-section. With the proper gradation of spacing, a flow with nearly uniform gradient of mean velocity is produced and the turbulent motion is moderately homogeneous. Measurements in such a flow have been made by Rose (1966, 1970) using cylinders of diameter 0.312 cm in a flow of average velocity 1540 cm s^{-1}. The total shear over the range of observation was about 2.7, and measurements were made of the intensities and Reynolds stress, and of spectra and correlations of the longitudinal component of the velocity fluctuation. The development of the Reynolds stress and the changes in intensity of the three components are shown in fig. 3.16, and Rose interprets the results as indicating an approach to an equilibrium state in which all these quantities are independent of decay-time. His measurements show that the Taylor dissipation length

$$\lambda = \left[\overline{u_1{}^2} \Big/ \overline{\left(\frac{\partial u_1}{\partial x_1}\right)^2} \right]^{1/2}$$

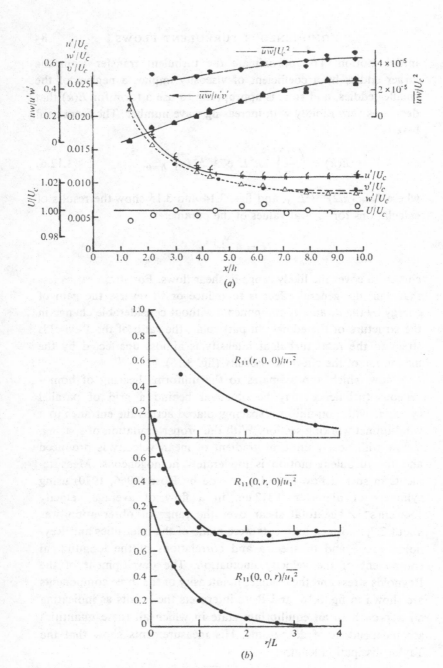

Fig. 3.16. Development of grid turbulence in a flow of uniform simple shearing. (a) Variation of intensities and Reynolds stress, $z/h = 0.5$, $y/h = 0$, $U_c = 50.5$ ft s^{-1}, $h = 12$ in. (b) Comparison of observed correlations with those calculated from the rapid distortion approximation (from Rose 1966).

increases steadily with decay-time and it is unlikely that the apparent energy equilibrium would persist. The possibility of structural equilibrium without energy equilibrium is open.

In comparing the results with the predictions of the rapid-distortion theory, it must be kept in mind that the initial turbulent flow cannot be controlled independently of the mean flow and that it may be far from isotropic, but the measurements immediately behind the grid suggest that the turbulence may be isotropic in the crude sense that the Reynold stress is small and the intensities of the three components nearly equal. Comparison of the intensity measurements with the theory shows that:

(1) The shear correlation coefficient, $-\overline{u_1 u_3}/[\overline{u_1^2 u_3^2}]^{1/2}$, increases initially nearly as 0.31β, compared with the rapid-distortion rate of 0.4β.

(2) The coefficient reaches a value near 0.45 near the end of the duct, considerably less than predicted values around 0.6.

(3) The ratio $-\overline{u_1 u_3}/\overline{q^2}$ approaches a value near 0.155, a little less than the theoretical values of 0.16 ± 0.01 for zero viscosity and 0.185 ± 0.01 for $N = 0.2$ and the range of β of 1.5–3.5.

(4) The intensities of the three components are in the predicted order, but the differences are less than is predicted by the theory.

To a considerable extent, the differences between the observed values and the predicted ones can be reconciled if the contributions of the smaller, more nearly isotropic eddies to the intensities is represented by adding equal amounts to the three component intensities. If one-quarter of the total energy resides in a group of isotropic smaller eddies and the remaining three-quarters has the form predicted by the theory for $N = 2$,

$$\frac{-\overline{u_1 u_3}}{[\overline{u_1^2 u_3^2}]^{1/2}} = 0.44 \quad \text{and} \quad \frac{-\overline{u_1 u_3}}{\overline{q^2}} = 0.141,$$

in good agreement. For likely values of the effective viscosity, the theory predicts very little variation of the coefficients over the range of the measurements, $\beta = 1.5 - 2.7$, possibly the cause of the apparent structural equilibrium.

The measurements of correlations and spectra are in good agreement with the predictions. Using a compressed specification of correlation form (see § 4.2), the measurements of Rose are compared with the theoretical predictions in table 3.1. The correspondence, both of shapes and of extent in r, is good and suggests strongly that the larger eddies of the turbulence have very nearly the structure predicted by the rapid-distortion theory.

TABLE 3.1 *Correlation shapes observed in sheared turbulence compared with predictions of the rapid-distortion theory*

	Theory ($\beta = 2$)		Rose ($\beta = 2.02$)		Rose ($\beta = 2.4$)	
	+	−	+	−	+	−
$R_{11}(r, 0, 0)$	4.2	−	4.4 in	−		
$R_{11}(0, r, 0)$	−	1.2 (0.08)		−	−	1.2 in (0.06)
$R_{11}(0, 0, r)$	2.2	−	1.86 in	−	2.10 in	−
$R_{13}(0, 0, r)$	3.8	−			3.3 in	−

Note: An entry in the '+' column indicates a nearly exponential form and gives the value of r for which $R(r)/R(0) = 0.05$. An entry in the '−' column gives the value of r for zero correlation and, in brackets, the extreme negative value of $R(r)/R(0)$.

The only serious divergence between the measurements and the predictions is that the intensity ratios are closer to one than is expected. Some of this is certainly caused by the presence of more nearly isotropic, smaller eddies which receive energy from the larger ones as well as from the mean flow. Generally, the measurements of the behaviour of turbulent fluid under either irrotational distortion or plane shearing show that the larger eddies, possessing most but not all of the energy, are persistent and coherent structures which interact with the mean flow in the 'rapid-distortion' manner and which lose energy and interact with the smaller eddies by processes that can be modelled by an effective viscosity.

3.13 Local isotropy and equilibrium of small eddies

The observations and conclusions of the previous sections concern mostly the aspects of the turbulent flow which depend on the energy-containing eddies, and one conclusion is that the rate of energy dissipation in turbulent flow is determined by the structure and intensity of these eddies. Since the actual conversion of mechanical energy to heat is carried out by much smaller eddies of almost negligible total energy, there is a strong suggestion that the energy is transferred to them by a cascade of instabilities, larger eddies breaking up to form eddies one size smaller that become unstable in their turn and break up into still smaller ones. If the rate of energy flow down the cascade of eddy sizes is limited by the capacity of the first instability, the smaller eddies must adjust their motion to pass on the imposed energy flow to the smallest eddies that dissipate it as heat. In 1941, A. N. Kolmogorov made explicit the consequences of the cascade hypothesis for the structure of the smaller eddies that contain only a small part of the whole turbulent energy. If the Reynolds number of the flow is sufficiently large, he showed that the smaller eddies must be in a state of absolute equilibrium in which the rate of receiving energy from larger eddies is very nearly equal to the sum of the rates of loss to smaller eddies by breaking up and by working against viscous stresses. Further, the motion of the smaller eddies is related only weakly to the, possibly, inhomogeneous and anisotropic large eddies and their motion should be nearly isotropic. On this basis, the theory of local isotropy asserts that the motion of the smaller eddies depends only on the energy flow from the energy-containing eddies and on the fluid properties, otherwise being the same for all kinds of turbulent flow. Accordingly, there is no need to study the smaller eddies in all the many kinds of turbulent flow. Once the energy loss from the large energy-containing eddies is known, the structure of the smaller ones is determined. The hypothesis of local isotropy and similarity is of central importance in the study of turbulent flow, and its validity provides the justification for later concentration of attention on the mean flow and the energy-containing eddies.

The theory of local isotropy applies to the small scale aspects of the turbulent motion, with eddies that are too small to contribute more than a small fraction to the total kinetic energy but are responsible for nearly the whole of the energy dissipation by viscous action. If the integrated spectrum function, $E_{ii}(k)$ (for convenience the suffices will be dropped), is used to define the energy associated with eddies of 'diameter' k^{-1}, the theory applies to wave numbers greater than k_e, where k_e is the smallest wave number satisfying the inequalities,

$$\int_{k_e}^{\infty} E(k)\, dk \ll \int_0^{\infty} E(k)\, dk = \overline{q^2}$$

and

$$\int_0^{k_e} k^2 E(k)\, dk \ll \int_0^{\infty} k^2 E(k)\, dk = \varepsilon/\nu.$$

(3.13.1)

The first condition is satisfied if k_e is large compared with the centroid of the spectrum function, in practice with L_ε^{-1} where L_ε is the dissipation length-scale of the turbulence. The second condition requires that k_e should be small compared with the centroid of the 'dissipation' spectrum function, $k^2 E(k)$. Both conditions can be satisfied if the Reynolds number of the turbulent motion, $(\overline{q^2})^{1/2} L_\varepsilon/\nu$, is large.

If the equilibrium eddies contain only a small fraction of the total energy, their diameters are small compared with those of the energy-containing eddies and with the scales of inhomogeneity of the mean flow and the turbulent motion, and their dynamics are nearly the dynamics of homogeneous turbulence. Then the equation for the spectrum function is

$$\frac{\partial E(k)}{\partial t} = -2\nu k^2 E(k) + 2\frac{\partial U_i}{\partial x_j} \int_{|\mathbf{k}|=k} \Phi_{ij}(\mathbf{k})\, dA(k)$$
$$- \frac{\partial U_i}{\partial x_j} \frac{\partial}{\partial k} \int_{|\mathbf{k}|=k} \frac{k_i k_j}{k} \Phi_{ll}(\mathbf{k})\, dA(k) - \frac{\partial S(k)}{\partial k}, \quad (3.13.2)$$

where the mean velocity gradient is the local value in the flow, and $S(k)$ is the rate of transfer of turbulent intensity from wave numbers smaller than k to wave numbers larger than k. The two terms involving the velocity gradient represent respectively transfer of energy from the mean flow to components of wave number k and the net effect of the redistribution of energy in wave-number space by the 'distortion' of equation (3.2.6). Integrating from k to infinity gives

$$S(k) - 2\nu \int_k^{\infty} k'^2 E(k')\, dk' = \frac{\partial}{\partial t} \int_k^{\infty} E(k')\, dk'$$
$$- 2\frac{\partial U_i}{\partial x_j} \int_{|\mathbf{k}|=k} \Phi_{ij}(\mathbf{k})\, d\mathbf{k}$$
$$- \frac{\partial U_i}{\partial x_j} \int_{|\mathbf{k}|=k} \frac{k_i k_j}{k} \Phi_{ll}(\mathbf{k})\, dA(k) \quad (3.13.3)$$

and the terms on the right are negligible unless either the mean velocity gradient or the logarithmic decay rate is large compared with the turbulent rate of shear, $(\overline{q^2})^{1/2}/L_\varepsilon$. For $k > k_e$,

$$S(k) = 2v \int_k^\infty k'^2 E(k') \, dk' \quad \text{and} \quad S(k_e) = 2\varepsilon. \tag{3.13.4}$$

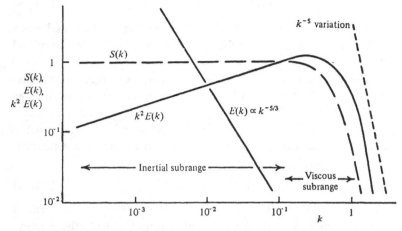

Fig. 3.17. Spectra of energy, dissipation and energy transfer in the Kolmogorov range of local similarity.

The spectral range, $k > k_e$, is nearly in a state of absolute equilibrium in the sense that all parts of it can adjust themselves to external changes at rates that are large compared with rates of change of the whole flow, i.e. rates of decay and rates of shear of the mean and turbulent motion. Then the structure of the equilibrium range of eddies is dependent only on the driving forces and on the fluid properties. The driving forces have their origin in the mean flow and in the eddies outside the equilibrium range, but if the energy reaching the range has passed through many intermediate stages of instability and breakdown, it is likely that the motion will be nearly independent of the directional bias and anisotropy of the large-scale motion. Without a full understanding of the mechanics of energy transfer between eddies, it is not possible to say how many stages are necessary, i.e. how large $k_e L_\varepsilon$ must be for effective isotropy, but some observations suggest that all eddies in the equilibrium range are effectively isotropic. Then the spectrum function is completely determined by the scalar, $E(k)$ which can depend only on the energy flow into the range and on the fluid properties because the decoupling introduced by the cascade of instabilities prevents special characteristics of the source of the energy flow having an influence on the motion. For an incompressible, Newtonian fluid, dimensional considerations lead to

$$E(k) = u_s{}^2 k_s{}^{-1} E^*(k/k_s), \tag{3.13.5}$$

where $u_s = (\varepsilon v)^{1/4}$, $k_s = \varepsilon^{1/4} v^{-3/4}$, are scales of velocity and wave number characteristic of the dissipating eddies, and E^* is a universal function.

If the Reynolds number of the turbulence is large, viscous dissipation is negligible over a large range of wave number with lower bound k_e, and viscous forces have no direct influence on the motion described by these components. In such an *inertial subrange* of the equilibrium spectrum, the spectrum function should be independent of the fluid viscosity and then

$$E(k) = C\varepsilon^{2/3} k^{-5/3}, \tag{3.13.6}$$

where C is an absolute constant of order one. The inertial form of the spectrum holds for eddies small compared with the energy-containing eddies, i.e. $kL_\varepsilon \gg 1$, but large compared with those responsible for viscous dissipation of energy. Energy is dissipated in the inertial subrange at a rate of

$$v \int_{ke}^{k_i} k^2 E(k)\, dk = \tfrac{3}{4} C v \varepsilon^{2/3}(k_i{}^{4/3} - k_e{}^{4/3}) = \tfrac{3}{4} C\left[\left(\frac{k_i}{k_s}\right)^{4/3} - \left(\frac{k_e}{k_s}\right)^{4/3}\right]\varepsilon,$$

where k_i is the upper limit of the range, and the rate fails to be small compared with the total rate ε if k_i is comparable with k_s. By definition of k_s and the dissipation length-scale L_ε,

$$k_s L_\varepsilon = \left(\frac{(\overline{q^2})^{1/2} L_\varepsilon}{v}\right)^{3/4} \tag{3.13.7}$$

and the condition for a recognisable inertial subrange is that $k_s L_\varepsilon \gg k_i L_\varepsilon \gg k_e L_\varepsilon$, and requires a very large value of the turbulence Reynolds number.

In practice, measurements are made of the one-dimensional spectrum functions $\phi_{11}(k_1)$ and $\phi_{22}(k_1)$ (where k_1 is in the direction of flow) or of the structure function. The one-dimensional functions are related to $E(k)$ by

$$\phi_{11}(k_1) = \frac{1}{2}\int_{k_1}^{\infty} E(k)\left(1 - \frac{k_1^2}{k^2}\right)\frac{dk}{k},$$

and
$$\phi_{22}(k_1) = \frac{1}{2}\left[\phi_{11}(k_1) + k_1\frac{d\phi_{11}(k_1)}{dk_1}\right],$$

(3.13.8)

since the motion in the equilibrium range is isotropic, and results similar to (3.13.5) follow:

$$\phi_{11}(k_1) = u_s^2 k_s^{-1}\phi_{11}^*(k_1/k_s).$$

(3.13.9)

Within the inertial subrange, equation (3.13.8) leads to

$$\phi_{11}(k_1) = \frac{9C}{55}\varepsilon^{2/3}k_1^{-5/3},$$

(3.13.10)

and

$$\phi_{22}(k_1) = \frac{12C}{55}\varepsilon^{2/3}k_1^{-5/3}.$$

(3.13.11)

The longitudinal structure function $B_{11}(r,0,0)$ is related to $\phi_{11}(k_1)$ by

$$B_{11}(r, 0, 0) = 2\int_0^{\infty}\phi_{11}(k_1)(1 - \cos k_1 r)\,dr$$

(3.13.12)

and, if $k_s^{-1} \ll r \ll L$, the integral is dominated by contributions from the inertial subrange of the spectrum. Then,

$$B_{11}(r, 0, 0) = \tfrac{68}{55}C\varepsilon^{2/3}r^{2/3}\int_0^{\infty}x^{-5/3}(1 - \cos x)\,dx$$
$$= \tfrac{81}{110}\tfrac{1}{3}!\,C\varepsilon^{2/3}r^{2/3}.$$

(3.13.13)

In the inertial subrange, the transverse structure function is related to the longitudinal function by

$$B_{22}(r, 0, 0) = B_{11}(r, 0, 0) + \tfrac{1}{2}r\,dB_{11}/dr$$
$$= \tfrac{4}{3}B_{11}(r, 0, 0).$$

(3.13.14)

3.14 Measurements of spectrum and structure functions

All arguments for the existence of an equilibrium range of locally isotropic eddies either give necessary conditions for its existence or depend on assumptions about the nature of energy transfer between eddies, and acceptance of the theory rests equally on plausibility of the arguments (which give confidence on its generality) and on comparison of its predictions with observations of real flows. The predictions concerning the spectrum function are:

(1) It has a universal form with scales of velocity and wave number determined by the energy dissipation and by the fluid viscosity.

(2) Within the inertial subrange of the equilibrium spectrum, the longitudinal spectrum function has the form

$$\phi_{11}(k_1) = C'\varepsilon^{2/3}k_1^{-5/3}. \qquad (3.14.1)$$

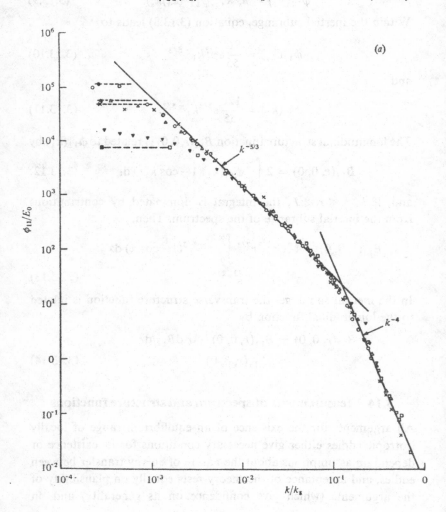

(3) The motion in the equilibrium range is locally isotropic and the longitudinal and transverse spectra are related by

$$\phi_{22} = \tfrac{1}{2}(\phi_{11} - k\phi_{11}').$$ (3.14.2)

Many measurements have been made which bear on the validity of these predictions but only a few can be considered here.

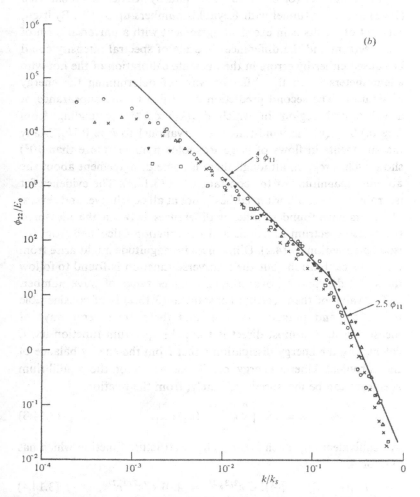

Fig. 3.18. One-dimensional spectra of grid turbulence at large Reynolds numbers compared with predictions of local similarity (from Kistler & Vrebalovich 1966). (a) Longitudinal component, (b) transverse component. Re_L: $\bigcirc$, 1.3 $\times 10^4$; $\triangle$, 7.5 $\times 10^3$; $\times$, 6.5 $\times 10^3$; $\square$, 1.5 $\times 10^3$; $\blacktriangledown$, 3.2 $\times 10^3$.

The evidence for the existence of a universal form for the spectrum function is good, even in flows with Reynolds numbers too small to satisfy the basic requirements for energy equilibrium. Figs. 3.18 and 3.19 show measurements of the longitudinal function made by Grant, Stewart & Moilliet (1962a) in a tidal flow of very large Reynolds number (of order 10^6–10^7), and by Kistler & Vrebalovich (1966) in a wind tunnel with Reynolds numbers up to 10^4. By itself, each set of results is in excellent agreement with a universal form of the spectrum and the differences in scale of spectral intensity could be caused either by errors in the absolute calibration of the hot-wire anemometers or in the different ways of determining the energy dissipation. The second prediction is verified by the appearance of a substantial region in which $\phi_{11}(k_1) \propto k^{-5/3}$ extending from $k \approx 0.1 L_\varepsilon^{-1}$ (in the wind-tunnel observations) to $k \approx 0.1 k_s$. Many measurements in flows of large Reynolds number (more than 10^3) show such a region, although there is some disagreement about the absolute magnitude of the constant C' in (3.14.1). The evidence for isotropy of the equilibrium eddies is not at all conclusive, and several observers have found significant differences between the measured transverse spectrum functions and the function calculated from the isotropic relation (3.14.2). Differences in magnitude might arise from errors in calibration, but the transverse function is found to follow the $k^{-5/3}$ distribution over a much smaller range of wave number.

The value of the spectrum constant in (3.13.1) is of considerable theoretical and practical interest, and there are several ways of measuring it. The most direct is from the spectrum function itself, calculating the energy dissipation either from the energy balance of the turbulent kinetic energy or, if the whole of the equilibrium spectrum can be measured accurately, from the relation,

$$\varepsilon = 15\nu \int_0^\infty k_1^2 \phi_{11}(k_1) \, dk_1. \tag{3.14.3}$$

An equivalent approach is through the structure function which has the form,

$$B_{11}(r, 0, 0) = \tfrac{9}{4}(\tfrac{1}{3})! \, C' \varepsilon^{2/3} r^{2/3} = 4.01 \, C' \varepsilon^{2/3} r^{2/3}, \tag{3.14.4}$$

for $k_s^{-1} \ll r \ll L_\varepsilon$ (from (3.13.10) & (3.13.13)). Structure functions are especially convenient for measurements in the atmospheric boundary layer but determination of the dissipation from the energy

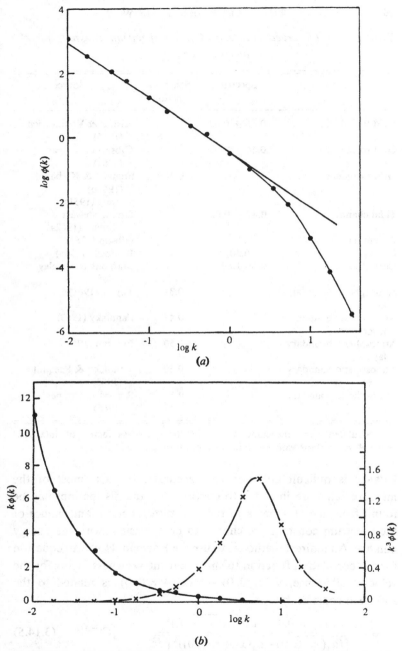

Fig. 3.19. One-dimensional spectra of turbulence in tidal flow compared with predictions of local similarity (from Grant, Stewart & Moilliet 1962a).

TABLE 3.2 *Observed values of C', the spectrum constant for the inertial subrange*

Flow	Spectrum	Structure function	Source
Grid turbulence	0.66 ± 0.03		Kistler & Vrebalovich (1966)
Grid turbulence	0.44 ± 0.02		Gibson & Schwarz (1963)
Grid turbulence		0.45 [a]	Frenkiel & Klebanoff (1967b) Stewart (1951)
Tidal channel	0.47 ± 0.02		Grant, Stewart & Moilliet (1962a)
Round jet	0.52		Gibson (1963)
Plane jet	0.50		Bradbury (1965)
Curved jet	0.68–1.80		Margolis & Lumley (1965)
Atmospheric boundary layer		0.31	Taylor (1961)
Atmospheric boundary layer		0.47	Panofsky (1963)
Atmospheric boundary layer		0.50	Frenzen (1965)
Atmospheric boundary layer		0.50	Panofsky & Pasquill (1963)
Atmospheric boundary layer		0.50	Record & Cramer (1966)

[a] Calculated from measured values of the skewness factor of $[u(x, y, z) - u(x + r, y, z)]$ for values of r in the inertial range.

balance is difficult and not very accurate. In fact, most of the measurements are intended to confirm that the dissipation may be found from measurements of the structure function, and values of the spectrum constant are chosen to be consistent with the energy balance. An indirect method is to use the Kármán–Howarth equation for the correlation function to show that the skewness factor for the velocity difference, $u_1(x_1, 0, 0) - u_i(x_1 + r, 0, 0)$, is related to the spectrum constant by

$$S_i = \frac{\overline{(u_1(x_1, 0, 0) - u_1(x_1 + r, 0, 0))^3}}{[\overline{(u_1(x_1, 0, 0) - u_1(x_1 + r, 0, 0))^2}]^{3/2}} = -\tfrac{4}{5} C'^{-3/2} \quad (3.14.5)$$

for values of r large compared with k_s^{-1} and small compared with the scale of the energy-containing eddies. Measurements of S_i by Stewart (1951) and by Frenkiel & Klebanoff (1965, 1967a, b) were made in grid turbulence of Reynolds number too small for the appearance of a distinct inertial subrange, and it is difficult to choose an appropriate value for S_i. Any value from -0.22 to -0.30 seems possible.

Table 3.2 summarises some of the information about the spectrum constant C'. From a consideration of the likely sources of error, the present state of knowledge is that

$$C' = 0.50 \pm 0.03. \tag{3.14.6}$$

3.15 Energy transfer in the inertial subrange

In view of our ignorance of the mechanisms that transfer energy between eddies of different sizes, the arguments for the existence of the equilibrium range and the similarity spectrum have been made as general as possible and the form of the spectrum derived by dimensional argument. The magnitude of the spectrum constant depends on the efficiency of the transfer process and an estimation of its value requires some assumptions about the nature of the process. The essence of the cascade theory of transfer is that a distinction may be made between eddies larger than a particular size and eddies smaller than that size, the larger ones losing energy to the smaller ones in an irreversible fashion. If the difference of size between the two groups of eddies were really large, the smaller eddies would develop Reynolds stresses in response to the distortion imposed by the larger ones in much the same way as the eddies of homogeneous turbulence respond to a mean flow distortion. Using the results of the previous discussion of distorted homogeneous turbulence, we know that the Reynolds stresses induced by mean velocity gradients $\partial U_i/\partial x_j$ are for small total strains,

$$-\overline{u_i u_j} = \tfrac{2}{5}\overline{u_1^2}\left(\frac{\partial U_i}{\partial x_j}+\frac{\partial U_j}{\partial x_i}\right)t_0, \tag{3.15.1}$$

where t_0 is the duration of the strain. Then the rate of working of the mean flow on the turbulence is

$$-\overline{u_i u_j}\,\frac{\partial U_i}{\partial x_j} = \tfrac{2}{5}\overline{u_1^2}\,\frac{\partial U_i}{\partial x_j}\frac{\partial U_i}{\partial x_j}\,t_0. \tag{3.15.2}$$

Now, both for distorted homogeneous turbulence and for ordinary

shear flows, the significant Reynolds stresses for normal conditions
are quite close to those given by (3.15.1) for a time sufficient to allow
unit strain in simple shearing, i.e. for

$$t_0 = \left(\frac{\partial U_i}{\partial x_j}\frac{\partial U_i}{\partial x_j}\right)^{-1/2}.$$

Then the average rate of working is nearly

$$-\overline{u_iu_j}\,\frac{\partial U_i}{\partial x_j} = \tfrac{4}{3}\overline{u_1^2}\left(\frac{\partial U_i}{\partial x_j}\frac{\partial U_i}{\partial x_j}\right)^{1/2}. \tag{3.15.3}$$

To apply these results to the spectrum transfer, a first approximation
is to suppose that components of wave number less than k provide
the mean velocity field and that components of wave numbers more
than k are the turbulence. Then the energy transfer from components
of wave number less than k to ones of wave number larger than k
is, in the inertial subrange,

$$\varepsilon = \tfrac{2}{15}\int_k^\infty E(k')\,\mathrm{d}k \times \left[\int_0^k k'^2E(k')\,\mathrm{d}k'\right]^{1/2}. \tag{3.15.4}$$

Substituting $E(k) = C\varepsilon^{2/3}k^{-5/3}$, we find that

$$C^{3/2} = 10/3^{1/2}, \qquad C = 3.2, \qquad C' = 0.27.$$

The value for the spectrum constant obtained by this argument
could be improved by abandoning a sharp distinction between large
eddies, i.e. wave numbers just less than k, and small eddies with
wave numbers just greater than k. Transfer of energy between such
components is unlikely because the Fourier components concerned
derive their amplitudes either from the same physical eddy or from
distinct eddies separated in space. In neither case is energy transfer
probable. If a zone of approximately half an octave were excluded
from the integrals in equation (3.15.4), the spectrum constant would
be equal to the observed value of about 0.50. The form for the
energy transfer (3.15.4) was first proposed by Obukhov (1941), and,
like other forms proposed by Heisenberg (1948) and by Kovasznay
(1948), assumes a separation of inertial and viscous effects that can
be valid only in the inertial subrange.

3.16 The equilibrium spectrum in the viscous subrange

No strong predictions can be made about the form of the spectrum
for wave numbers in the viscous range, which depends on the exact

nature of the interactions that transfer energy between eddies. Several attacks on the problem have been made, all of them unsatisfactory in some ways. The best known assume that $S(k)$, the transfer term in the spectrum equation, is determined by the form of the spectrum function in the neighbourhood of k – in accordance with the cascade hypothesis – but also that it is independent of the fluid viscosity. The latter assumption is reasonable in the inertial range but must be suspect in the viscous range where more energy is lost by working against viscous forces than is passed to smaller eddies. The first formulation is due to Heisenberg (1948) who argued that the effect of the smaller eddies, i.e. components with wave numbers more than k, on the larger ones is equivalent to an eddy viscosity. His form is

$$S(k) = \tfrac{32}{9}C^{-3/2} \int_0^k E(k')k'^2 \, dk' \times \int_k^\infty [E(k')/k'^3]^{1/2} \, dk', \quad (3.16.1)$$

where C is the spectrum constant in equation (3.13.6). The second, due to Obukhov (1941), assumes that the smaller eddies provide a Reynolds stress for the larger ones to work on, and that the Reynolds stress is proportional to the intensity of the smaller eddies. Then,

$$S(k) = \frac{8C^{-3/2}}{3\sqrt{3}} \left[\int_0^k E(k')k'^2 \, dk' \right]^{1/2} \int_k^\infty E(k') \, dk'. \quad (3.16.2)$$

The third, due to Kovasznay (1948), is the simplest and assumes the transfer to be determined by the local spectral intensity $E(k)$. Then,

$$S(k) = 2C^{-3/2}[E(k)]^{3/2}k^{5/2}. \quad (3.16.3)$$

Within the inertial subrange, each form leads to the standard result, $E(k) = C\varepsilon^{2/3}k^{-5/3}$, but each leads to anomalous results in the viscous range, viz.,

Heisenberg: $E(k) \propto k^{-7}$ for $k \ll k_s$, i.e. $\partial^n u/\partial x^n$ is unbounded if $n \geqslant 3$, contrary to the Navier–Stokes equations.

Obukhov: $E(k)$ increases indefinitely for large k/k_s.

Kovasznay:

$$E(k) = C\varepsilon^{2/3}k^{-5/3}\left[1 - \tfrac{1}{4}C\left(\frac{k}{k_s}\right)^{4/3}\right]^2 \quad (3.16.4)$$

becoming zero at $k/k_s = (4/C)^{3/4}$.

Another approach to the spectrum of the dissipating eddies, different but also unsatisfactory, supposes that the diffusive action of turbulence extends vortex filaments and concentrates vorticity into sheets whose thicknesses are governed by a balance between outward viscous diffusion and the lateral compression associated with the stretching. Assuming the turbulent rates of strain provide the lateral compression and resemble locally plane straining, it can be shown that the spectrum function in the dissipation range is

$$E(k) = \left(\frac{8}{\pi}\right)^{1/2} \varepsilon^{1/4} \nu^{5/4} \left(\frac{k}{k_s}\right)^{-2} \exp\left[-2\left(\frac{k}{k_s}\right)^2\right] \qquad (3.16.5)$$

(Townsend 1951b). A serious objection is that the vorticity of the sheets and the lateral compressions are parts of the same motion

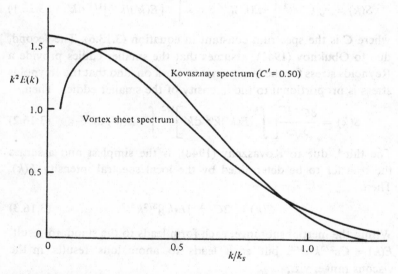

Fig. 3.20. Comparison of spectral distribution of energy dissipation, as given by the vortex-sheet model and the transfer assumption of Kovasznay (1948).

and are not independent of each other, but the assumed motion does satisfy the equations of motion and the behaviour of the spectrum function for large k/k_s is good. Naturally, no account is taken of the inertial subrange.

Measurements of the spectrum function in the viscous range are in fair agreement with the forms derived from the Heisenberg,

Kovasznay and vortex-sheet models. All three forms satisfy the constraint that

$$\int_0^\infty k^2 E(k) \, dk = \varepsilon/\nu \qquad (3.16.6)$$

and the first two must give the same inertial spectrum for small k/k_s. A close resemblance of form is therefore not surprising. The similarity of the vortex-sheet spectrum indicates that the vortex sheets, if they exist, have nearly the average thickness calculated from the crude assumptions of the theory. Fig. 3.20 shows the predicted dissipation spectra, $k^2 E(k)$, in the non-dimensional form of the similarity theory for the Kovasznay and vortex-sheet models. The Heisenberg spectrum is hardly distinguishable from the Kovasznay spectrum in the form used here.

3.17 Local isotropy in non-Newtonian fluids

If distortion of the fluid leads to stresses which cannot be described by a Newtonian coefficient of viscosity, some modification of the theory is necessary but the results for the inertial subrange should be valid. The fundamental property of the inertial subrange is that molecular stresses exert a negligible influence on the motion of the eddies in the range, and so the nature of stresses should be irrelevant. If the instantaneous molecular stresses depend only on the present and past rates of deformation and rotation of the fluid particle now at the position of measurement, the stresses are determined by quantities dependent on the smallest eddies of the turbulent motion, and the principle of Reynolds number similarity will be valid if Reynolds numbers based on the coefficients analogous to viscosity are large. For really large Reynolds numbers, the condition for an inertial subrange can be satisfied and the ordinary $k^{-5/3}$ spectrum should describe the motion. Naturally, the motion in the viscous subrange must depend on the exact rheological properties of the fluid, but a rapid decrease in spectral intensity is inevitable.

Another kind of non-Newtonian behaviour occurs if the fluid contains long-chain molecules or fibres of lengths comparable with scales of the turbulent motion. The effective stresses then depend on the flow deformation over volumes of diameter comparable with the length of the molecule or fibre, and all motions on that scale or less must be affected by molecular stresses. The condition for Reynolds number similarity or an inertial subrange is that the overall lengths of the molecules should be small compared with the scales of motion, a condition not depending directly on a Reynolds number. The condition is still more complicated if the molecules exist in unstrained fluid as rolled-up chains of diameter about 10 nm but can be unrolled by shearing to lengths of order 100 μm. Behaviour of this kind is found in dilute solutions of polymers.

CHAPTER 4

INHOMOGENEOUS SHEAR FLOW

4.1 Large eddies and the main turbulent motion

Both the ideal 'complete' flows and the more complicated flows of practical importance are strongly inhomogeneous, and the nature and consequences of the inhomogeneity are similar for all flows of a single class. For example, many aspects of the turbulent motion are similar in all free turbulent flows, the differences between wakes and jets being far less than the differences between jets and any examples of wall turbulence or of convective turbulence. In free turbulence, inhomogeneity arises from spreading of the flow into the ambient non-turbulent fluid, while the inhomogeneity of wall turbulence is caused by the physical restriction of the solid boundary. It could be argued that the classification of turbulent flows depends on the kind of inhomogeneity imposed by the flow boundaries and on the mode of energy release to the turbulent motion.

The Kolmogorov theory of local isotropy, supported by observation, asserts that all fully turbulent flows share a common structure of the smaller eddies which contribute little to the total energy. Eddies of sizes too large to be included in the range of local equilibrium and isotropy and containing a substantial fraction of the energy may still be moderately small compared with the scales of inhomogeneity of the flow, and may be confined to a region within which the mean velocity gradient is nearly uniform. Larger eddies have sizes comparable with scales of inhomogeneity and their motion cannot be discussed without taking account of the spatial variation of the mean velocity gradient. Whether a clear distinction can be drawn between quasi-homogeneous eddies of a *main turbulent motion* and a possibly weaker set of *large eddies* is debatable, but it is convenient to discuss turbulent motion as if the distinction were valid. Then the flow is composed of:

(1) the mean velocity field $U(x)$,

(2) the large-eddy motion $\mathbf{u}'(\mathbf{x})$, and

(3) the main turbulent motion $\mathbf{u}''(\mathbf{x})$.

Locally, the influence of the mean velocities and the large-eddy motion on the main turbulent motion is approximately that of a velocity field of uniform gradient, composed of:

(1) a bulk convection of the motion with velocity $\mathbf{U} + \mathbf{u}'$ and

(2) a distortion by the nearly uniform velocity gradient,

$$\frac{\partial U_i}{\partial x_j} + \frac{\partial u_i'}{\partial x_j}.$$

In most turbulent flows, variations of mean velocity are considerably greater than the turbulent fluctuations and the large-eddy contribution to the velocity gradient is a small perturbation of the mean flow gradient. Further, the mean velocity distribution imposes plane shearing in all relevant parts of the flow, and we may expect that the main turbulent motion is dominated by the eddy structures that are selected and amplified by plane shearing. In that event, the main turbulent motion should be structurally similar in all shear flows and might be distinguishable from the large-eddy motion by the universality of its structure. Evidence for structural similarity of the main turbulent motion is examined in the next section.

4.2   Structural similarity of the main turbulent motion

Both the main turbulent motion and the large eddies contribute to the velocity fluctuations, but the large eddy contribution varies from one part of a flow to another and from flow to flow while variations in the contribution from the main motion should be limited nearly to changes in the scales of velocity and length. In principle, measurements of correlation functions could be analysed to extract the common structure of the main turbulent motion, but the inaccuracy and incompleteness of the available measurements make it preferable to compare observations with correlation functions calculated from the rapid-distortion theory. The functions to be used are calculated for initially isotropic turbulence with the 'exponential' correlation function

$$R_{ij}(\mathbf{r}) = [\delta_{ij}(1 - \tfrac{1}{2}r/L_0) + \tfrac{1}{2}r_i r_j/(rL_0)]\, \mathrm{e}^{-r/L_0} \qquad (4.2.1)$$

and the spectrum function,

$$\Phi_{ij}(\mathbf{k}) = (\delta_{ij} - k_i k_j / k^2) \frac{2}{\pi^2} \frac{k^2 L_0^5}{(1 + k^2 L_0^2)^3} \qquad (4.2.2)$$

(L_0 is the integral length-scale of the turbulence) that has been subjected to a plane shear with a total strain ratio of two. The correlation function (4.2.1) is a fair approximation to that observed in grid turbulence, and a strain ratio of two gives a degree of anisotropy comparable with that found in common shear flows.

The simplest index of structure is the relative values of the components of the Reynolds stress tensor, and ratios found in a number

TABLE 4.1 *Reynolds stress tensor in shear flow*

Flow	$\tau/\overline{u^2}$	$\tau/\overline{w^2}$	$2\tau/(\overline{u^2} + \overline{w^2})$	$\tau/\overline{q^2}$	$\overline{u^2}/\overline{w^2}$	$\overline{u^2}/\overline{v^2}$
Boundary layer						
$a = 0$						
Outer layer	0.24	0.48	0.32	0.13	1.95	1.37
Inner layer	0.20	1.2	0.35	0.12	6.3	2.16
$a = -0.15$						
Outer layer	0.27	0.47	0.34	0.13	2.4	2.1
Inner layer	0.17	0.57	0.23	0.095	3.6	2.0
$a = -0.255$						
Outer layer	0.29	0.60	0.39	0.14	2.1	1.3
Inner layer	0.13	0.48	0.21	0.07	3.4	1.4
Cylindrical mixing layer	0.51	0.60	0.54	0.18	1.2	1.0
Plane jet	0.38	0.53	0.45	0.16	1.4	1.45
Plane wake	0.36	0.48	0.41	0.15	1.3	1.6
Pipe						
$r = \frac{1}{2}R$	0.30	0.61	0.40	0.14	2.1	1.5
Near wall	0.21	1.00	0.34	0.115	4.8	1.3
Channel						
$z = 0.1d$	0.33	1.22	0.52	0.205	3.7	3.0
$z = \frac{1}{2}D$	0.275	0.80	0.41	0.16	2.9	2.65
Plane mixing layer	0.38	0.66	0.43	0.17	1.75	–

Notes: (1) In the boundary-layer data, a is the exponent of the power-law describing the variation of free stream velocity.

(2) Sources: Boundary layer, Klebanoff (1955) and from Bradshaw (1967b); cylindrical mixing layer, Bradshaw, Ferriss & Johnson (1964); channel, Laufer (1951); pipe, Laufer (1955); plane jet, Bradbury (1965); plane wake, Townsend (1949a); plane mixing layer, Liepmann & Laufer (1947).

of shear flows are shown in table 4.1. Contributions from the large eddies and errors of measurement combine to produce a large scatter of the ratios, but the dispersion is less if measurements close to walls are omitted. Table 4.2 contains values of $-\overline{u_1 u_3}/\overline{u_1}^2$ and $-\overline{u_1 u_3}/\overline{q^2}$ as averages over several groups of flows.

TABLE 4.2 Variability of stress coefficients

Group of flows	$-\overline{u_1 u_3}/\overline{u_1}^2$ $=\tau/\overline{u^2}$	S.D.	$-\overline{u_1 u_3}/\overline{u_3}^2$ $=\tau/\overline{w^2}$	S.D.	$-\overline{u_1 u_3}/\overline{q^2}$ $=\tau/\overline{q^2}$	S.D.
All	0.29	0.10	0.69	0.25	0.13	0.03
No wall layers	0.34	0.08	0.58	0.14	0.15	0.02
Wall layers with small pressure gradients	0.25	0.06	1.14	0.17	0.15	0.04
Wall layers with large pressure gradients	0.15	0.02	0.52	0.1	0.08	0.02

Details of the eddy structure are provided by the correlation function, and the experimental errors are less for measurements of ratios than for measurements of turbulent intensities. Most of the published work concerns the 'principal' correlation functions,

$$R_{11}(r, 0, 0), \qquad R_{11}(0, r, 0), \qquad R_{11}(0, 0, r),$$
$$R_{22}(r, 0, 0), \qquad R_{22}(0, r, 0), \qquad R_{22}(0, 0, r),$$
$$R_{33}(r, 0, 0), \qquad R_{33}(0, r, 0), \qquad R_{33}(0, 0, r),$$

i.e. those with separations parallel to the conventional axes of reference. In the initial isotropic turbulence, principal correlations have the form, $\exp(-r/L_0)$, if the velocity components are parallel to the separation (a *longitudinal* correlation), and the form,

$$(1 - \tfrac{1}{2}r/L_0) \exp(-r/L_0),$$

if the components are at right angles to the separation (a *transverse* correlation). Rapid distortion changes both the length scales of the correlations and their forms, as shown in fig. 4.1. In particular, the transverse correlation $R_{11}(0, 0, r)$ assumes the 'longitudinal' form, while the longitudinal correlation $R_{22}(0, r, 0)$ assumes the 'transverse' form and develops negative values. In general, the rapid-distortion correlations are remarkably similar to those observed by

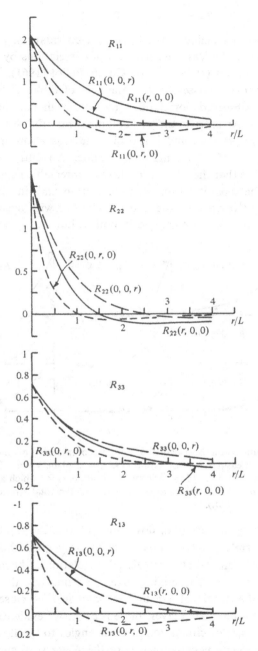

Fig. 4.1. Normal components of the correlation function, calculated for a strain ratio of two (from Townsend 1970).

Grant (1958) in a cylinder wake and in a boundary layer (figs. 4.2, 4.3, 4.4) and to the less extensive sets of correlations by Bradshaw, Ferriss & Johnson (1964) and by Comte-Bellot (1961). The degree of similarity can be assessed with the help of table 4.3 which summarises the observed correlations by entries in columns labelled '+' and '−'. An entry in the '+' column means that the correlation is everywhere positive, and it specifies the separation at which the correlation is 5% of its maximum value. An entry in the '−' column means that the correlation takes appreciable negative values and gives the separation for zero correlation and, in brackets, the extreme negative value. Note that correlations with negative values not exceeding −0.01 are entered in both columns.

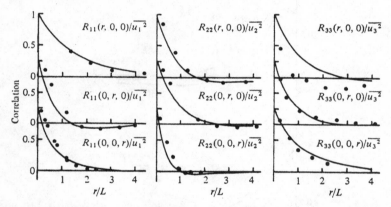

Fig. 4.2. Comparison of calculated normal components of the correlation function with measurements by Grant (1958) in a plane wake. ($U_1 d/v = 1300$, $x/d = 533$, fixed wire at $z/d = 4$ except for $R_{33}(0, 0, r)$ for which $z/d = 2.8$. The 'initial' integral scale is taken to be $L = 3d$ and the total strain ratio is two (from Townsend 1970)).

Considering the variety of flows and authors, the similarity of the observed correlation functions is remarkably good, both between different flows and with the calculated correlations. The most notable discrepancies occur for mixing layers, particularly the cylindrical layer near the nozzle of a round jet, and in flow close to a solid boundary. In the mixing layer, there is extensive evidence of regular eddies with axes of circulation at right angles to the flow, and they cause the strongly negative values of the $R_{11}(r, 0, 0)$ correlation. In an equilibrium layer near a wall, the motion is composed of attached eddies with a wide range of size and the observed correlation function is an average over all the eddy sizes. One consequence is that the $R_{22}(r, 0, 0)$ correlation does not take negative values as it does in flows remote from walls (see § 5.8). Another feature of wall turbulence is the abnormal extent of the $R_{11}(r, 0, 0)$ correlation

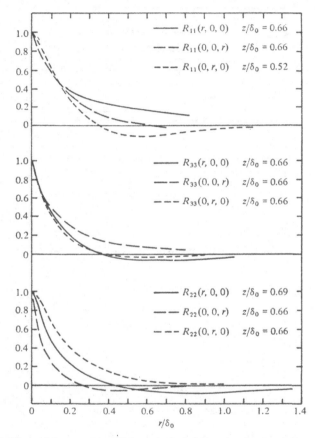

Fig. 4.3. Normal components of the correlation function in the outer part of a boundary layer (after Grant 1958).

compared with all the other components, possibly arising from transient patterns of turbulent-induced secondary flow (see § 7.21).

Similarity between calculated and observed correlations is confirmed by the less extensive measurements that have been made of correlations for displacements inclined to the flow axes, particularly in the $x_1 O x_3$ plane. Fig. 4.5 shows equal-value contours for R_{11}, R_{22}, R_{33}, calculated from the rapid-distortion theory, and they may be compared with Grant's measurements in a wake (fig. 4.6). The most striking feature is that the contours for R_{11} and R_{22} resemble in

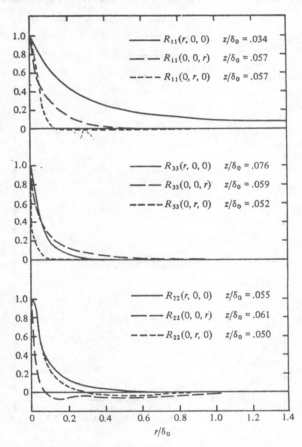

Fig. 4.4. Normal components of the correlation function in the inner part of a boundary layer (after Grant 1958).

shape ellipses with long axes in the positive quadrant at an angle of about 30° to the Or_1 axis, while those of R_{33} are much less elongated and are nearly symmetrical about the flow axes. These features are reproduced in the boundary layer for R_{33}, and in the wake for all three correlations if both sensors are in the same half of the flow.

TABLE 4.3 Normal correlation functions in shear flows

Component	Rapid distortion		Homogeneous shear		Wake		Cylindrical mixing layer		Inner boundary layer		Outer boundary layer		Channel	
	+	–	+	–	+	–	+	–	+	–	+	–	+	–
$R_{11}(r,0,0)$	4.2	–	5.1	–	6	–	–	8 (0.05)	30	–	5	–	–	–
$R_{11}(0,r,0)$	–	1.2 (0.08)	–	1.2 (0.06)	–	2.3 (0.08)	–	4.2 (0.15)	–	2.6 (0.02)	–	1.7 (0.12)	–	1.4 ( )
$R_{11}(0,0,r)$	3.5 (0.01)	–	2.1	–	5	–	7	–	–	–	2.9 (0.03)	–	2.2	–
or	2.2													
$R_{22}(r,0,0)$	–	1.6 (0.10)	?	?	–	2.3 (0.08)	–	4 (0.18)	–	5 (0.02)	–	2.2 (0.08)	?	?
$R_{22}(0,r,0)$	–	2.5 (0.035)	?	?	–	3.1 (0.06)	9	–	–	5 (0.02)	3.0	–	–	1.5
$R_{22}(0,0,r)$	–	1.0 (0.045)	?	?	–	1.8 (0.03)	–	4 (0.14)	–	1.3 (0.06)	–	1.3 (0.06)	–	1.2 ( )
$R_{33}(r,0,0)$	–	2.9 (0.04)	?	?	–	1.5 (0.16)	–	6 (0.02)	?	?	–	1.9 (0.06)	?	?
$R_{33}(0,r,0)$	–	3.5 (0.002)	?	?	4	–	4.4	–	?	?	–	2 (0.02)	?	?
or	1.9													
$R_{33}(0,0,r)$	–	–	?	?	?	?	1.4	–	?	?	?	?	–	1.0 ( )
$R_{13}(r,0,0)$	3.8	–	?	?	?	?	?	?	4	–	?	?	?	?
$R_{13}(0,0,r)$	3.8	–	?	?	?	?	?	?	4	–	?	?	?	?
$R_{13}(0,r,0)$	–	1.05 (0.12)	?	?	?	?	?	?	–	3 (0.20)	–	0.8 (0.18)	?	?
$R_{13}(0,0,r)$	3.2	–	3.1	–	?	?	?	?	–	–	3	–	?	?

Notes: (1) For each flow the scale of length has been chosen convenient but preserving the ratios of scales in each flow. No information is denoted by '?'.

(2) Sources: Homogeneous shear – Rose (1966); wake, inner and outer boundary layer – Grant (1958); cylindrical mixing layer – Bradshaw et al. (1964); two-dimensional channel – Comte-Bellot (1961).

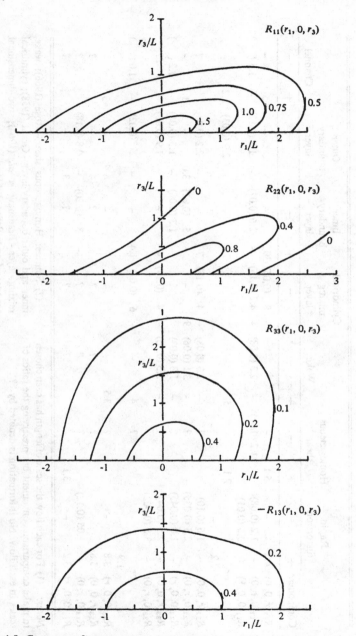

Fig. 4.5. Contours of constant correlation for separations n the plane $r_2 = 0$, for a strain ratio of two (from Townsend 1970).

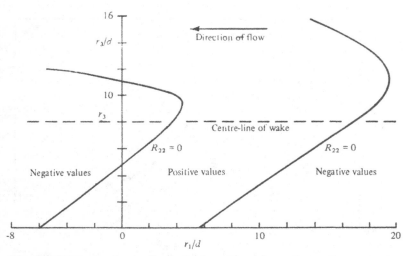

Fig. 4.6. Contours of zero $R_{22}(r_1, 0, r_3)$ measured in a plane wake by Grant (1958). To simplify comparison with fig. 4.5, the co-ordinate directions are arranged so that the fixed wire is at $x_3/d = -7.6$ and the direction of r_1 is opposite to the direction of flow (from Townsend 1970).

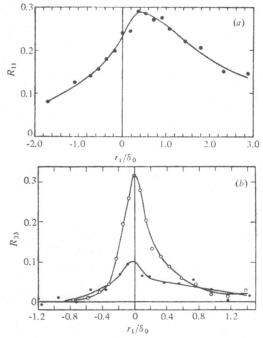

Fig. 4.7. Correlations in a boundary layer with fixed separation normal to the wall (a) $R_{11}(r_1, 0, r_3)$ with $z/\delta_0 = 0.159$, $r_3/\delta_0 = 0.295$; (b) $R_{33}(r_1, 0, r_3)$ with $z/\delta_0 = 0.140$, r_3/δ_0: $\bigcirc = 0.145$; $\bullet = 0.29$ (from Tritton 1967).

5

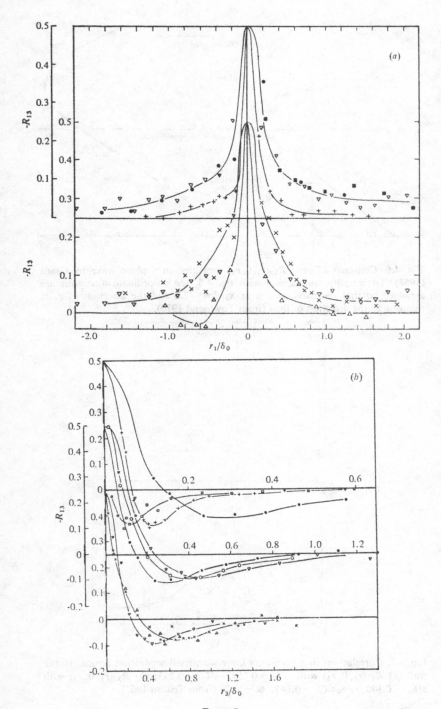

Fig. 4.8.

The shape of the correlation contours means that correlations with constant displacement in the direction of shear, i.e. for constant r_3, and variable displacement in the stream direction reach their maximum values for non-zero r_1, roughly for $r_1/r_3 = 1.2$ for R_{11}

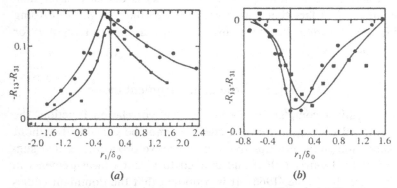

(a) (b)

Fig. 4.9. Measurements of R_{13} in a boundary layer by Tritton (1967): (a) with fixed separation normal to the wall and small separation in the Ox_2 direction, ●, $-R_{13}$ for $r_3/\delta_0 = 0.295$ and $x_3/\delta_0 = 0.158$, ■, $-R_{31}$ for $r_3/\delta_0 = 0.28$ and $x_3/\delta_0 = 0.163$, (b) with fixed separations in both transverse directions, ●, $-R_{13}$ for $r_2/\delta_0 = 0.325$ and $r_3/\delta_0 = 0.22$, ■, $-R_{31}$ for $r_2/\delta_0 = 0.325$ and $r_3/\delta_0 = 0.21$, both for $x_3/\delta_0 = 0.20$.

and R_{22}, compared with the much smaller value $r_1/r_3 = 0.17$ for R_{33}. Measurements by Favre, Gaviglio & Dumas (1957), by Grant (1958) and by Tritton (1967) in boundary layers, and by Bowden (1962) and by Bowden & Howe (1963) in tidal channels show the effect clearly for R_{11} and R_{22} but not for R_{33} (fig. 4.7).

Lastly, Tritton (1967) has measured the correlation $R_{13}(\mathbf{r})$ with a variety of displacement directions in a boundary layer. He finds that the behaviour of $R_{13}(r_1, 0, r_3)$ is generally similar to that of $R_{33}(r_1, 0, r_3)$ in agreement with the calculated correlations (figs 4.8, 4.9). His observations agree with the calculations that the correlation $R_{13}(0, r, 0)$ changes sign and attains relatively large positive values ($R_{13}(0)$ is negative).

Fig. 4.8. Measurements of R_{13} in a boundary layer by Tritton (1967). (a) $-R_{13}(r, 0, 0)$; $+$, $z_3/\delta_0 = 0.048$; ●, $z_3/\delta_0 = 0.156$; ■, $z_3/\delta_0 = 0.171$; ∇, $z_3/\delta_0 = 0.51$; $\times$, $z_3/\delta_0 = 0.79$; $\triangle$, $z_3/\delta_0 = 1.15$. (b) $-R_{13}(0, r, 0)$; $\square$, $z_3/\delta_0 = 0.017$; $+$, $z_3/\delta_0 = 0.049$; ●, $z_3/\delta_0 = 0.155$; $\bigcirc$, $z_3/\delta_0 = 0.29$; ∇, $z_3/\delta_0 = 0.51$; $\times$, $z_3/\delta_0 = 0.785$; $\triangle$, $z_3/\delta_0 = 1.15$.

In each of the flows, the effect of the flow inhomogeneity on the motion is to produce features of the correlation function that are not duplicated in other flows, but the basic similarity of the correlation functions is more striking than the differences. Further, the common structure implied by the similarity is one of considerable complexity, a fact that adds significance to the possibility of using rapidly distorted isotropic turbulence as a model structure for turbulent flows.

4.3   Nature of the main turbulent motion

Even with a complete knowledge of the correlation function, the form and distribution of the eddies cannot be calculated without some ambiguity. Perhaps the most objective procedure is that suggested by Lumley (1965) and applied to Grant's measurements in the wake by Payne (1966). It is proposed that the dominant eddies of the turbulence have the same velocity pattern as the first eigenfunction of the equation,

$$\int R_{ij}(\mathbf{x}, \mathbf{r})\phi_j^{(n)}(\mathbf{x}+\mathbf{r})\, d\mathbf{r} = \lambda^{(n)}\phi_i^{(n)}(\mathbf{x}). \tag{4.3.1}$$

If the velocity field is decomposed into a sum of orthogonal functions, the first eigenfunction of (4.3.1) is the best approximation to it obtainable with a single term and is, in this sense, an optimal representation. In practice, the method encounters two difficulties, first the construction of a complete correlation function from the incomplete measurements and, secondly, the indeterminacy of relative phases in the directions of homogeneity, to say nothing of the computational problem. Fig. 4.10 is a sketch of the form of the dominant eddy form in a wake, obtained by Payne. Notice the double-roller form and the tendency for the axes of the rollers to lie against the direction of shear.

A less elegant but simpler method is to attempt to construct the simplest eddy form whose presence in the motion would lead to the observed correlations. Referring to the calculated correlations of figs. 4.1 and 4.5, the following features may be considered relevant:

(1) $R_{11}(0, r, 0)$ becomes negative but $R_{11}(0, 0, r)$ does not, implying that return flow of the u_1 component is concentrated in the Ox_2 direction.

(2) $R_{22}(0, r, 0)$ is persistently negative for large values of r, indicating that convergence or divergence of flow in the Ox_2 direction is common.

(3) $R_{22}(r, 0, 0)$ takes large negative values which persist, showing that return flow of the u_2 component is mostly in the Ox_1 direction.

(4) $R_{13}(0, r, 0)$ is initially negative but changes sign and attains comparatively large positive values.

(5) $R_{33}(r, 0, 0)$ becomes negative but $R_{33}(0, r, 0)$ does not, showing return flow of u_3 to be mostly in the Ox_1 direction.

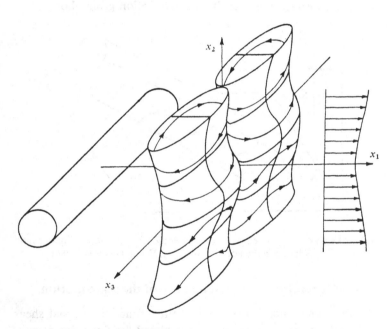

Fig. 4.10. Sketch of the velocity pattern of the eddy giving best fit to the correlation measurements of Grant, obtained by Lumley & Payne.

All these features but the last are consistent in a general way with the presence of pairs of roller eddies with axes inclined at about 30° to Ox_1 (fig. 4.11). Near the centre of the double roller, the plane of circulation is inclined at a moderately small, negative angle to the x_1Ox_2 plane. The eddy form bears some resemblance to the motions observed near the wall in boundary-layer flow by Kline,

Reynolds, Schraub & Runstadler (1967). Further additions to the eddy structure could be made to account for the last feature and others, but it is really not likely that the larger eddies are all of one type, size and orientation. Consideration of the rapid-strain equations suggest that prolonged strain of any simple eddy converts it to an elongated double roller with motion nearly confined to the $x_1 O x_2$ plane, and the observed motion may be a combination of eddies that have reached this stage and others with appreciable motion in the $O x_3$ direction. Even remembering the warning, it is probable that the dominant eddies of shear flow are double-roller eddies whose average member fits the description given above.

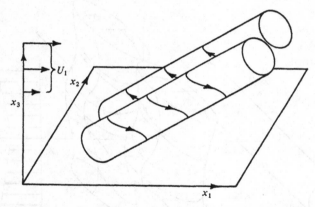

Fig. 4.11. Sketch of an inclined double-roller eddy. (Arrows on the lines around the cylinders indicate the eddy streamlines. From Townsend 1970.)

4.4   Generation and maintenance of the main motion

The close resemblance between correlation functions in real shear flows and the correlation functions calculated for finite rapid shear of initially isotropic turbulence is interesting but poses some problems. Using the simple argument that similar results imply similar origins, we might suppose that turbulent fluid begins as unstrained, unorganised turbulence which undergoes plane shearing until the total strain ratio is about four when it is converted back into unstrained fluid. At a particular point in the flow, parcels of fluid will be observed at all stages of the cycle and the observed correlations will be an average of the correlations and resemble the calculated correlations for a strain ratio of two. The basic difficulty with a cyclic straining model is to provide a plausible way of converting strained turbulent fluid back to its original condition of small Reynolds shear stress.

Alternatively, it may be that the observed structure is an equilibrium structure, attained after sufficient shearing and depending on a balance between the elongation of the roller eddies by the shear and diffusion of them by interaction with the smaller eddies. That interaction with the smaller eddies does reduce the anisotropy of the strained turbulence is clear from the observations of distorted homogeneous turbulence, but no evidence of an equilibrium structure has been found for irrotational strain. For plane shearing, the measurements of Rose (1966) may be consistent with an equilibrium structure but the results are hardly decisive. Theoretically, the effect of the smaller eddies on the larger ones may be modelled by an eddy viscosity, and then the work of Pearson (1959) would show that the energy of the larger eddies (and the Reynolds stress) decays after a large total strain. Another objection to the simple hypothesis of a universal equilibrium structure is that, although the dominant eddies seem to have much the same form in most flows, there are consistent differences between different flows and particularly between flows of different classes.

The differences between flows suggests that the reasons for the presence of eddies with the rapid-strain structure may not be the same for all flows or, at least, that the reasons are different for free turbulence and wall turbulence. In free turbulent shear flows, fluid is entrained continuously at the sharp interface dividing the ambient and the turbulent fluid, and parcels of fluid have really undergone finite and not very large total strains. For example, near the positions of maximum shear, the total strain since entrainment is in the range 3–15 for common flows assuming that the fluid parcel has followed a mean streamline since entrainment. Two factors may reduce the effective value of the total strain. The first has been mentioned, the influence of the smaller eddies which reduces the anisotropy and gives the appearance of a lower total strain. The second is the random movement of eddies across the flow by the velocity fields set up by their neighbours. Qualitatively, the effect is to diffuse the effective strain, i.e. the average total strain of parcels at a particular point, and it tends to reduce effective strains near the flow centre and to increase it near the edges.

For pipe flow and for wall flow in general, the total strain following a fluid parcel may be very large, even after allowance for diffusion, but the presence of the wall sets a limit to the maximum size of an eddy with its 'centre' at any position. It follows that eddies which are near the maximum size at one distance from the wall would become relatively small if they were transported to a position much further from the wall. The random movement of eddies normal to the wall has the effect of importing relatively small, fully strained eddies from nearer the wall (which were large in their original position), and relatively large, unstrained eddies from positions further from the wall. The small eddies join the comparatively isotropic smaller eddies and cease to contribute to the main structure, while the imported large eddies begin to be strained and to acquire the rapid-strain structure. In return, small, unstrained eddies are exported inwards and large, strained eddies outwards. Then the effective total strain at any point is the product of the residence time of an eddy in the layer and the rate of shear, and it is finite.

While these arguments may make it plausible that the main structure in shear flow resembles that produced by a finite rapid distortion, they imply that the effective strain varies from flow to flow, and from point to point in a single flow. The variations exist, but the principal changes in the dominant eddies are in the slopes of the axes of the rollers and in the inclinations of the planes of circulation. In particular, the calculated ratio of Reynolds stress to total intensity, $a_1 = |-\overline{uw}|/\overline{q^2}$, is comparatively insensitive to the value of the effective strain (see fig. 3.14), and tables 4.1 and 4.2 show that the measured values show no clear correlation with effective strain. However, it is probable that the ratio is less in flows with large effective strains.

4.5 Flow inhomogeneity and the large eddies

It is common to discuss turbulent shear flows in terms of a main turbulent motion of scale rather less than the scale of lateral inhomogeneity and a group of large eddies of lateral scale comparable with the width of the flow, that is to say, in terms of two groups of eddies distinct in size and structure. Recent experimental work has cast doubt on the existence of large eddies in this simple sense, particularly in boundary layers and in channel flow, but it is useful to consider separately components of the motion existing in and formed by nearly uniform shear and components whose existence and form are determined by the inhomogeneity of the flow.

Large eddies as a distinct group were first postulated to account for the nature of energy diffusion and entrainment in turbulent wakes, and the work of Grant (1958) and of Keffer (1965) has confirmed their identity and shown that they are connected with the folding of the interface between turbulent and non-turbulent fluid. In wakes, though possibly not in other free turbulent flows, their circulation lies nearly in the $x_1 O x_3$ plane and they are noticeably periodic in the $O x_1$ direction. Being part of the entrainment mechanism, their centres are located near the outer edge of the flow but their size is such that they influence most parts of the flow. In jets and boundary layers, similar eddies may exist but their size is less and their velocity field is confined to a narrow region near the mean position of the interface. Here they may be distinguished by their quasi-periodicity in the $O x_1$ direction and by their plane of circulation.

A different kind of motion dependent on flow inhomogeneity is found near a wall, and depends on the inhomogeneity of eddy sizes, not as in free turbulence on inhomogeneity in velocity gradient and turbulent intensity. An eddy with its centre at distance L from the wall may be of size not larger than L, and its velocity field must be such that motion is parallel to the wall at distances small compared with L. Consequently, the contribution of this eddy to the Reynolds stress is zero at the wall, reaches a maximum value near the eddy centre and then decreases. On the other hand, its contribution to intensities of components parallel to the wall is finite at the wall. The consequence is that superposition of the velocity fields of eddies with a wide range of L, necessary to produce (say) a uniform distribution of Reynolds stress, produce at the same time an 'inactive' swirling motion near the wall whose magnitude depends on the total thickness of the flow.†

† 'Inactive' motion is particularly intense in boundary layers in adverse pressure gradients. The low values of the ratios in table 4.2 are a consequence of its presence.

Another modification of the flow depends on the variation of effective strain across the flow. The primary effect of the variation is to cause changes in the average inclination of the roller axes but there is a tendency for adjacent rollers to align themselves and form a composite double roller extending across the entire width of the flow (see fig. 4.10). The reason may be that this configuration leads to the least velocity gradients for the larger eddies and so to the least energy transfer to the smaller eddies. The favoured configuration is probably more like a linked chain of the unit double rollers than a closely organised structure extending over the whole width of the flow.

The appearance of double-roller eddies has been attributed to their size being small enough for them to develop in a uniform gradient of mean velocity. Smallness probably depends on the roller diameter being small compared with the flow width, and reference to fig. 4.5 shows that the diameter is about $6d$, rather less than the flow width of about $12d$ (d is the cylinder diameter).

4.6 The dependence of Reynolds stress on mean velocity

A central problem in the theory of turbulent shear flow is the establishment of a second relation between the Reynolds stress and the mean velocity field, one that is independent of the Reynolds equation for the mean velocity. Many relations have been proposed but none of the simpler ones are generally valid. The reasons can be made clear by considering the equation for the turbulent kinetic energy,

$$U_k \frac{\partial(\frac{1}{2}\overline{q^2})}{\partial x_k} + \overline{u_i u_j} \frac{\partial U_i}{\partial x_j} + \frac{\partial}{\partial x_k}(\overline{pu_k} + \frac{1}{2}\overline{q^2 u_k}) + \varepsilon = 0, \qquad (4.6.1)$$

and the assumption of structural similarity of the turbulent motion.

If the motion has everywhere the same geometrical structure, it can be specified completely by its intensity $\overline{q^2}$ and by its length scale. Conveniently, we use the dissipation length-scale L_ε, so that

$$\varepsilon = (\overline{q^2})^{3/2}/L_\varepsilon = a_1^{-3/2}\tau^{3/2}/L_\varepsilon. \qquad (4.6.2)$$

For almost unidirectional mean flow, the energy equation becomes

$$\frac{1}{2a_1}\left(U \frac{\partial\tau}{\partial x} + W \frac{\partial\tau}{\partial z}\right) - \tau \frac{\partial U}{\partial z} + \frac{\partial}{\partial z}(\overline{pw} + \frac{1}{2}\overline{q^2 w}) + a_1^{-3/2}\tau^{3/2}/L_\varepsilon = 0, \quad (4.6.3)$$

where $a_1 = -\overline{|uw|}/\overline{q^2}$ is a constant. In this way, all but one of the terms has been expressed in terms of the mean velocity field and the Reynolds stresses. In order, they describe the effects on the energy balance of (1) advection of turbulent energy from upstream, (2) production of turbulent energy by working of Reynolds stresses against the mean flow, (3) lateral diffusion of turbulent energy by diffusive movements and by flow down pressure gradients, and (4) dissipation of turbulent energy.

To use (4.6.3) as a second relation between mean velocity and Reynolds stress, it is necessary to assign values to the term representing lateral transport of energy. The direction of the energy flux, $\overline{pu_i} + \frac{1}{2}\overline{q^2u_i}$, is expected to be broadly from regions of large intensity to regions of smaller intensity and, consistently with an assumption of structural similarity, a possibility is that

$$\overline{pw} + \tfrac{1}{2}\overline{q^2w} = -C(\overline{q^2})^{3/2}\,\mathrm{sgn}(\partial\overline{q^2}/\partial z). \qquad (4.6.4)$$

Observations show that the transfer depends mostly on flow in eddies large enough to span most of the flow, taking the form either of convection of smaller eddies or transfer of energy from one part of a large eddy to another. In that case, the transport may be better represented by a pattern of *convection velocity* $\mathscr{V}$, directed from high intensity regions and defined so that

$$\overline{pw} + \tfrac{1}{2}\overline{q^2w} = \tfrac{1}{2}\mathscr{V}\overline{q^2}. \qquad (4.6.5)$$

The magnitude of the convection velocity is characteristic of the whole section of the flow, and Bradshaw, Ferriss & Atwell (1967) have shown that satisfactory predictions of boundary-layer development can be made using (4.6.3) and a universal distribution of convection velocity across the layer.

In general, all the terms in the energy equation are of comparable magnitude but, in some circumstances, only two of them may be significant. The most important case is wall turbulence. Near a rigid boundary, advection and lateral diffusion of energy are both negligible, and the equation takes the form,

$$\frac{\partial U}{\partial z} = a_1^{-3/2}\tau^{1/2}/L_\varepsilon, \qquad (4.6.6)$$

essentially the basic equation of the momentum-transfer version of the mixing-length theory. If a value can be assigned to the length scale, by dimensional reasoning or otherwise, the mean flow problem can be solved.

Use of the energy equation in the form (4.6.3) is not appropriate if the Reynolds stress or the mean velocity gradient changes sign since the shear coefficient a_1 cannot be constant in a region of small or zero stress. An alternative is to assume a distribution of eddy viscosity, i.e. that

$$-\overline{uw} = v_T \, \partial U/\partial z \qquad (4.6.7)$$

where the eddy viscosity v_T may vary with position but remains positive. Although use of the eddy viscosity concept does lead to useful results particularly for flows developing in slowly changing external conditions, no clear justification has been produced for a general situation and its physical basis is far less sound than that of structural similarity.

4.7 Statistical distribution of velocity fluctuations

The correlation or the spectrum function may be a complete specification of a turbulent flow only if the probability distribution function for the velocity field is joint-normal, that is, the probability of having velocities $\mathbf{u}$, $\mathbf{u'}$, $\mathbf{u''}$, etc. at the points $\mathbf{x}$, $\mathbf{x'}$, $\mathbf{x''}$, etc. can be expressed in the form,

$$P(\mathbf{u}, \mathbf{u'}, \mathbf{u''}, \ldots) = C \exp[-(\text{quadratic in } \mathbf{u}, \mathbf{u'}, \mathbf{u''}, \text{etc.})].$$

A joint-normal distribution implies spatial homogeneity of the flow and cannot describe inhomogeneous flow, but departures from the normal distribution are an essential feature of any turbulent motion. For example, the skewness factor for the velocity gradient $\partial u_1/\partial x_1$,

$$S_1 = \overline{\left(\frac{\partial u_1}{\partial x_1}\right)^3} \Big/ \left[\overline{\left(\frac{\partial u_1}{\partial x_1}\right)^2}\right]^{3/2}, \qquad (4.7.1)$$

appears in the equation for the mean square vorticity of the turbulence (Batchelor 1953),

$$\frac{d\overline{\omega_1^2}}{dt} = -\frac{7}{3\sqrt{5}} S_1 (\overline{\omega_1^2})^{3/2} - 10v \overline{\left(\frac{\partial \omega_1}{\partial x_1}\right)^2}, \qquad (4.7.2)$$

as a factor in the term describing the rate of production of vorticity by stretching of vortex lines. If it were zero, as it would be with a normal distribution, turbulent vorticity would be rapidly destroyed by viscous effects and the high rates of energy dissipation that are characteristic of turbulent flow could not be maintained.

The reason for the departures of the statistical distribution from the joint-normal form is to be found in the non-linearity of the equations of motion, and the nature of the departures is a large subject not yet fully explored. For practical purposes, only the single-point distribution is of much interest and the following examples of departures from the normal distribution,

$$P(\mathbf{u}) = C \exp(-\tfrac{1}{2}a_{ij}u_iu_j) \qquad (4.7.3)$$

($a_{ij}u_iu_j$ is always positive), have been chosen for their relevance to turbulent shear flows.

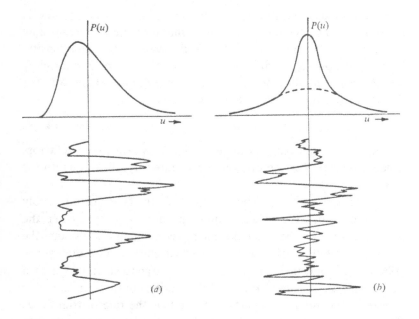

Fig. 4.12. Probability distribution functions for one component of the velocity fluctuation. (a) Skewed distribution typical of flows with strong gradients of turbulent intensity. (b) Two-component distribution, typical of irregular, non-uniform distribution of intensity in space.

The probability distribution function for velocity fluctuations at a point is very nearly normal in homogeneous turbulence, less so near the centre of a shear flow, and very far from normal near the edge of a free turbulent flow such as a jet. The departures are of two kinds, (1) skewness of the distributions so that odd moments are non-zero, and (2) distortion of the distribution leading to abnormally large values for high-order even moments (fig. 4.12).

The skewness of the distributions is connected with the convection of turbulent energy by turbulent movements, a process that, as a rule, transfers energy from regions of large intensity to regions of smaller intensity. In free turbulent flows, the lateral convective flux of energy, $\frac{1}{2}[\overline{w(u^2 + v^2 + w^2)}]$, is of considerable magnitude except near a plane of symmetry. Further, since the velocity components, u and w, are strongly correlated in regions of appreciable Reynolds stress, the longitudinal flux $\frac{1}{2}[\overline{u(u^2 + v^2 + w^2)}]$, is also non-zero.

Near the edges of free turbulent flows, fully turbulent flow is intermittent and the motion at a point alternates between slow fluctuations of low intensity and rapid fluctuations of high intensity. During each phase, the distributions are distributed nearly normally but the effect of inhomogeneity of turbulent intensity makes the flatness factor of a velocity component, say

$$\beta_4(u) = \overline{u^4}/(\overline{u^2})^2 \tag{4.7.4}$$

exceed its 'normal' value of three. Abnormally large flatness factors usually indicate that the distribution of intensity of the quantity is spotty.

The small-scale components of the velocity field are noticeably intermittent, and the effect is more pronounced the greater the difference in sizes between the energy-containing eddies and the viscous eddies that dissipate the turbulent energy. As Kolmogorov (1962) points out, transfer of energy from large to small eddies by a cascade process means that spatial fluctuations in the rate of energy transfer from eddies of a particular size bias the rate of transfer at the next step in the cascade so that the fluctuation not only persists but is amplified. The ultimate result is that some parts of the flow are (instantaneously) regions of large dissipation while others have a very low rate of dissipation. The consequent 'spottiness' of dissipation in space has been observed in grid turbulence by Batchelor & Townsend (1949), in boundary layers by Sandborn (1959) and in the lower atmosphere by Stewart, Wilson & Burling (1970), by Gibson, Stegen & Williams (1970) and many others. In principle, it means that the form of the equilibrium spectrum of turbulence depends on the Reynolds number.

CHAPTER 5

TURBULENT FLOW IN PIPES AND CHANNELS

5.1 Introduction

Turbulent flows bounded by rigid – or, at any rate, nearly im-movable – walls make up an important class of basic flows, with members including the most homogeneous and symmetrical of all turbulent shear flows. In approximate order of interest and of knowledge, they are:

(1) pressure flow along straight pipes of circular section,

(2) pressure flow between parallel planes,

(3) flow of a liquid down a wide open channel with uniform depth,

(4) plane Couette flow between parallel planes, with or without a pressure gradient,

(5) pressure flow in the annular space between concentric cylinders.

In all these flows, the mean flow is everywhere in the same direction, and all mean values (except the pressure) are functions of a single co-ordinate measuring distance from the flow centre.

The structural feature that distinguishes all shear flows confined by a solid boundary is the existence of an *equilibrium layer* of *wall turbulence* near the boundary. In equilibrium layers, the rates of generation and of dissipation of turbulent energy are exceptionally large compared with those in other parts of the flow, and the motion is, to a considerable extent, determined by local conditions, par-ticularly by the stress distribution in the layer. We shall see that many properties of pipe and channel flows depend largely on the nature and properties of the equilibrium layers and very little on the flow outside them.

5.2   Equations of motion for unidirectional mean flow

Consider the flow between smooth rigid walls situated at the planes, $z = 0$ and $z = 2D$, with mean flow velocity U everywhere parallel to Ox and with all mean values functions of z alone. The equations for the mean velocity reduce to

$$\left. \begin{array}{c} \dfrac{\partial \overline{uw}}{\partial z} = -\dfrac{\partial P}{\partial x} + \nu \dfrac{\partial^2 U}{\partial z^2}, \\[4mm] 0 = -\dfrac{\partial P}{\partial y}, \\[4mm] \dfrac{\partial \overline{w^2}}{\partial z} = -\dfrac{\partial P}{\partial z}. \end{array} \right\} \qquad (5.2.1)$$

The second and third equations may be integrated to give

$$P + \overline{w^2} = P_0(x) \qquad (5.2.2)$$

and, since the velocity fluctuation is zero at a rigid surface, $P_0(x)$ is the pressure measured at either wall. Then the first equation may be integrated to give

$$\tau = \nu \frac{\partial U}{\partial z} - \overline{uw} = \tau_0 + \frac{dP_0}{dx} z, \qquad (5.2.3)$$

where $\tau_0 = \nu[\partial U/\partial z]_{z=0}$ is the shear stress at the wall at $z = 0$. It follows that the shear stress at the wall $z = 2D$ is

$$\tau_1 = \nu \left[\frac{\partial U}{\partial z} \right]_{z=2D} = \tau_0 + 2 \frac{dP_0}{dx} D. \qquad (5.2.4)$$

So far the translation velocities of the two walls have been left unspecified. For channel flow between stationary walls, the flow is symmetrical about the central plane $z = D$, and then

$$\tau_0 = -\tau_1 = -\left(\frac{dP_0}{dx} \right) D. \qquad (5.2.5)$$

As the mean pressure gradient and the wall stress are related in this way, it is possible and convenient to use τ_0, D and ν as the parameters defining the flow, and to define the Reynolds number of the flow as

$$R_\tau = \tau_0^{1/2} D/\nu. \qquad (5.2.6)$$

If the two boundaries are not moving at the same speed, the flow is only (anti-) symmetrical if the pressure gradient is zero, and it is necessary to use two parameters besides D and v to specify the flow. Again it is better to use the wall stresses, τ_0 and τ_1, rather than the relative velocities of the walls and the pressure gradient.

Flows homogeneous over concentric cylindrical surfaces are also statistically axisymmetric and are described using cylindrical polar co-ordinates, with Ox measured along the axis which is the direction of mean flow, r distance from the axis, and u, v, w the components of the velocity fluctuation in the axial, circumferential and radial directions. If the outer boundary of the flow is a smooth circular cylinder of radius R, the equations of mean flow are

$$
\left.
\begin{aligned}
\frac{1}{r}\frac{\partial \overline{uw}r}{\partial r} &= -\frac{\partial P}{\partial x} + \frac{v}{r}\frac{\partial}{\partial r}\left(r\frac{\partial U}{\partial r}\right), \\
0 &= -\frac{1}{r}\frac{\partial P}{\partial \theta}, \\
\frac{\partial \overline{w^2}}{\partial r} + \frac{\overline{w^2}-\overline{v^2}}{r} &= -\frac{\partial P}{\partial r},
\end{aligned}
\right\}
\tag{5.2.7}
$$

and, as before, integrate to

$$
P + \overline{w^2} + \int_R^r \frac{\overline{w^2}-\overline{v^2}}{r}\, dr = P_0(x)
\tag{5.2.8}
$$

and to

$$
\tau = v\frac{\partial U}{\partial r} - \overline{uw} = -\tau_0\frac{R}{r} + \frac{1}{2}\frac{dP_0}{dx}\left(r - \frac{R^2}{r}\right),
\tag{5.2.9}
$$

where $-\tau_0$ and $P_0(x)$ are the shear stress and pressure measured at the boundary. If the flow is between the outer cylinder and an inner one of radius R_1, the wall stresses are related by

$$
\tau_1 = -\tau_0\frac{R}{R_1} + \frac{1}{2}\frac{dP_0}{dx}\frac{R_1{}^2 - R^2}{R_1}.
\tag{5.2.10}
$$

If there is no inner cylinder, the stress is necessarily zero at the axis and

$$
\left.
\begin{aligned}
\tau_0 &= -\tfrac{1}{2}R\frac{dP_0}{dx} \\
\tau &- -\tau_{0'}/R.
\end{aligned}
\right\}
\tag{5.2.11}
$$

and

Again the parameters τ_0, R and ν may be used to define the flow and to define the Reynolds number of the flow as

$$R_\tau = \tau_0^{1/2} R/\nu. \qquad (5.2.12)$$

5.3   Reynolds number similarity in pipe and channel flow

For small Reynolds numbers of flow, the flow in a channel is laminar and the velocity distribution is

$$U = \frac{\tau_0 z}{\nu}\left[1 - \frac{z}{2D}\right], \qquad (5.3.1)$$

the well-known parabolic profile. At higher Reynolds numbers, the flow becomes turbulent and it is expected that the Reynolds stresses will be large compared with the direct viscous stresses over most of the flow, the ratio increasing with increasing Reynolds number. Since the stresses are certainly of the same sign and the total stress is determined by (5.2.3), the velocity gradients in the fully turbulent parts of the flow must be small compared with their values in a laminar flow with the same wall stress. At the wall, the velocity fluctuations and the Reynolds stress are zero, and so

$$\left[\frac{\partial U}{\partial z}\right]_{z=0} = \tau_0/\nu \qquad (5.3.2)$$

just as in the laminar flow. It follows that the turbulent velocity profile must be of the flattened form indicated in fig. 5.1, the flattening increasing with the Reynolds number of the flow.

The necessity for negligible Reynolds stress in the immediate neighbourhood of the wall means that there is a finite region of flow in which the mean viscous stresses are not small compared with the turbulent stresses, distinguished as the *viscous layer*. Outside the viscous layers lies the region of fully turbulent flow where the viscous stresses are small compared with the Reynolds stresses, and this region usually occupies most of the channel.

The existence of the viscous layer shows that the whole flow cannot be independent of the fluid viscosity, and the principle of Reynolds number similarity can be applied only to the fully turbulent region in which the direct effects of the viscous stresses are negligible. If the viscous layers are thin compared with the channel width, the defining parameters for the fully turbulent region are the channel width and the velocity conditions and stresses at the boundaries of the region. The energy input to the region is determined entirely by the pressure gradient and by its internal flow and so, disregarding the remote possibility that the very thin viscous layers may influence

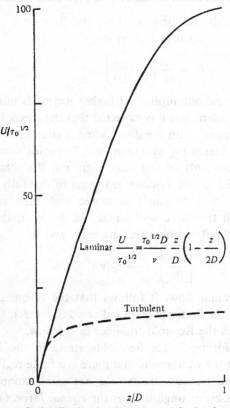

Fig. 5.1. Mean velocity distribution in a channel for laminar and turbulent flow at the same Reynolds number, $\tau_0^{1/2}D/\nu = 200$.

the rest of the flow, it follows that the fully turbulent flow depends only on the width and on the wall stresses. Remembering that a velocity of translation U_t does not affect the dynamics of the motion, independence of the fluid viscosity implies that the mean velocity is

$$U = U_t + \tau_0^{1/2}f(z/D, \tau_1/\tau_0) \qquad (5.3.3)$$

for the general channel flow, or

$$U = U_t + \tau_0^{1/2}f(z/D) \qquad (5.3.4)$$

for flow between walls not in relative motion. Similar results exist for turbulent parameters not appreciably influenced by eddies responsible for the viscous dissipation of energy, e.g. the correlation function is of the form

$$R_{ij}(z; \mathbf{r}) = \tau_0 R_{ij}*[z/D; \mathbf{r}/D]. \qquad (5.3.5)$$

In particular, the stress distribution of (5.2.3) is of the similarity form – naturally.

The translation velocity U_t appearing in (5.3.4, 5.3.5) depends on conditions outside the fully turbulent region. In a smooth-walled channel, its ratio to $\tau_0^{1/2}$, the *friction velocity* which is the velocity scale of the motion in the region, is a function of the flow Reynolds number. If the walls are rough with roughness elements of dimensions small compared with the channel width, the motion in the fully turbulent region is still determined by the width and by the pressure gradient, but the translation velocity will depend on the nature of the flow around the roughnesses. In general, whatever the wall conditions, provided that their influence on the central flow is limited to a transfer of stress from the walls, the relative motion in the fully turbulent region depends only on the wall stresses and on the channel width.

With flow between stationary walls, it is convenient to choose U_t as the velocity at the channel centre $U(D)$, and equation (5.3.4) is written as a *defect law*,

$$\frac{U(D) - U}{\tau_0^{1/2}} = f(z/D), \qquad (5.3.6)$$

valid for rough and smooth channels. This and previous predictions from the principle of Reynolds number similarity have their obvious counterparts for flow through cylindrical pipes.

5.4 Wall similarity in the region of constant stress

In channel flow and indeed in most turbulent flows with solid boundaries, there is a region adjacent to the wall within which the total shear stress is nearly constant and whose motion is determined to a considerable extent by the shear stress and by the nature of the wall. This property is a consequence of the effect of turbulent flow in reducing mean velocity gradients over the greater part of the channel and concentrating the large velocity gradients in the region close to the wall. In the equation for the turbulent energy in channel flow,

$$-\overline{uw}\,\frac{\partial U}{\partial z} - \frac{\partial}{\partial z}(\overline{pw} + \tfrac{1}{2}\overline{q^2 w}) = \varepsilon, \qquad (5.4.1)$$

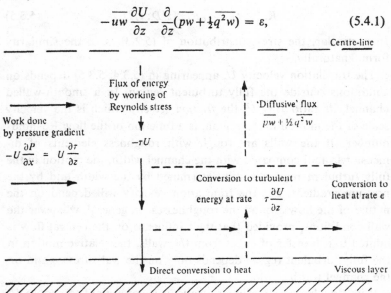

Fig. 5.2. Flow and transformations of energy in a channel. *Note*: (1) Horizontal arrows indicate conversion processes, vertical arrows indicate lateral fluxes. (2) Full lines refer to energy of the mean flow, broken lines refer to turbulent kinetic energy.

the terms involving streamwise gradients have disappeared and the remaining terms represent (1) production of turbulent energy by working of mean velocity gradients against the Reynolds stresses, (2) net energy gain from the lateral flow of energy by fluctuations of velocity and pressure, and (3) viscous energy dissipation. The form of the turbulent velocity profile is such that $\partial U/\partial z$ increases sharply as the wall is approached and so the rate of production of turbulent energy, $-\overline{uw}\,\partial U/\partial z$, reaches a maximum value just outside the viscous layer (see fig. 5.2).

Consider now the motion in a layer next the wall with thickness z_1 small compared with the channel width but large enough to extend into the fully turbulent part of the flow. Inside the layer, the rates of energy production and, presumably, of dissipation are much larger than those anywhere else except in the corresponding layer on the far wall, and it is a plausible assumption that the vigorous motion within the layer is nearly independent of the weaker motion around the channel centre and the vigorous but remote motion near the far wall. Then, the motion within the layer is determined by the boundary conditions at the plane $z = z_1$ and at the wall, and, since the mean velocity distribution and the turbulent motion are related by equation (5.4.1), the conditions at the edge concern only the turbulent motion. For aspects of the motion concerned with the energy balance, the edge conditions should be fully described by the Reynolds stress and by the lateral flux of turbulent energy. In the thin layer under consideration, the Reynolds stress at the edge is very nearly equal to the wall stress which specifies the boundary conditions at the wall.

Ignoring (for the moment) the lateral flux of energy, we conclude that the aspects of the motion relevant to the processes of energy production and dissipation depend on the three parameters, τ_0, z_1 and ν. Dimensional considerations show then that

$$
\left.
\begin{aligned}
\frac{\partial U}{\partial z} &= \frac{\tau_0^{1/2}}{kz} f\left(\frac{\tau_0^{1/2} z}{\nu}, \frac{z}{z_1}\right), \\
-\overline{uw} &= \tau_0 g\left(\frac{\tau_0^{1/2} z}{\nu}, \frac{z}{z_1}\right), \\
\varepsilon &= \frac{\tau_0^{3/2}}{kz} h\left(\frac{\tau_0^{1/2} z}{\nu}, \frac{z}{z_1}\right),
\end{aligned}
\right\}
\qquad (5.4.2)
$$

where k is a constant. Now the only restrictions placed on the choice of z_1 were that it should be small compared with D but large enough for the edge to lie within the fully turbulent flow. Hence, the non-dimensional functions are independent of the value of z_1 (if one exists) and the motion in the layer is determined by τ_0 and ν alone, i.e.

$$
\left.
\begin{aligned}
\frac{\partial U}{\partial z} &= \frac{\tau_0^{1/2}}{kz} f\left(\frac{\tau_0^{1/2} z}{\nu}\right), \\
-\overline{uw} &= \tau_0 g\left(\frac{\tau_0^{1/2} z}{\nu}\right), \\
\varepsilon &= \frac{\tau_0^{3/2}}{kz} h\left(\frac{\tau_0^{1/2} z}{\nu}\right).
\end{aligned}
\right\}
\qquad (5.4.3)
$$

These equations are the formal expression of the principle of wall similarity for a constant-stress layer on a smooth wall.

Well clear of the viscous layer, the motion is fully turbulent and independent of the magnitude of the fluid viscosity and so, for large values of $\tau_0^{1/2}z/v$,

$$
\left.
\begin{aligned}
-\overline{uw} &= \tau_0, \\
\varepsilon &= \frac{\tau_0^{3/2}}{kz}
\end{aligned}
\right\}
\tag{5.4.4}
$$

with a suitable choice of the *Kármán constant* k. Neglecting the effect of any lateral flux of turbulent energy, the energy equation shows that

$$
\frac{\partial U}{\partial z} = \frac{\tau_0^{1/2}}{kz}
\tag{5.4.5}
$$

for large $\tau_0^{1/2}z/v$.

Several comments should be made on the preceding argument for the existence of wall similarity in a constant stress layer. First, the neglect of the net effect of lateral energy flux may be defended by pointing out that Reynolds stress is nearly independent of distance from the wall and so that, since Reynolds stress and turbulent energy depend on the same eddies, gradients of turbulent energy are likewise small and the diffusive flux is unlikely to be appreciable. Secondly, the largest eddies in the layer are probably comparable in size with its thickness and, unless the layer thickness is small compared with the channel width, they would certainly be influenced by the presence of the far wall. An essential property of any equilibrium layer is that its thickness should be a small fraction of the total width of the flow. Lastly, care has been taken to emphasise that the conclusions apply only to those aspects of the motion that are concerned with the formation of the Reynolds stress and with the energy dissipation. It is known that many important aspects of the motion contribute little to these quantities and do not conform to wall similarity.

By integrating (5.4.5), we obtain

$$
U = \frac{\tau_0^{1/2}}{k}\left[\log\frac{\tau_0^{1/2}z}{v} + A\right],
\tag{5.4.6}
$$

the *logarithmic velocity distribution* or 'law of the wall' for a smooth surface, where $k \simeq 0.41$ and $A \simeq 2.3$. It has been derived in many ways, by using the Prandtl mixing-length theory, by the von Kármán similarity theory, and by the especially attractive method of requiring an overlap between the ranges of validity of wall similarity and the defect law, to mention but a few. The cautious and somewhat pedantic approach used here is justified by the ample evidence that the turbulent motion in a constant-stress layer is too complicated to conform with sweeping assumptions of complete similarity.

Although the arguments have been presented for pressure flow in a two-dimensional channel, it should be clear that they can be applied in any wall flow in which an *equilibrium layer* can be distinguished with the properties:

(1) Its thickness is a small fraction of the total width of the turbulent flow.

(2) Terms in the turbulent energy equation that describe advection of turbulent energy by the mean flow are small compared with the production term, so that the complete equation may be approximated by the 'homogeneous' form of (5.4.1).

(3) The variation of shear stress across the equilibrium layer is small compared with the wall stress.

5.5   Flow over rough walls

In practice, the bounding walls of a turbulent flow may have corrugations or other irregularities distributed uniformly over the surface with heights small compared with the flow width, and then the shear stresses are transmitted to the walls by combinations of tangential viscous stresses and normal pressures generated by the flow near the roughness elements. At distances from the wall large compared with the extent of the flow patterns set up by individual roughness elements, the turbulent flow is unlikely to be affected by the exact nature of the roughness and, as with a smooth wall, it will be determined by the averaged wall stresses, the channel width and the fluid viscosity. Then the arguments of the previous section show that the velocity distribution in the constant-stress layer is given by

$$\frac{dU}{dz} = \frac{\tau_0^{1/2}}{kz} \qquad (5.5.1)$$

but only for z considerably larger than z_r, the characteristic scale of the roughness elements. If $\tau_0^{1/2}z_r/v$, the Reynolds number of flow around the roughness elements, is very small, (5.5.1) remains valid for values of z for which the mean velocity and Reynolds stress are negligible, and the smooth-wall distribution is recovered. If the roughness Reynolds number is not small, the integrated equation contains a constant of integration which depends on the flow around the roughness elements and it can be written in the form,

$$U = \frac{\tau_0^{1/2}}{k} \log z/z_0 \qquad (5.5.2)$$

valid for $z \gg z_r$, and large $\tau_0^{1/2} z/v$. z_0 is the *roughness length* of the surface and, together with the friction velocity $\tau_0^{1/2}$, it defines a 'slip velocity' between the wall and a standard location in the constant-stress layer (figs. 5.3.)

The magnitude of the roughness length (or of the slip velocity) depends on the flow around the roughness elements, which is determined by the velocities and pressures near the inner edge of the equilibrium layer, and two kinds of roughness behaviour can be distinguished, depending on whether the relevant fields of velocity and pressure are specified by motion in the equilibrium layer alone or by motions outside. The first kind, 'k' type roughness in the nomenclature of Perry, Schofield & Joubert (1969), is shown by surfaces with irregular corrugations and protuberances, e.g. sandpaper. On such surfaces, the flow around the roughness elements is determined almost entirely by the velocity in the innermost part of the equilibrium layer, and it responds very quickly to changes of velocity. For geometrically similar roughnesses, the flow depends only on the size of the roughness elements, on the friction velocity and on the viscosity, so that

$$z_0 = z_r F_r(\tau_0^{1/2} z_r/v). \qquad (5.5.3)$$

For large values of the roughness Reynolds number, the flow around the elements will be independent of the viscosity and then $z_0 \propto z_r$. For small values, the flow at the wall is aerodynamically smooth and the roughness length takes the value for a smooth wall, i.e.

$$z_0 = e^{-A} v/\tau_0^{1/2}. \qquad (5.5.4)$$

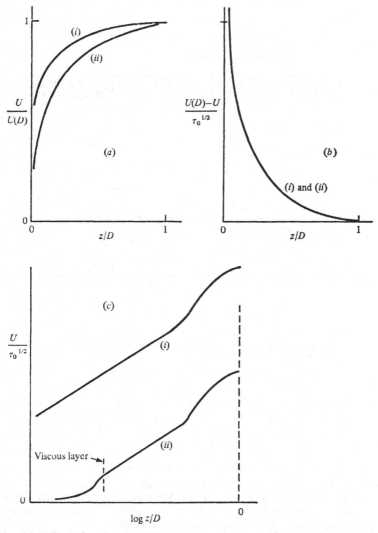

Fig. 5.3. Effect of a change of Reynolds number or surface roughness on the mean velocity distribution in a channel. (*a*) Velocity profiles as fractions of the central velocity. (*b*) Velocity defects as fractions of the friction velocity. (*c*) Velocity as a fraction of the friction velocity, showing change of the translation velocity.

The relative magnitudes of the roughness length and the mean height of the roughness elements are of importance for the lower

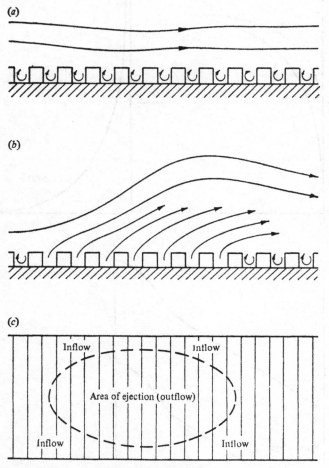

Fig. 5.4. Possible mechanism for 'd' type roughness behaviour. (a) Quiescent flow with stable vortices between the roughness elements. (b) Ejection of fluid induced by a transient region of adverse pressure gradient of extent comparable with the layer thickness. (c) Plan of the region of outflow.

limit to the validity of the logarithmic velocity distribution. For effectively smooth walls, equation (5.5.4) shows that $z_0 \approx 0.1 \, \nu/\tau_0^{1/2}$, while the logarithmic distribution is valid for (roughly) $\tau_0^{1/2} z/\nu$ greater than thirty or for z/z_0 greater than 300. For fully rough flow at large values of the Reynolds number, z_0 is commonly about one-tenth of the average height of the roughness elements and the logarithmic distribution can be valid only at heights considerably larger than that, say for z/z_0 greater than fifty. At both extremes, the minimum value of z/z_0 is quite large.

The second type of roughness behaviour is found when the shape of the roughness elements is such that the flow around them is almost unstable and can be disturbed violently by small fluctuations of large scale in the outside flow. Large-scale disturbances are characteristic of flow near the channel centre and, in the 'd' type roughness, the effect of the rough surface on the turbulent flow is dependent on the channel width and not only on the dimensions of the roughness elements. In that case, the roughness length becomes a fixed fraction of the channel width. The 'd' type behaviour is shown by rough surfaces composed of similar rectangular bars laid transversely to the direction of flow with spacing equal to the width of the bars. The 'normal' flow consists of stable vortices in the channels between the bars with relatively undisturbed flow over them, but it seems probable that large-scale pressure fluctuations can lead to simultaneous ejection of the stagnant fluid over areas comparable with the flow width and with normal velocities comparable with the friction velocity. Fig. 5.4 attempts to sketch a likely situation and to show that the penetration of the ejected fluid is likely to be a small but constant fraction of the extent of the area of ejection. If the spacing of the elements is small compared with the width of the ejection area, that width determines the roughness length.

It may be observed that the two kinds of roughness are extreme versions and that intermediate forms can exist.

5.6   Mean flow in the central region

The arguments used to derive velocity distributions in the equilibrium layers cannot be used in the central regions of the flows, but we do know that the velocity distributions has the form prescribed by Reynolds number similarity and that the velocity gradients there are small compared with those in the wall layers. It follows that the relations between volume flow, or relative velocities of the boundaries and the boundary stresses are only weakly dependent on the precise form of the velocity profile in the central region.

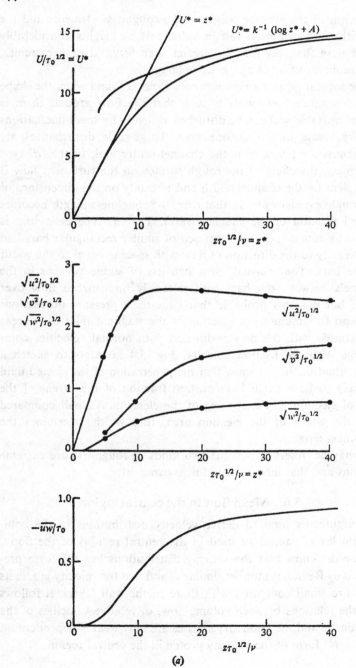

(a)

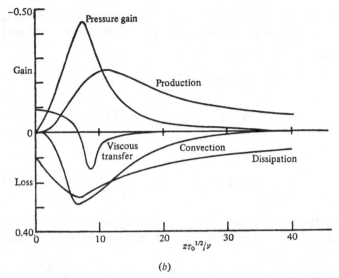

(b)

Fig. 5.5. Measurements within the constant-stress layer (from Laufer 1955). (a) Mean flow, turbulent intensities and Reynolds stress. (b) Terms in the equation for turbulent kinetic energy.

Identifying the velocity of translation in the similarity equation (5.3.3) with $U(D)$ the velocity at the channel centre, it becomes

$$U = U(D) + \tau_0^{1/2} f(z/D, \tau_1/\tau_0). \tag{5.6.1}$$

If constant-stress layers exist on both walls, the function must be such that the appropriate velocity distributions are reproduced near the walls, that is,

$$U = \frac{\tau_0^{1/2}}{k} \log z/z_0 \quad \text{for small } z/D,$$

and

$$U = U(D) - \frac{\tau_1}{k|\tau_1|^{1/2}} \log[(2D-z)/z_1] \quad \text{for small } (2D-z)/D. \tag{5.6.2}$$

The behaviour is possible if

$$f(\eta) \sim \frac{1}{k}[\log \eta + C_0] \quad \text{if } \eta \ll 1,$$

and

$$f(\eta) \sim \frac{\tau_1}{|\tau_1\tau_0|^{1/2}k}[\log(2-\eta) + C_1] \quad \text{if } (2-\eta) \ll 1, \tag{5.6.3}$$

where the quantities C_0 and C_1 depend only on the ratio τ_1/τ_0 and have magnitudes comparable with one. Comparison of (5.6.1) and (5.6.3) then shows that

$$U(D)/\tau_0^{1/2} = (1/k)(\log D/z_0 - C_0)$$

$$= U_1/\tau_0^{1/2} - \frac{\tau_1}{k|\tau_1\tau_0|^{1/2}}(\log D/z_1 - C_1) \qquad (5.6.4)$$

and

$$kU_1/\tau_0^{1/2} = (\log D/z_0 - C_0) + \frac{\tau_1}{|\tau_1\tau_0|^{1/2}}(\log D/z_1 - C_1) \quad (5.6.5)$$

giving the central flow velocity and the relative velocity of the bounding surfaces as functions of the surface stresses and the roughness lengths.

If the total flow in the viscous layer (or among the roughness elements) is small, the bulk flow velocity, defined as the volume flow divided by the channel area, can be expressed in terms of the similarity function of equation (5.6.1) as

$$U_m = U(D) + \tfrac{1}{2}\int_0^2 f(\eta)\,d\eta.$$

Then, from (5.6.4),

$$U_m/\tau_0^{1/2} = 1/k(\log D/z_0 - C_0) + \tfrac{1}{2}\int_0^2 f(\eta)\,d\eta. \qquad (5.6.6)$$

For flow over aerodynamically smooth surfaces,

$$z_0 = \nu/\tau_0^{1/2}\exp(-A), \qquad z_1 = \nu/|\tau_1|^{1/2}\exp(-A),$$

and equations (5.6.5, 5.6.6) provide a complete solution to the mean flow problem if the wall constants, k and A, and the shape parameters of the similarity function, C_0, C_1 and $\tfrac{1}{2}\int_0^2 f(\eta)d\eta$, are known.

For pressure flow,

$$U_m/\tau_0^{1/2} = 1/k(\log \tau_0^{1/2}D/\nu + \text{constant})$$

and the most accurate determination of the Kármán constant is by measuring the dependence of the friction coefficient, $c_f = \tau_0/U_m^2$, on the flow Reynolds number, $R_m = U_m D/v$, which are related by

$$c_f^{-1/2} = 1/k \log(R_m c_f^{1/2}) + \text{constant.}$$

Velocity distributions in the central region are conveniently discussed in terms of the *eddy viscosity*, defined as

$$v_T = \tau \left/ \frac{\partial U}{\partial z} \right. . \tag{5.6.7}$$

Within the constant-stress layers, it varies linearly with distance from the boundary,

$$v_T = k\tau_0^{1/2} z \quad \text{and} \quad v_T = k|\tau_0|^{1/2}(2D - z) \tag{5.6.8}$$

and it must have a maximum value near the centre of the flow. The very simple assumption that the eddy viscosity is constant outside the equilibrium layers describes well the velocity distributions in the three simple flows for $\tau_1/\tau_0 = -1, 0, +1$, the limits of the wall layers being at the positions where the eddy viscosity given by (5.6.8) equals the central value, i.e. at

$$z_e = (1/k)(v_T/\tau_0^{1/2}) = (1/k)R_s^{-1}D. \tag{5.6.9}$$

The assumption of a continuous distribution implies a smooth velocity distribution but has no deeper justification, and small discontinuities have little effect on the results.

For pressure flow in a channel ($\tau_1/\tau_0 = -1$), the assumption leads to the composite velocity distribution,

$$\left. \begin{aligned} \frac{U - U(D)}{\tau_0^{1/2}} &= -\tfrac{1}{2}R_s(1 - z/D)^2 \quad \text{for } \alpha < z/D < 2 - \alpha, \\ &= -\tfrac{1}{2}R_s(1 - \alpha)^2 - \frac{1}{k}\log \alpha + \frac{1}{k}\log z/D \\ &\qquad \text{for } 0 < z/D < \alpha, \end{aligned} \right\} \tag{5.6.10}$$

where

$$\frac{U(D)}{\tau_0^{1/2}} = \frac{1}{k}\log \frac{\alpha D}{z_0} + \tfrac{1}{2}R_s(1 - \alpha)^2,$$

$$\alpha = \frac{z_e}{D} = (kR_s)^{-1},$$

6

and

$$
\left.\begin{aligned}
C_0 &= C_1 = -\frac{1}{2}\frac{(1-\alpha)^2}{\alpha} - \log\alpha, \\
\tfrac{1}{2}\int_0^2 f(\eta)\,d\eta &= -\frac{1}{k}(\tfrac{1}{6}\alpha^{-1}+\tfrac{1}{2}\alpha+\tfrac{1}{3}\alpha^2).
\end{aligned}\right\}
\tag{5.6.11}
$$

A free-surface flow with $\tau_1 = 0$ has a constant-stress equilibrium layer on the lower surface and a 'zero-stress' layer on the upper surface (see § 5.15). It is quite difficult to detect the presence of the zero-stress layer from the mean velocity profile which is nearly the same as in one-half of a channel flow with width $4D$ and a pressure gradient of θg (θ is the slope of the channel). The bottom stress is $\tau_0 = 2\theta g D$, and the composite velocity distribution is

$$
\left.\begin{aligned}
\frac{U-U_1}{\tau_0^{1/2}} &= -\tfrac{1}{4}R_s\left(2-\frac{z}{D}\right)^2 \quad \text{for } \alpha < z/D < 2, \\
&= -\tfrac{1}{4}R_s(2-\alpha)^2 - \frac{1}{k}\log\alpha + \frac{1}{k}\log Z/D \\
&\qquad\qquad\qquad\qquad \text{for } 0 < z/D < \alpha,
\end{aligned}\right\}
\tag{5.6.12}
$$

where U_1 is an effective surface velocity given by

$$
kU_1/\tau_0^{1/2} = \log\frac{\alpha D}{z_0} + \tfrac{1}{4}kR_s(2-\alpha)^2.
\tag{5.6.13}
$$

The relevant constants are

$$
\left.\begin{aligned}
C_0 &= -\tfrac{1}{4}(2-\alpha)^2/\alpha - \log\alpha + \frac{1}{4\alpha}, \\
\tfrac{1}{2}\int_0^2 f(\eta)\,d\eta &= -\frac{1}{k}(\tfrac{1}{12}\alpha^{-1}+\tfrac{1}{4}\alpha+\tfrac{1}{12}\alpha^2).
\end{aligned}\right\}
\tag{5.6.14}
$$

For plane Couette flow ($\tau_1 = \tau_0$), the composite distribution is

$$
\begin{aligned}
\frac{U-U(D)}{\tau_0^{1/2}} &= R_s\left(\frac{z}{D}-1\right) \\
&= -R_s(1-\alpha) - \frac{1}{k}\log\alpha + \frac{1}{k}\log z/D,
\end{aligned}
\tag{5.6.15}
$$

and

$$U_1 = 2U(D) = 2\frac{\tau_0^{1/2}}{k}\left[\log\frac{\alpha D}{z_0} + R_s(1-\alpha)\right]. \quad (5.6.16)$$

The constants are

$$\left. \begin{array}{l} C_0 = C_1 = -\dfrac{1-\alpha}{\alpha} - \log\alpha, \\[2ex] \dfrac{1}{2}\displaystyle\int_0^2 f(\eta)\,d\eta = 0. \end{array} \right\} \quad (5.6.17)$$

For axisymmetric flow along a pipe of circular section, the velocity distribution is

$$\left. \begin{array}{l} \dfrac{U-U_0}{\tau_0^{1/2}} = -\tfrac{1}{2}R_s(r/R)^2 \quad \text{for } r/R < (1-\alpha), \\[2ex] \qquad = -\tfrac{1}{2}R_s(1-\alpha)^2 - \dfrac{1}{k}\log\alpha + \dfrac{1}{k}\log\left(1-\dfrac{r}{R}\right) \\[2ex] \qquad\qquad \text{for } 1 > R/r > (1-\alpha) \end{array} \right\} \quad (5.6.18)$$

and the axial velocity is given by

$$kU(0)/\tau_0^{1/2} = \log\frac{\alpha R}{z_0} + \tfrac{1}{4}kR_s(1-\alpha)^2. \quad (5.6.19)$$

The constants are

$$\left. \begin{array}{l} C_0 = -\log\alpha - \tfrac{1}{2}kR_s(1-\alpha)^2, \\[2ex] 2\displaystyle\int_0^1 \eta f(\eta)\,d\eta = -\dfrac{1}{k}(\tfrac{1}{4}\alpha^{-1} + \alpha + \tfrac{1}{2}\alpha^2 + \tfrac{1}{4}\alpha^3). \end{array} \right\} \quad (5.6.20)$$

Values of the flow constant, R_s, and of the parameters α, C_0, C_1, $\frac{1}{2}\int^2 \eta f(\eta)\,d\eta$, are given in table 5.1, based on measurements of the velocity distribution in the flows. The results for pipe flow can be compared with the accurate measurements of the dependence of pressure gradient on flow rate. The bulk flow velocity measurements are represented very well by

$$kU_m/\tau_0^{1/2} = \log(\tau_0^{1/2}R/v) + 0.81 \quad (5.6.21)$$

(Schlichting 1955, p. 413) with $k = 0.407$. From the table, the relation should be

$$kU_m/\tau_0^{1/2} = \log(\tau_0^{1/2}R/\nu) + A - 1.35,$$

where A is the additive constant in the logarithmic distribution (5.4.6), approximately 2.2. Also, the difference between the bulk flow velocity and the axial velocity is given by Goldstein (1938) as

$$(U(0) - U_m)/\tau_0^{1/2} = 4.07 \tag{5.6.22}$$

compared with a value of 4.21 using the value $\alpha = 0.16$ in (5.6.20).

TABLE 5.1   *Flow constants for pipe and channel flows*

Flow	τ_1/τ_0	U_mD/ν	R_s	α	C_0	C_1	$\dfrac{(U_1-U_m)}{\tau_0^{1/2}}$
2-D channel [a]	−1	61,600	12.9	0.19	−0.78	0.78	−2.4
Plane Couette	1	18,000	7–9 [b]				
				0.29	−1.23	−1.23	0
		18,000	8.5 [c]				
Open channel [d]	0	50,000	13.0	0.19	−0.78	–	−2.4
Pipe [e]	–	300,000	15.2	0.16	−0.36	–	−4.21

[a] From Laufer (1951).
[b] From Reichardt (1956).
[c] From Robertson (1959).
[d] From Finley *et al.* (1966). Note that D here is equal to the flow depth, *not* half the flow depth.
[e] From Laufer (1955).

5.7   The turbulent motion in constant-stress equilibrium layers

Assuming that local values of Reynolds stress, energy dissipation and lateral flux of turbulent energy depend only on wall stress and distance from the wall places few restrictions on the nature of the turbulent motion, but the assumption leads to a definite form of the mean velocity distribution. By measuring the terms in the equation for the turbulent kinetic energy, the assumptions may be tested and the existence of local energy equilibrium confirmed. For example, Laufer (1955) has made comprehensive measurements in pipe flow with results shown in figs 5.5, 5.6. A better reason for faith in the similarity hypothesis is that the predicted distribution of mean velocity, the 'law of the wall', is found in many different kinds of

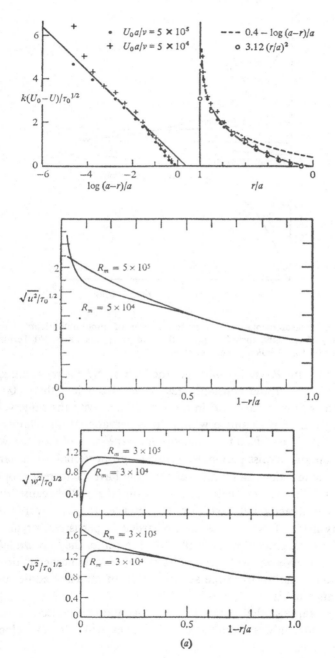

(a)

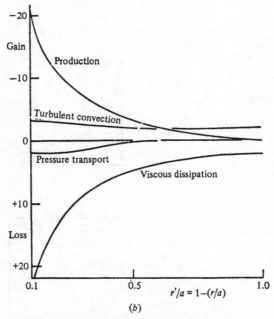

Fig. 5.6. Measurements in the central region of pipe flow (from Laufer 1955). (a) Mean flow, turbulent intensities and Reynolds stress. (b) Terms in the equation for turbulent kinetic energy.

wall flow with Reynolds numbers (defined as $\tau_0^{1/2} z_c/\nu$, where z_c is the thickness of the constant-stress layer) from less than 100 in laboratory flows to over 10^6 in the atmosphere, with the same value of the Kármán constant to within the observational uncertainty of about 2%. It is difficult to imagine how the presence of the wall could impose a dissipation length-scale proportional to distance from it unless the main eddies of the flow have diameters proportional to distance of their 'centres' from the wall because their motion is directly influenced by its presence. In other words, the velocity fields of the main eddies, regarded as persistent, organised flow patterns, extend to the wall and, in a sense, they are *attached* to the wall. We proceed to consider the observed characteristics of a motion made up from the superposition of attached eddies of a wide range of sizes.

Let us suppose that the main, energy-containing motion is made up of contributions from 'attached' eddies with similar velocity distributions,

$$u_i(\mathbf{x}) = u_0 f_i[(\mathbf{x} - \mathbf{x}_a)/z_a], \tag{5.7.1}$$

where $\mathbf{x}_a = (x_a, y_a, z_a)$ is the centre of a particular eddy, and u_0 is its velocity scale. Then the correlation function for a flow built from a random superposition of the eddies is given by

$$R_{ij}(\mathbf{r}; z) = \int \int P(u_0, \mathbf{x}_a) u_0^2 f_i \left[\frac{\mathbf{x} - \mathbf{x}_a}{z_a} \right] f_j \left[\frac{\mathbf{x} + \mathbf{r} - \mathbf{x}_a}{z_a} \right] du_0 \, d\mathbf{x}_a$$

$$= \int_{l_0}^{L_0} N(z_a) I_{ij}(\mathbf{r}/z_a; z/z_a) \, dz_a/z_a, \tag{5.7.2}$$

where $P(u_0, \mathbf{x}_a)$ is the probability distribution function for eddy velocity u_0 and centre position $\mathbf{x}_a$, which, by the flow homogeneity, must be independent of x_a and y_a, l_0 and L_0 are the lower and upper limits of z_a,

$$N(z_a) = \int_0^\infty u_0^2 z_a^3 P(u_0, \mathbf{x}_a) \, du_0$$

is the intensity of eddies with centres distant z_a from the wall, and

$$I_{ij}(\mathbf{r}/z_a; z/z_a) = \int \int_{-\infty}^\infty f_i(\mathbf{x}^*) f_j(\mathbf{x}^* + \mathbf{r}/z_a) \, dx^* \, dy^* \tag{5.7.3}$$

measures the contribution to the correlation function from eddies with centres distant z_a from the wall. The basic requirements of wall similarity will be met if the distribution function $I_{ij}(0; z/z_a)$ permits uniform Reynolds stress for $l_0 \ll z \ll L_0$ deep within the fully turbulent layer, and if energy input and loss for each eddy are concentrated near its centre.

The presence of the wall means that the normal velocity is zero on it and so $f_3(0) = 0$. For small values of z/z_a, $f_3 \approx z/z_a \, fn(x, y)$ while f_1 and f_2 are not restricted. It follows that the behaviour of the integrals $I_{ij}(0; z^*)$ for small values of z^* is

$$I_{13} \propto z^*, \quad I_{33} \propto z^{*2},$$

$$\left. \right\} \tag{5.7.4}$$

I_{11} and I_{22} approach non-zero values.

Since the component eddies are finite structures, all the f_i and the I_{ij} become very small for large values of z^*. Possible forms for the components of $I_{ij}(0; z^*)$ are shown in fig. 5.7, and it is clearly possible that the contribution of eddies of one size to the Reynolds stress can be concentrated in the general neighbourhood of the distance from the wall of their centres. From the conditions (5.7.4), the integral,

$$\int_{z/L_0}^{z/l_0} I_{13}(0; z^*) \, dz^*/z^*,$$

is nearly independent of z if $l_0 \ll z \ll L_0$, and constant Reynolds stress is possible if

$$N(z_a) = \text{constant.} \tag{5.7.5}$$

Using the same distribution of eddy intensity, inspection of the integral in (5.7.2) shows that

$$\left.\begin{aligned} \overline{u^2}/\tau_0 &= C_1 + D_1 \log L_0/z, \\ \overline{v^2}/\tau_0 &= C_2 + D_2 \log L_0/z, \\ \overline{w^2}/\tau_0 &= C_3, \qquad -\overline{uw}/\tau_0 = 1, \end{aligned}\right\} \tag{5.7.6}$$

all for $l_0 \ll z \ll L_0$. The constants depend on the forms of the attached eddies.

It now appears that simple similarity of the motion is not possible with attached eddies and, in particular, that the stress–intensity ratio, $-\overline{uw}/\overline{q^2}$, depends to some extent on position in the layer. The variation of the ratio does not invalidate the previous similarity analysis because the 'non-similar' logarithmic terms in the expressions for $\overline{u^2}$ and $\overline{v^2}$ represent motions which are large-scale swirling in planes parallel to the wall and do not extract energy from the mean flow or affect the rate of energy transfer to smaller eddies for viscous dissipation. Swirling motions contribute little to the Reynolds stress, and their effect on that part of the layer between the point of observation and the wall is one of slow random variations of 'mean velocity' which cause corresponding variations of wall stress. It is possible and useful to regard the 'swirl' component of the local motion as an *inactive* component which may be ignored in any discussion of the local flow, for example when using similarity assumptions to interpret the turbulent energy equation as an equation for the Reynolds stress. On the other hand, the inactive flow at one level is an essential part of active flow at other higher levels, and the distinction should be regarded as a device for reconciling descriptions based on local parameters (Bradshaw, Ferris &

Atwell 1967) with the reality of large eddies extending to the wall.

Because the time scale of the swirling, inactive motion is much longer than the adjustment time of the eddies with centres between

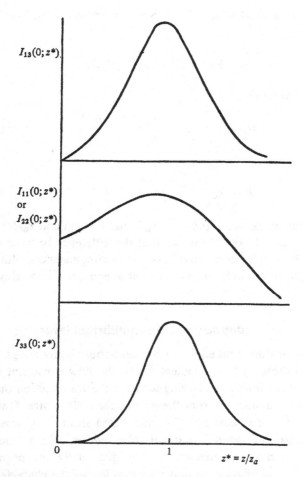

Fig. 5.7. Contributions to Reynolds stress and turbulent intensities from a single attached eddy.

an observation point and the wall, the intervening motion will be nearly in a state of wall equilibrium, adjusted to give the current velocity at the point. So, if $\mathbf{u}'$ is the velocity fluctuation caused by inactive motion, the current velocity $U + u'$ is related to the current wall stress τ by

$$|U+u'| = \frac{\tau^{1/2}}{k_0} \log z/z_0, \tag{5.7.7}$$

where k_0 is a local value of the Kármán constant. Squaring and averaging,

$$U^2 + \overline{u'^2} + \overline{v'^2} = \frac{\bar{\tau}}{k_0^2} \log^2 z/z_0$$

and, approximately,

$$U = \frac{\bar{\tau}^{1/2}}{k_0} \log z/z_0 \left(1 - \frac{\overline{u'^2} + \overline{v'^2}}{2U^2}\right) \tag{5.7.8}$$

or,

$$k = k_0 \left(1 + \frac{\overline{u'^2} + \overline{v'^2}}{2U^2}\right). \tag{5.7.9}$$

In constant-stress layers, $(\overline{u^2} + \overline{v^2})/\tau_0$ is about seven and τ_0/U^2 is unlikely to exceed 0.005. It follows that the difference between k and k_0 is unlikely to be detectable in ordinary circumstances, although, in principle, it would become important at extremely large Reynolds numbers.

5.8   Eddy structure in equilibrium layers

Although the dominant eddies in constant-stress layers are generally similar to those in free turbulent flows, the flow constraint at the wall prevents eddies from having scales in the Oz direction that are larger than the distance from the wall of the eddy centre. It will be recalled that the typical eddy in unrestricted shear flows resembles closely a pair of inclined, parallel roller eddies, which would be restricted by a solid boundary both longitudinally and normal to the wall. One configuration that preserves locally the characteristics of a double-roller eddy while satisfying the wall constraint is a 'double-cone' eddy, sketched in fig. 5.8. In the double-cone eddy, circulation takes place around the surfaces of two adjacent cones diverging from a common vertex on the wall, and, since the eddy has contact with the wall over its whole length, it is an attached eddy. Other possibilities exist, e.g. a distribution of shorter double-roller eddies of various sizes each with lateral extent comparable with distance from the wall, but the flow visualisation studies of Kline *et al.* (1967) show clearly flow structures resembling double cones and it is useful to compare measurements with a model based on double-cone eddies.

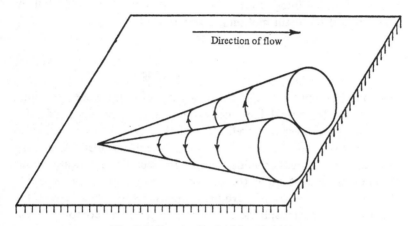

Fig. 5.8. Sketch of a double-cone eddy.

A conical eddy has a velocity distribution of the form,

$$u_i(\mathbf{x}) = u_0(x - x_0)f_i[(y - y_0)/z_a, z/z_a], \qquad (5.8.1)$$

where $u_0(x - x_0)$ is the velocity scale at a section distant $x - x_0$ from the cone vertex at $(x_0, y_0, 0)$, and $z_a(x - x_0)$ measures the cone diameter at that section. It could be described as an organised assembly of simple attached eddies of the whole range of sizes, and the main motion of the equilibrium layer is assumed to be the superposition of similar conical eddies with vertices distributed over the wall with number density n. Then the correlation function is

$$R_{ij}(r; z) = n \int_{l_0}^{L_0} \int_{-\infty}^{\infty} u_0(x)u_0(x+r_1)f_i(y^*, z^*)f_j(y^*+r_2^*, z^*+r_3^*)$$

$$\times (dx/dz_a)z_a \, dz_a \, dy^* \qquad (5.8.2)$$

and the condition of constant Reynolds stress for $l_0 \ll z \ll L_0$ is satisfied if

$$u_0^2(x)z_a^2 \, dx/dz_a = \text{constant.} \qquad (5.8.3)$$

A simple form of double-cone eddy is specified by

$$
\left.
\begin{aligned}
-af_1 = f_3 &= z^*(1 - y^{*2}) \exp[-\tfrac{1}{2}(y^{*2} + z^{*2})] \\
f_2 &= -y^*(1 - z^{*2}) \exp[-\tfrac{1}{2}(y^{*2} + z^{*2})].
\end{aligned}
\right\}
\tag{5.8.4}
$$

(The wall condition is satisfied, and $\partial v/\partial y + \partial w/\partial z = 0$, i.e. the contribution of $\partial u/\partial x$ to the velocity divergence is neglected.) Carrying out the integration in (5.8.2), it is found that the turbulent intensities of the main motion are given by

$$
\left.
\begin{aligned}
\overline{u^2} = a^{-1}\tau_0, \qquad \overline{w^2} &= a\tau_0, \\
\overline{v^2} = (4/3)a\tau_0(-1.58 &+ 2\log L_0/z).
\end{aligned}
\right\}
\tag{5.8.5}
$$

It should be emphasised that the invariance of $\overline{u^2}$ is for the particular eddy form. In typical laboratory flows, measurements inside the constant-stress layer and well clear of the viscous layer are in the range $z = 0.05$ to $0.1D$, and it is only in the atmospheric boundary layer that much smaller values of z/D are accessible, typically about 0.003 (D is understood as the effective thickness of the flow). Table 5.2 lists some ratios of turbulent intensities to shear stress for laboratory and atmospheric equilibrium layers, and compares them with values calculated from equations (5.8.5) for relevant values of z/L_0. It is clear that the difference of the ratios $\overline{v^2}/\tau_0$ is similar to that given by the model, and that the rather scattered measurements of $\overline{w^2}/\tau_0$ are similar in both conditions. Significant variation of the ratio $\overline{u^2}/\tau_0$ occurs, not predicted by the model, but it is less than the variation of $\overline{v^2}/\tau_0$.

TABLE 5.2   *Intensity–stress ratios in laboratory and atmospheric equilibrium layers*

	Observed		Equations (5.8.5) $a = 0.4$	
Ratio	Laboratory	Atmosphere	$z/L_0 = 0.05$	$z/L_0 = 0.01$
$\overline{u^2}/\tau_0$	4.8	6.2	3.3	3.3
$\overline{v^2}/\tau_0$	2.2	4.5	2.3	4.1
$\overline{w^2}/\tau_0$	1.1	0.7–1.3	0.4	0.4

The observed correlation function departs considerably from the simple similarity form,

$$
R_{ij}(\mathbf{r}; z) = \tau_0 \mathcal{R}(\mathbf{r}/z)
\tag{5.8.6}
$$

TABLE 5.3   Comparison of correlations measured in an equilibrium layer with double-cone, double roller and rapid-distortion models

	Observed		Double cone				Double roller		Rapid distortion	
			$z/L = 0.1$		$z/L = 0.05$					
	+	−	+	−	+	−	+	−	+	−
$R_{11}(r, 0, 0)$	30	—	4.2	—	4.2	—	6.92	—	4.2	—
$R_{11}(0, r, 0)$	—	2.6 (0.02)	—	1.4 (0.18)	—	1.4 (0.18)	—	1.10 (0.54)	—	1.2 (0.08)
$R_{11}(0, 0, r)$	7	—	13.3	—	13.3	—	4.00	—	2.2 *or*	3.5 (0.01)
$R_{22}(r, 0, 0)$	7	—	~20	—	~30	—	—	2.83 (0.13)	—	1.6 (0.10)
$R_{22}(0, r, 0)$	—	5 (0.02)	—	7.4 (0.19)	—	13.4 (0.21)	—	1.41 (0.13)	—	2.5 (0.035)
$R_{22}(0, 0, r)$	—	1.3 (0.06)	—	4.0 (0.20)	—	8.3 (0.21)	—	1.63 (0.13)	—	1.0 (0.045)
$R_{33}(r, 0, 0)$	4.4	—	4.2	—	4.2	—	6.92	—	—	2.9 (0.04)
$R_{33}(0, r, 0)$	1.4	—	—	1.4 (0.18)	—	1.4 (0.18)	—	1.10 (0.54)	1.9 *or*	3.5 (0.002)
$R_{33}(0, 0, r)$	7	—	13.3	—	13.3	—	4.00	—	3.8	—

Note: Double roller has a slope of 30°. For conventions see page 110.

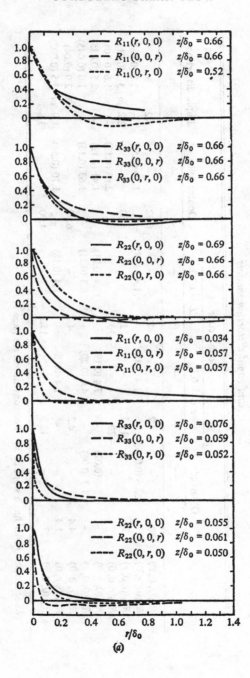

(a)

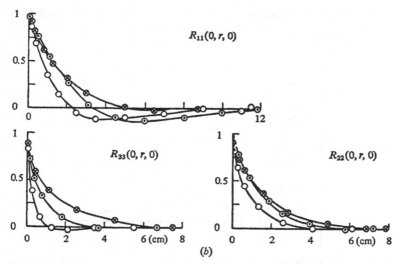

Fig. 5.9. Correlation functions for wall flow; (a) Measurements by Grant (1958) in a boundary layer; (b) Measurements by Comte-Bellot (1961) in a two-dimensional channel. z/D: $\otimes$ = 1.0, $\odot$ = 0.44, $\bigcirc$ = 0.11.

and appears to have two 'components', a 'wall' component scaling with distance from the wall and an 'inactive' component varying slowly with separation and dependent on the whole flow. Here, the two components do not indicate two distinct groups of eddies. Fig. 5.9 shows correlations measured by Grant (1958) in a boundary layer and by Comte-Bellot (1961) in a channel. They may be compared with correlations calculated for double-cone eddies of the form (5.8.4) for $z/L_0 = 0.1$, a value typical of laboratory measurements (fig. 5.10). Table 5.3 compares the observations with the calculated rapid-distortion structure and correlations for a simple inclined double-roller model as well, using the numerical description of § 5.4. Considering the crudity of the model, the agreement is satisfactory. Notice that the change of type in the component $R_{22}(r, 0, 0)$ from the rapid-distortion form found in homogeneous flows is correctly predicted. The serious discrepancy is with the longitudinal component $R_{11}(r, 0, 0)$ and may be traced to use of too simple a model.†

† The component $R_{11}(r, 0, 0)$ is very sensitive to transitory transverse flow patterns, driven by 'non-Newtonian' instability of the mean flow (see § 7.21).

The simple double-cone form of the main eddies gives a qualitative description of the forms of the instantaneous correlation function,

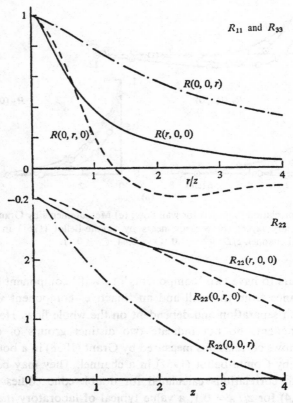

Fig. 5.10. Correlation functions calculated from a double-cone model.

but measurements of convection velocities from space–time correlations show that different parts of the cone move with different velocities. Changes in shape of the cone must occur and any particular form represents an average over the lifetime of individual eddies.

5.9   Motion in the viscous layer next the wall

The boundary conditions that must be satisfied at a smooth wall are

$$u = v = w = 0 \quad \text{for } z = 0,$$

and continuity requires that $\partial w/\partial z = 0$. From these conditions and the equations of motion a number of simple relations may be derived. Neglecting the variation of total stress in the viscous layer,

$$\tau_0 = -\overline{uw} + \nu \frac{\partial U}{\partial z}$$

and it follows that

$$\left.\begin{array}{l}
\left(\dfrac{\partial U}{\partial z}\right)_{z=0} = \dfrac{\tau_0}{\nu}, \\[2mm]
\left(\dfrac{\partial^2 U}{\partial z^2}\right)_{z=0} = \dfrac{1}{\nu}\left(\dfrac{\partial \overline{uw}}{\partial z}\right)_{z=0} = 0, \\[2mm]
\left(\dfrac{\partial^3 U}{\partial z^3}\right)_{z=0} = \dfrac{1}{\nu}\left(\dfrac{\partial^2 \overline{uw}}{\partial z^2}\right)_{z=0} = 0, \\[2mm]
\left(\dfrac{\partial^4 U}{\partial z^4}\right)_{z=0} = \dfrac{1}{\nu}\left(\dfrac{\partial^3 \overline{uw}}{\partial z^3}\right)_{z=0} = \dfrac{3}{\nu}\left[\dfrac{\partial u}{\partial z}\dfrac{\partial^2 w}{\partial z^2}\right]_{z=0}.
\end{array}\right\} \quad (5.9.1)$$

Since the second and third derivatives of mean velocity are zero at the wall, the velocity variation will be closely linear for an appreciable range, and the departure from linearity, when it comes, will be rather abrupt. A rough approximation is to neglect the transition between the linear distribution and the fully turbulent logarithmic distribution and to represent the whole velocity variation by

$$\left.\begin{array}{l}
U = \dfrac{\tau_0}{\nu}z \quad \text{for} \quad \dfrac{{\tau_0}^{1/2}z}{\nu} < a_0, \\[3mm]
U = \dfrac{{\tau_0}^{1/2}}{k}\left[\log \dfrac{{\tau_0}^{1/2}z}{\nu} + A\right] \quad \text{for} \quad \dfrac{{\tau_0}^{1/2}z}{\nu} > a_0.
\end{array}\right\} \quad (5.9.2)$$

Two conditions must be satisfied, that U is continuous and that the change of slope at the junction is sufficient that the viscous stress in the fully turbulent part should be small compared with τ_0, that is to say,

$$\left.\begin{array}{l}
ka_0 = \log a_0 + A, \\[1mm]
ka_0 \gg 1.
\end{array}\right\} \quad (5.9.3)$$

The Kármán constant k is characteristic of the fully turbulent flow and is nearly 0.41. Then, replacing the symbol $\gg$ by $=5\times$, it is found that

$$A = 2.5, \qquad a_0 = 12.2$$

close to the observed values.

Within the viscous layer, dissipation of turbulent energy greatly exceeds production by working against Reynolds stresses and the motion is driven by the outside fully turbulent flow through viscous stresses and pressure gradients. To a first approximation comparable with that for the composite mean velocity profile of (5.9.2), the u and v components of the fluctuation always increase linearly with distance from the wall and so are determined by the motion at the edge of the viscous layer. Their correlation function in the xOy plane should be very little different from the same correlation just outside the layer, and might imply a double roller or perhaps a linear jet. Measurements by Elswick (1967) have been analysed by Bakewell & Lumley (1967) to give the structure of fig. 5.11.

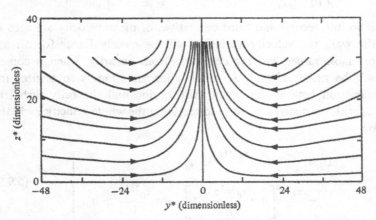

Fig. 5.11. Sketch of typical eddy motion in the viscous layer of channel flow (from Bakewell & Lumley 1967).

A better description of the details of the motion is provided by the theory of Schubert & Corcos (1967) which uses observations of the pressure and velocity fields at the layer edge as boundary conditions for linearised equations of motion. It provides a good description of the variations of intensity and Reynolds stress in the layer.

5.10 Fluctuations of pressure and shear stress on a wall

The magnitudes and characters of the pressure and stress fluctuations on the wall under a turbulent flow have considerable practical interest for problems of noise production and for deformation of the wall material, and their measurement is possible without appreciable disturbance of the flow, unlike that arising from the use of hot-wire anemometers in the viscous layer. Fluctuations of shear stress may be measured from the fluctuations of transfer rates of heat or matter from a small source on the wall (Bellhouse & Schultz 1968, Mitchell & Hanratty 1966). Mitchell & Hanratty give measurements of the longitudinal and lateral correlations for $\partial u/\partial z$, and time spectra of $\partial u/\partial z$ at a fixed point, measured in a circular pipe at Reynolds numbers from 7,500 to 23,800. They show:

(1) that the longitudinal correlation extends to distances comparable with pipe diameter,

(2) that the transverse correlation becomes negative for separations comparable with the thickness of the viscous layer, and may have a long negative tail,

(3) the time spectra scale with pipe diameter and mean flow velocity rather than with friction velocity and fluid viscosity. All the features are compatible with attached eddies with axes in the direction of flow.

Measurements of wall pressure do not give information about the viscous layer alone since the pressure at a point is related to the velocity field by the Poisson equation,

$$-\nabla^2 p = 2 \frac{\partial U}{\partial z} \frac{\partial w}{\partial x} + \frac{\partial^2}{\partial x_i \, \partial x_j} (u_i u_j - \overline{u_i u_j}). \tag{5.10.1}$$

If we consider Fourier components of the pressure fluctuation over the wall, the equation indicates that the amplitude will be influenced by velocity fluctuations within a distance comparable with the wavelength of the component. Then, eddies of the main motion with centres distant z from the wall have diameters comparable with z and are the main contributors to wave numbers around z^{-1}. They are expected to move with the mean velocity at z, and we expect that the convection velocity of pressure components of wave number k will be determined by the mean velocity at height ak^{-1} (a is a constant near one). The measurements, in particular those of Willmarth & Woolridge (1962) and of Bull (1967) confirm the expectation, the convection velocity of the largest wave-number components being $0.59\,U_1$ and of the smallest $0.90\,U_1$ (the observations were made in unaccelerated boundary layers with free stream velocity U_1), i.e. nearly the mean velocities just clear of the viscous layer and in the middle of the outer flow.

The correlation function for the wall pressure is dominated by the large-scale fluctuations at considerable distances from the wall,

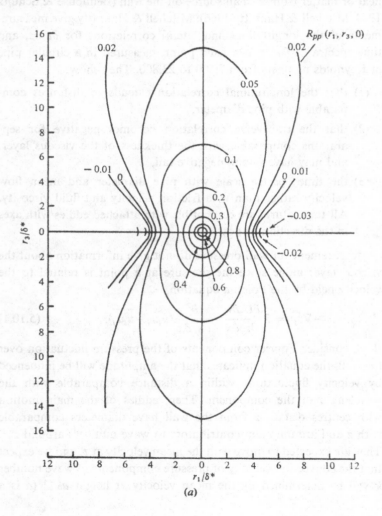

(a)

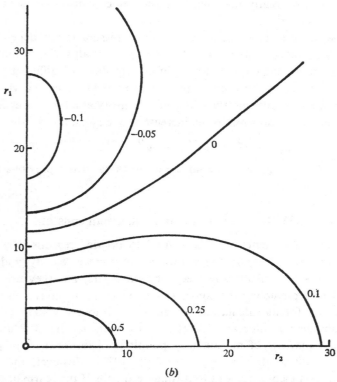

(b)

Fig. 5.12. Pressure correlations at the wall; (a) Contours of constant correlation by Bull (1967); (b) Contours calculated from

$$\nabla^2 p = -2 \frac{\partial U}{\partial z} \frac{\partial w}{\partial x}.$$

and the forms observed by Bull are in reasonable agreement with calculations based on neglect of source terms other than $2(\partial U/\partial z) \times (\partial w/\partial x)$ in equation (5.10.1). A comparison between measurements and the calculations for attached eddies with the (partial) velocity distributions,

$$w \propto z(x - x_0)(y - y_0) \exp -\frac{1}{2}\left(\frac{(x - x_0)^2 + (y - y_0)^2 + (z - z_0)^2}{z_0^2}\right)$$

$$(5.10.2)$$

is shown in fig. 5.12. The main features arise from the $\partial w/\partial x$ factor and are not greatly influenced by the velocity distribution in the attached eddies.

The mean square fluctuation of wall pressure is not dependent simply on the velocity scale of the flow because all scales of motion make equal contributions to it. Supposing flows of different Reynolds number differ only in the relative thickness of the viscous layer, it is obvious that the additional intensity at a larger Reynolds number is proportional to the increase in $\log L_0/l_0$, and so that

$$\overline{p^2} = \tau_0^2(C \log L_0/l_0 + D), \tag{5.10.3}$$

where L_0 and l_0 have the same meanings as in the previous sections.

5.11 The magnitude of the Kármán constant

Similarity arguments have been used to provide reasons why the law of the wall describes the relation between mean velocity gradient and stress in constant-stress equilibrium layers, but they are not capable of predicting the constant. At the moment, no satisfactory procedure for its calculation has emerged although Malkus (1956) has obtained a value and recent progress with numerical solutions of the equations of motion may permit its determination in particular flow conditions (e.g. Deardorff 1970). However, the following argument does lead to a rough estimate of the Kármán constant.

Consider the history of an attached eddy (or of a section of a double-cone eddy). It appears, probably as a result of the shearing of random motion, grows to full strength and then is broken up and dispersed. For much of its life, it is contributing to the local Reynolds stress, extracting energy from the mean flow and passing the energy on to the cascade for dissipation, and it must be near a condition of energy equilibrium for much of its life. Assuming a velocity distribution,

$$\left. \begin{aligned} -au = w &= u_0 z^*(1-y^{*2}) \exp{-\tfrac{1}{2}(y^{*2}+z^{*2})}, \\ v &= -u_0 y^*(1-z^{*2}) \exp{-\tfrac{1}{2}(y^{*2}+z^{*2})}, \end{aligned} \right\} \tag{5.11.1}$$

where $y^* = y/z_0$, $z^* = z/z_0$, and z_0 is the position of the eddy centre, it may be shown that the rate of transfer of energy from the mean flow is

$$T = \frac{3\pi^{1/2}}{8} \frac{\tau_0^{1/2} z_0}{ka} u_0^2. \tag{5.11.2}$$

In a constant-stress layer, the eddy viscosity for the mean flow is $k\tau_0^{1/2}z$, and, if the energy loss from the attached eddy can be described by an eddy viscosity, it will be $k'\tau_0^{1/2}z$, where k' is another constant. Using this eddy viscosity, the rate of energy loss to smaller eddies is calculated to be

$$E = \frac{3\pi^{1/2}}{16} k'\tau_0^{1/2} z_0 u_0^2 (13 + 7a^{-2}) \tag{5.11.3}$$

and equating the two rates leads to

$$kk' = \frac{2}{13a + 7a^{-1}}. \tag{5.11.4}$$

If the eddy viscosity for the mean flow were equal to the transfer viscosity for the attached eddy, (5.11.3) leads to a value of $k = 0.32$ for the assumed form of the attached eddy.† The effective value of the transfer eddy viscosity is likely to equal the mean flow value close to the wall where the scale of the attached eddy is large compared with the scale of the local motion, but it will fall below it in the outermost parts where the scale of the eddy motion is comparatively small. The value is therefore a lower limit, but it is as close as can be expected from arguments of this kind.

5.12 Turbulent flow and flow constants

The assumption of constant eddy viscosity outside the wall layers is a simple procedure that describes the mean velocity distributions for the stress ratios, 1, 0 and -1, with fair accuracy, and values of the flow constant are available for the three ratios. Values for other ratios are of importance for lubrication flows and Reynolds (1963) has proposed an interpolation, devised solely on grounds of convenience. It is perhaps more satisfying though not necessarily more accurate to interpolate from considerations of the turbulent flow.

In pressure flow between parallel planes, the turbulent motion in

† The factor a measures the inclination of the planes of circulation in the attached eddy. From observation, it is probably in the range 0.5–0.7, but the expression (5.11.4) is not sensitive to its value in this range.

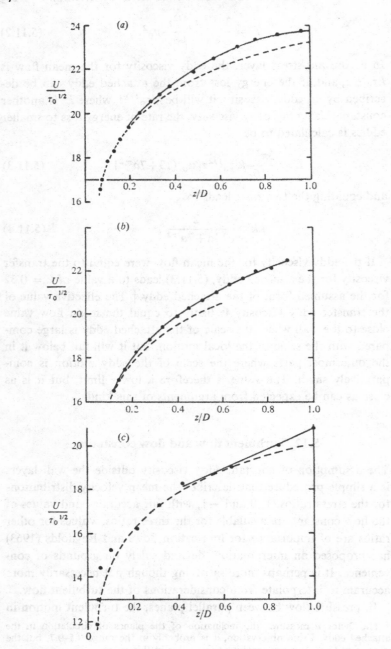

the central region is consistent with the dominance of inclined, double-roller eddies which are inclined in opposite directions in the two halves of the channel and which possibly may have a tendency to form hairpin eddies of the kind observed by Grant (1958) in a wake. However this may be, the turbulent intensity is small near the central plane and it seems likely that the motion in one half may be nearly uninfluenced by that in the other half. Support for this view is provided by the observation that the flow constant for pressure flow is very nearly twice that for open channel flow with zero stress on one boundary. If we ignore the weak, 'zero-stress' equilibrium layer on the free surface, the velocity and stress distributions are identical with those in one-half of pressure flow with the same wall stress. Evidently, fluid movements across the central plane have little effect since the replacement of the plane by an impermeable free surface induces no major change in the flow.

If a plane of zero Reynolds stress is an effective boundary to the turbulent motion, a channel with negative stress ratio can be considered as made up of two parts, separated by the plane,

$$z/D = \frac{2\tau_0}{\tau_0 - \tau_1} = \eta_b \qquad (5.12.1)$$

on which $\tau = 0$. On either side, the eddy viscosity is specified by the relevant wall stress, the layer thickness and the flow constant for open channel flow R_{s0}, i.e.

$$
\left.
\begin{aligned}
v_T &= R_{s0}^{-1}\tau_0^{1/2}D\frac{2\tau_0}{\tau_0 - \tau_1} \quad \text{for } z/D < \eta_b, \\
&= R_{s0}^{-1}\tau_0^{1/2}D\frac{-2\tau_1}{\tau_0 - \tau_1} \quad \text{for } z/D > \eta_b.
\end{aligned}
\right\} \qquad (5.12.2)
$$

Fig. 5.13. Mean velocity profiles for channel flow; (a) two-dimensional channel flow (Stevenson 1958),

$$\frac{\tau_0^{1/2}D}{\nu} = 1540, \ldots \frac{1}{k}\log\frac{z}{D} + 23.3;$$

(b) free surface flow (Finley et al. 1966),

$$\frac{U_1 D}{\nu} = 3 \times 10^4, \ldots \frac{1}{k}\log\frac{z}{D} + 21.75;$$

(c) Plane Couette flow (Robertson 1959),

$$\frac{U_1 D}{\nu} = 10^4, \ldots \frac{1}{k}\log\frac{z}{D} + 20.$$

The stress on the dividing plane is zero and, although the eddy viscosity changes discontinuously across it, the two velocity profiles join smoothly.

For positive values of the stress ratio, the division into two parts is not possible. The difference between the flow constants for free surface and plane Couette flow may be regarded as a consequence of the greater relative thickness of the wall equilibrium layer in the free surface flow (fig. 5.13). Now the limit to the extent of an equilibrium layer in parallel flow is set by the condition, that the effect of the far boundary should be small both as a constraint on eddy size and as a source of energy. In free surface flow, little energy is available from the zero-stress equilibrium layer, unlike the Couette flow, and it should not be surprising that the equilibrium layer is relatively thinner in Couette flow. In the present state of knowledge, the effect might be represented by the simple interpolations,

$$\alpha_1 = \frac{z_e}{D} = \alpha_f - \frac{\tau_1}{\tau_0}(\alpha_f - \alpha_c),$$

and

$$R_s = 6.5 + 2\tau_1/\tau_0$$

(5.12.3)

using observed values.

5.13 Similarity flows in channels and pipes of varying widths

Fully developed flows along uniform pipes and channels are homogeneous in the direction of flow, and the distributions of mean values are not only similar but identical at all sections. If the channel section changes in size but not in shape, it is still possible for the distributions to be similar at all sections. First, if the section changes very slowly, the flow is always a close approximation to that in a uniform channel of the same width and the imposed volume flow, and the flow at different sections is described by the defect law,

$$U = U_1 + \tau_0^{1/2} f(z/D)$$

and by

$$\overline{u_i u_j} = \tau_0 g_{ij}(z/D)$$

and similar expressions. The only difference from the uniform channel is that τ_0 and D are functions of x, related by a condition for conservation of mass,

$$U_m D = U_1 D + \tau_0^{1/2} D \int_0^1 f(x)\, \mathrm{d}x$$
$$= Q \text{ (a constant)} \tag{5.13.1}$$

and the uniform channel friction law (5.6.6).

Necessary conditions for the quasi-stationary kind of similarity development are that the terms in the equations for mean velocity and turbulent kinetic energy should approximate to those for flow in a uniform channel. The critical requirement is that the term representing advection of mean flow momentum should be small compared with the lateral stress gradient,

$$\left| U \frac{\partial U}{\partial x} + W \frac{\partial U}{\partial z} \right| \ll \left| \frac{\partial \tau}{\partial z} \right|. \tag{5.13.2}$$

If the approximation of quasi-stationary equilibrium is satisfied, the friction coefficient τ_0/U^2 is nearly constant, and, at the channel centre,

$$U \frac{\partial U}{\partial x} + W \frac{\partial U}{\partial z} = U_1 \frac{\mathrm{d}U_1}{\mathrm{d}x} \approx \frac{U_1^2}{D} \frac{\mathrm{d}D}{\mathrm{d}x} \tag{5.13.3}$$

since U_1 and U_m differ only by a small amount. The condition is so that the angle of divergence should be everywhere small compared with the friction coefficient, which is commonly of order 0.0015. In short, the angle must be extremely small.

In the second kind of similar development, advection terms in the mean value equations are not negligible and the flow similarity depends on a moving equilibrium with the flow at any section developing from flow patterns of different scales and intensities at upstream sections. Its attainment depends first on the existence of external flow conditions that make the *self-preserving* development possible and secondly on a sufficiently long period of development to allow establishment of the moving equilibrium. Self-preserving flow, homogeneous in the Oy direction and with mean flow in the xOz plane, exists if

$$\left. \begin{aligned} U &= U_1 + u_0 f(z/l_0), \\ \overline{uw} &= q_0^2 g_{13}(z/l_0), \\ \overline{q^2} &= q_0^2 g(z/l_0), \\ \varepsilon &= q_0^3/l_0\, h(z/l_0), \\ \overline{pw} + \tfrac{1}{2}\overline{q^2 w} &= q_0^3 j(z/l_0), \text{ etc.} \end{aligned} \right\} \tag{5.13.4}$$

Necessary conditions for the flow are that the equations for the mean velocity and the turbulent energy can be satisfied with these forms and with U_1, u_0, q_0 and l_0 functions only of x. In terms of the non-dimensional functions, the equation for the mean velocity is

$$\frac{dU_1}{dx}(U_1 - u_0\eta f' + u_0 f) + U_1\left(\frac{du_0}{dx}f - \frac{u_0}{l_0}\frac{dl_0}{dx}\eta f'\right) + u_0\frac{du_0}{dx}f^2$$

$$-\frac{u_0}{l_0}\frac{d(u_0 l_0)}{dx}f'\int_0^{} f\,d\eta + \frac{q_0^2}{l_0}g'_{13} = -\frac{dP_0}{dx} + \frac{vu_0}{l_0^2}f'' \qquad (5.13.5)$$

and the equation for the turbulent energy is

$$-\tfrac{1}{2}q_0^2\frac{dU_1}{dx}\eta g' + \tfrac{1}{2}U_1\left(\frac{dq_0^2}{dx}g - \frac{q_0^2}{l_0}\frac{dl_0}{dx}\eta g'\right) + \tfrac{1}{2}u_0\frac{dq_0^2}{dx}fg$$

$$-\frac{1}{2}\frac{q_0^2}{l_0}\frac{du_0 l_0}{dx}g'\int_0^{} f\,d\eta + \frac{q_0^3}{l_0}j' + \frac{q_0^3}{l_0}h + \frac{u_0 q_0^2}{l_0}f'g_{13} = 0. \qquad (5.13.6)$$

The equations can be satisfied by non-dimensional functions independent of x only if the ratios of the coefficients are independent of x, which requires that

(1) $\tau_0^{1/2}$, u_0, q_0 and U_1 preserve a constant ratio,

(2) dl_0/dx = constant,

(3) $dP_0/dx \propto u_0^2/D$, and

(4) in general, that only the fully turbulent part of the flow is included.

Taking into account the boundary conditions, the conditions are satisfied for a two-dimensional channel converging or diverging at a constant angle, and the equations become

$$\frac{D}{x}\left(1 + \frac{u_0}{U_1}f\right)^2 - \frac{u_0^2}{U_1^2}g_{13}' = \frac{D}{U_1^2}\frac{dP_0}{dx} - \frac{vu_0}{U_1^2 D}f'' \qquad (5.13.7)$$

and

$$\frac{D}{x}g\left(1 + \frac{u_0}{U_1}f\right) - f'g_{13} - \frac{u_0}{U_1}(j' + h) = 0 \qquad (5.13.8)$$

after equating $\tau_0^{1/2}$, u_0 and q_0, and l_0 to D. (Note that the origin of $\eta = z/D$ is now the channel centre, not one wall.) Overall conservation of momentum requires that

$$\frac{d}{dx} \int_0^D U^2 \, dz = -\frac{dP_0}{dx} D - \tau_0 \qquad (5.13.9)$$

or, in terms of the similarity function,

$$\pi = -\frac{D}{U_1^2} \frac{dP_0}{dx} = \frac{\tau_0}{U_1^2} - \theta_0 \int_0^1 \left(1 + \frac{u_0}{U_1} f\right)^2 d\eta, \qquad (5.13.10)$$

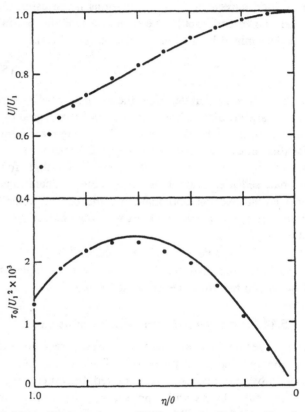

Fig. 5.14. Mean velocity and stress distributions for flow in a diverging channel (● from measurements by Ruetenik 1954). Full line is calculated from data for a uniform channel.

where $\theta_0 = D/x$ is the semi-angle of divergence. The character of the flow depends on the relative magnitudes of the terms, and it may be analysed in much the same way as ordinary channel flow by considering the flow to have two parts, a central flow with nearly constant eddy viscosity and equilibrium layers at the walls. An important difference is that the wall stress is less than the maximum stress for appreciable angles of divergence, and the approximation of nearly constant stress in the equilibrium layer may not be valid.

Detailed measurements of the flow in a diverging channel with a semi-angle of one degree have been made by Ruetenik (1954, see also Ruetenik & Corrsin 1955) and by Craya & Milliat (1955), and they are in good agreement with calculations made using the two-layer model (Townsend 1960). The magnitude of the central eddy viscosity is determined by a Clauser relation (see § 7.4),

$$v_T = R_0^{-1} \int_0^D |U - U_1| \, dz \qquad (5.13.11)$$

using the value $R_0 = 28$ that describes the mean flow between parallel planes. There are no disposable constants and the degree of agreement can be seen in the fig. 5.14. Notice the strong maximum in the stress distribution even for a small angle of divergence.

A point of some interest is to find the maximum angle for which unidirectional, self-preserving flow is possible. Calculations using the same Clauser relation and matching the central velocity distribution to that in an equilibrium layer with zero wall stress indicate a critical angle of

$$\theta_0 = 0.075^c = 4.3°.$$

Milliat (1957) has reported steady flow for a semi-angle of three degrees, not much less than the calculated value.

5.14 Equilibrium layers with variable stress

Equilibrium layers with nearly constant shear stress are the most common kind but the basic properties of energy equilibrium and independence of the outer flow are shared by other equilibrium layers, in particular layers with appreciable variation of stress. A general definition of an equilibrium layer is one satisfying the first two conditions of § 5.4, (1) that the advection terms in the turbulent energy equation are small compared with the production and dissipation terms, and (2) that the thickness is a small fraction of the flow thickness, and with properties determined by the local variation of stress with distance from the wall. Except insofar as the stress distribution depends on the whole flow, the motion is independent of conditions upstream or outside the equilibrium layer.

It has been asserted that the properties of the constant-stress layer depend on the existence of attached eddies which make up the main turbulent motion and contribute the greater part of the Reynolds stress, and it has been shown that a constant-stress flow may be constructed from a suitable size distribution of the eddies. Referring to § 5.7, the relation between Reynolds stress and the size distribution of attached eddies is

$$-\overline{u_1 u_3} = -\int_{l_0}^{L_0} N(z_a) I_{13}(z/z_a; 0)\, dz_a/z_a \qquad (5.14.1)$$

where $N(z_a)$ is positive and measures the intensity of attached eddies of size z_a. Since I_{13} varies in general as z/z_a for small values of z/z_a, not all stress distributions can be constructed from a particular variety of attached eddy and, if the condition of strong dissipation is met by having a wide range of sizes, the limit occurs for a linear variation of $N(z_a)$ with z_a. If $N(z_a) \propto (1 + \alpha z_a)$,

$$-\overline{u_1 u_3} \propto \alpha z(\log L_0/z + B) + A \qquad (5.14.2)$$

and, for the special form of attached eddy with

$$I_{13}(x; 0) \propto x \exp(-\tfrac{1}{2} x^2),$$

$$-\overline{u_1 u_3} \propto \left(\frac{\pi}{2}\right)^{1/2} + \alpha z(\log L_0/z - 0.06), \qquad (5.14.3)$$

both for $l_0 \ll z \ll L_0$. The stress distribution (5.14.3) is an approximation to a linear distribution of stress but it and all stress distributions constructed from similar attached eddies must have negative curvature if the stress gradient is positive (fig. 5.15).

Supposing that a broad distribution of attached eddies can produce the required stresses, we may attempt to find the mean velocity profile by considering the balance of turbulent energy. The simple approach, equivalent to using the mixing-length theory, is to neglect lateral transport of turbulent energy so that the energy equation is

$$\tau \frac{\partial U}{\partial z} = \varepsilon \qquad (5.14.4)$$

and to argue that the local dissipation of energy is mostly from attached eddies with centres nearby. Then their rate of energy loss is determined by their scale and intensity, i.e. $\varepsilon = \tau^{3/2}/(kz)$, and

$$\frac{\partial U}{\partial z} = \frac{\tau^{1/2}}{kz}. \qquad (5.14.5)$$

However, it seems unlikely that lateral transport of energy can be neglected, and its magnitude must be considered.

Our concept of an attached eddy is of a flow pattern which is finite in size, mechanically coherent and resistant to disintegration. If energy is supplied or removed in one part of the pattern, energy flows from or to that part to preserve the structure, in much the same way as a rotating flywheel responds to frictional or driving forces. Even in a constant-stress layer, it is possible that the centroid of energy transfer from the mean flow does not coincide with the centroid of energy transfer to smaller eddies in a typical attached eddy, implying a uniform flux of energy through the layer which has no effect on the energy balance. If the distribution of mean velocity or the friction of the smaller eddies are not those of a constant-stress layer, the distributions of energy gain and loss will be changed and the overall result will be a non-uniform flux of energy dependent on the stress distribution.

Consider the energy flux normal to the wall and averaged over planes parallel to the wall for a single attached eddy, $u_0^3 Q(z/z_a)$, where u_0 measures the intensity of eddies with centres near z_a and depends on the local stress. Then the overall flow in an assembly of attached eddies has the form,

$$\overline{pw} + \tfrac{1}{2}\overline{q^2 w} = \int \int N(z_a, u_0) u_0^3 Q(z/z_a)\, du_0\, dz_a/z_a. \qquad (5.14.6)$$

If the distribution of u_0 is similar at all heights, the overall flow is independent of z for a constant-stress layer, but two factors lead to a change with variable stress. The first is a direct consequence of the changed size distribution of the eddies, not dependent on a change in $Q(z/z_a)$. By itself, it would give

$$\overline{pw} + \tfrac{1}{2}\overline{q^2 w} = c \int [\tau(z_a)]^{3/2} Q_0(z/z_a) \frac{dz_a}{z_a}, \qquad (5.14.7)$$

where $Q_0(z/z_a)$ is the flux for attached eddies in constant-stress conditions. The second contribution comes from the redistribution of energy gain and loss in individual eddies. If stress increases with distance from the wall, it is exposed to velocity gradients that are abnormally large for large z/z_a and abnormally small for small z/z_a. Supposing energy loss to smaller eddies to be little influenced by

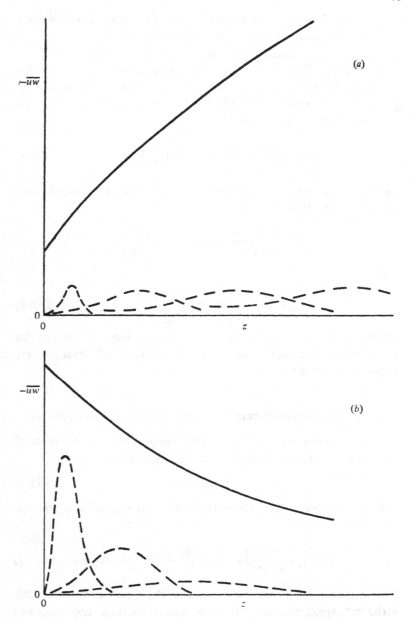

Fig. 5.15. Building of (a) positive and (b) negative stress gradients by superposition of attached eddies of similar forms.

7

the pattern of energy gain, energy flux inwards is increased in each attached eddy with a similar effect on the overall flux.

For small deviations from constant stress, both effects lead to overall energy flux proportional to stress gradient, but the dependence for large gradients is open to speculation. In strong gradients, it is possible that the flux reaches a limiting value proportional to $\tau^{3/2}$, and possibly

$$\overline{pw} + \tfrac{1}{2}\overline{q^2 w} = -B'\tau^{3/2} \operatorname{sgn}\left(\frac{\partial \tau}{\partial z}\right). \qquad (5.14.8)$$

Supposing the energy loss to smaller eddies to be unchanged, substitution in the energy equation,

$$\tau \frac{\partial U}{\partial z} - \frac{\partial}{\partial z}(\overline{pw} + \tfrac{1}{2}\overline{q^2 w}) = \varepsilon = \frac{\tau^{3/2}}{kz},$$

gives

$$\frac{\partial U}{\partial z} = \frac{\tau^{1/2}}{kz}\left[1 - B\frac{z}{\tau}\left|\frac{\partial \tau}{\partial z}\right|\right]. \qquad (5.14.9)$$

Other forms can be devised without difficulty, but this one has the advantage of simplicity, and, since the effects of diffusion are not large, it may be adequate.

5.15 Equilibrium layers with linear distributions of stress

In many equilibrium layers on impermeable walls, the variation of stress may be approximated by a linear distribution,

$$\tau = \tau_0 + \alpha z \qquad (5.15.1)$$

and, substituting in the relation between stress and velocity gradient (5.14.9),

$$U = \frac{\tau_0^{1/2}}{k} \log\left|\frac{\tau^{1/2} - \tau_0^{1/2}}{\tau^{1/2} + \tau_0^{1/2}}\right| + \frac{2(1 - B\operatorname{sgn}(\alpha))}{k}\tau^{1/2} + \text{constant.} \quad (5.15.2)$$

The constants of integration depend on the flow between the fully turbulent region and the wall. If the wall stress is not too small and the Reynolds number is large, the stress is almost the same at the wall and at the inner edge of the fully turbulent region, and the profiles should reduce to the constant-stress form for small values of z. That, is,

$$U = \frac{\tau_0^{1/2}}{k} \log\left[\frac{(\tau_0+\alpha z)^{1/2}-\tau_0^{1/2}}{(\tau_0+\alpha z)^{1/2}+\tau_0^{1/2}} \frac{4\tau_0}{\alpha z}\right] +$$

$$+ \frac{2(1-B\,\mathrm{sgn}(\alpha))}{k}[(\tau_0+\alpha z)^{1/2}-\tau_0^{1/2}] \qquad (5.15.3)\dagger$$

a form reducing to $U = \tau_0^{1/2}/k \log z/z_0$ if $|\alpha z| \ll \tau_0$. In practice, the variation of stress across an equilibrium layer can be large only with positive stress gradients, that is for flows in adverse pressure gradients.

Flows with zero wall stress are of particular interest, and the constant of integration in (5.15.2) is, for a smooth wall, dependent on α and ν, i.e.

$$U = \frac{2(1-B)}{k}(\alpha z)^{1/2} + A_0(\alpha\nu)^{1/3}. \qquad (5.15.4)$$

Using the arguments of § 5.8 that the whole velocity distribution may be approximated by joining the viscous profile,

$$U = \frac{1}{2}\frac{\alpha}{\nu}z^2 \qquad (5.15.5)$$

to the fully turbulent profile with a sudden, large change of slope at the junction, it is found that

$$A_0 = r^{1/3}(\tfrac{1}{2}r - 2)k_0^{-4/3}, \qquad (5.15.6)$$

where r is the ratio of the slopes on either side of the junction and $k_0 = k/(1 - B)$. For $r = 5$, $A_0 = 2.2$, but the viscous term is usually negligible in flows of moderate or large Reynolds numbers.

The approximation of linear variation of stress may be tested by substituting the calculated profile in the equation of mean motion,

$$\frac{\partial \tau}{\partial z} = U \frac{\partial U}{\partial x} + w \frac{\partial U}{\partial z} + \frac{dP_0}{dx}. \qquad (5.15.7)$$

If the wall stress is not small, most of the velocity variation takes place close to the wall and the first term on the left varies very little

† The mixing-length form of the linear-stress profile with $B = 0$ was derived and tested against observations by Szablewski (1960). The special form for zero wall stress, $U = 2(1 - B)(\alpha z)^{1/2}/k$, was found by Stratford (1959a) in an investigation of boundary-layer development with zero wall stress, and by Ellison (1960) for channel flow with a free surface.

over most of the layer. For example, substitution of the smooth-wall profile for constant stress leads to the result,

$$\tau = \tau_0 + \frac{dP_0}{dx}z + \frac{\tau_0^{1/2}}{k^2}\frac{d\tau_0^{1/2}}{dx}\left[\left(\log\frac{\tau_0^{1/2}z}{\nu}+A-1\right)^2+1\right]z \quad (5.15.8)$$

and, if $\log(\tau_0^{1/2}z/\nu) + A = kU/\tau_0^{1/2}$ is large at the centre of the equilibrium layer, the stress variation is substantially linear over the whole layer.

At the other extreme, if the stress on the wall is zero, substitution of $U = 2/k_0(\alpha z)^{1/2}$ leads to

$$\tau = \tau_0 + \frac{dP_0}{dx}z + \frac{2}{3k_0}\frac{d\alpha}{dx}z^2. \quad (5.15.9)$$

In this case, the contribution of the flow acceleration to the stress variation is far from linear but it is always small compared with the contribution from the pressure gradient. In boundary layers for example, $dP_0/dx = -U_1\,dU_1/dx$ and acceleration has a negligible influence on the stress distribution if

$$\frac{2}{3k_0^2}(1-2a)\frac{z}{x} \ll 1, \quad (5.15.10)$$

where $a = d(\log U_1)/d(\log x)$, a condition which is strongly satisfied in any flow conforming to the boundary layer approximation.

In general, the stress distribution in equilibrium layers is linear to a good approximation with an effective gradient given by

$$\alpha = U_e\frac{dU_e}{dx} + \frac{dP_0}{dx} \quad (5.15.11)$$

where U_e is an average value, say the velocity at the centre of the equilibrium layer (the choice is not critical). Except in zero-stress layers, the pressure gradient is not a good approximation to the stress gradient.

The velocity distribution (5.15.3) is in good agreement with observations in boundary layers in adverse pressure gradients, describing the velocity distribution over a much greater thickness than the simple logarithmic distribution (Szablewski 1960, Townsend 1961a). For large values of $\alpha z/\tau_0$, the distribution is

$$U = U_t + \frac{2}{k_0}(\alpha z)^{1/2} - \frac{1+B}{k}\frac{\tau_0}{(\alpha z)^{1/2}}, \quad (5.15.12)$$

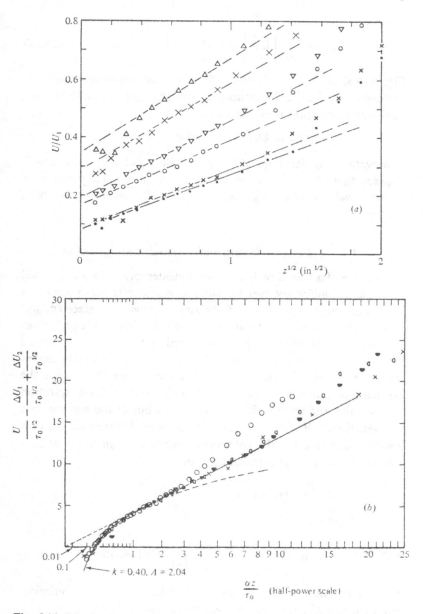

Fig. 5.16. Distributions of mean velocity in boundary layers in strong adverse pressure gradients. (a) From Townsend (1961a), (b) from Perry, Schofield & Jonbert (1969).

where

$$U_t = \frac{\tau_0^{1/2}}{k}\left[\log\frac{4\tau_0}{\alpha z_0} - 2(1-B)\right]. \qquad (5.15.13)$$

Fig. 5.16 shows the result of plotting some boundary layer profiles as functions of $z^{1/2}$. By estimating the gradients in (5.15.11), the linear-stress constant is estimated to be

$$k_0 = k/(1-B) = 0.48 \pm 0.03.$$

Observations by Stratford (1959a) indicate a value of 0.50, and it appears that the diffusion parameter B is about 0.2 which is a plausible value with the sign that indicates diffusion down the stress gradient.

5.16 Equilibrium layers with surface transpiration

The nature of a wall flow is changed considerably if fluid is injected or extracted uniformly over the surface, and considerable work has been done on the possibility of controlling flows by surface suction or blowing. As with other wall flows, the more important properties of flows with transpiration are determined by the equilibrium layers next the walls, but the presence of a normal mean velocity makes for some changes in the analysis. It is convenient to separate the normal component of the mean velocity into the transpiration velocity W_0, the volume flow per unit area out of the surface, and the remainder W, zero at the surface and determined by flow accelerations through the continuity equation. Then the equations for mean velocity and turbulent kinetic energy are

$$U\frac{\partial U}{\partial x} + (W + W_0)\frac{\partial U}{\partial z} = -\frac{dP_0}{dx} + \frac{\partial \tau}{\partial z} \qquad (5.16.1)$$

and

$$U\frac{\partial(\tfrac{1}{2}\overline{q^2})}{\partial x} + (W + W_0)\frac{\partial(\tfrac{1}{2}\overline{q^2})}{\partial z} + \overline{uw}\frac{\partial U}{\partial z} + \frac{\partial}{\partial z}(\overline{pw} + \tfrac{1}{2}\overline{q^2 w}) + \varepsilon = 0 \quad (5.16.2)$$

and an equilibrium layer may exist if the balance of turbulent energy is not affected by advection from upstream.

Consider first flow in which the pressure gradient is not sufficient to induce appreciable flow acceleration within the equilibrium layer, i.e. $|dP_0/dx|z \ll \tau_0$ the wall stress. Then the equation for the mean velocity integrates to

$$\tau = \tau_0 + W_0 U, \qquad (5.16.3)$$

expressing constancy of total momentum flux. The energy equation is then

$$W_0 \frac{\partial \frac{1}{2}\overline{q^2}}{\partial z} + \overline{uw}\,\frac{\partial U}{\partial z} + \frac{\partial}{\partial z}(\overline{pw} + \frac{1}{2}\overline{q^2 w}) + \varepsilon = 0 \qquad (5.16.4)$$

and, after making the usual similarity assumptions that $\frac{1}{2}\overline{q^2} = a_1 \tau_0$ and that $\varepsilon = \tau^{3/2}/(kz)$, it may be integrated in the form,

$$\frac{2\tau^{1/2}}{W_0} + \frac{2a_1 W_0}{\tau_0^{1/2}} + B\,\text{sgn}\,W_0\log\tau = \frac{1}{k}\log z + \text{const.} \qquad (5.16.5)$$

In practice, the transpiration ratio $W_0/\tau_0^{1/2}$ is small and the second and third terms on the left are small. For a smooth surface, the constant of integration depends on the viscosity, the surface stress and the transpiration ratio. A form of (5.16.5) that points up the relation to the similarity layer without transpiration is

$$\frac{2}{W_0}[(\tau_0 + U W_0)^{1/2} - \tau_0^{1/2}] = \frac{1}{k}\left[\log\frac{\tau_0^{1/2} z}{\nu} + A(W_0/\tau_0^{1/2})\right], \qquad (5.16.6)$$

where the constant of integration is a function of $W_0/\tau_0^{1/2}$ and approaches the additive constant A for small values of $W_0/\tau_0^{1/2}$. Equation (5.16.6) was obtained by Dorrance & Dore (1954) directly from the mixing-length theory.

The very large changes in velocity profile produced by transpiration can be seen from the distributions in fig. 5.17. With suction, the maximum velocity cannot exceed $-\tau_0/W_0$ and the region of appreciable velocity variation may be kept very small. With blowing, the stress increases with distance from the wall and, usually, the wall stress is small and the flow thick.

Appreciable variations of total momentum flux in an equilibrium layer are likely to occur only with blowing (positive W_0). If the variation is

$$\tau - W_0 U = \tau_0 + \alpha z$$

the variation of stress is given, to the mixing-length approximation, by

$$\frac{1}{W_0}\frac{d\tau}{dz} - \frac{\alpha}{W_0^2}\tau = \frac{\tau^{1/2}}{kz}. \qquad (5.16.7)$$

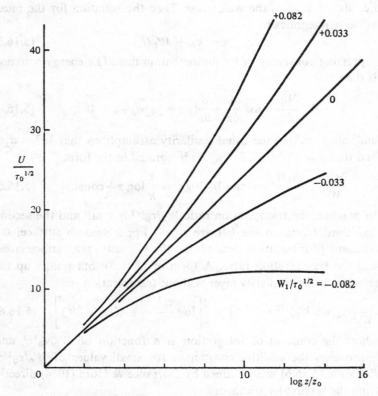

Fig. 5.17. Distributions of mean velocity near a porous wall with uniform suction or transpiration.

5.17 Equilibrium layers with variable direction of flow

So far, only flow that is nearly unidirectional has been considered, but there is no more difficulty in extending the idea of an equilibrium layer to flows in which the flow direction changes with distance from the surface. In the first place, by assembling attached eddies so that eddies with centres at different heights point in different directions, it is possible to construct a flow in which the stress vector τ_{i3} changes direction with height in a reasonably smooth manner. Then the attached eddies are in some form of equilibrium with the mean flow gradients and the friction of the smaller eddies, an equilibrium that should be expressed in some generalisation of the basic equation (5.14.9). A natural assumption is that the stress vector is everywhere in the direction of the velocity gradient vector $\partial U_i/\partial x_3$, and then energy arguments lead to

$$\frac{\partial U_i}{\partial x_3} = \frac{\tau_{i3}/\tau^{1/2}}{kz}\left[1 - B\frac{z}{\tau}\left|\frac{\partial \tau}{\partial z}\right|\right],$$ (5.17.1)

where $\tau = (\tau_{13}{}^2 + \tau_{23}{}^2)^{1/2}$.

Near identity of the directions could be a result either of the process of generation or of a mechanism for alignment of the developed eddies. To the extent that the active portion of each eddy is to some degree concentrated near its centre, the generation process should lead to a normal double-roller eddy with shear stress and axis aligned with the velocity gradient. Further, the inclined and trailing form of the double roller means that fully developed eddies will be kept in alignment by the velocity gradient. Naturally, if directions and magnitudes of shear change rapidly with height, eddies may be distorted or not aligned with the local shear, but the effect is essentially similar to that of energy diffusion in an attached eddy and is expected to be quite small.

CHAPTER 6

FREE TURBULENT SHEAR FLOWS

6.1   General properties of free turbulence

Free turbulent flows are bounded on at least one side by ambient fluid of nearly the same density which is not turbulent and is usually in irrotational flow. Their most striking characteristic is that, at any moment, the fluid in turbulent, vortical motion is divided from the fluid in irrotational motion by a fairly well-defined *intermittency surface*. The surface is indented on a scale comparable with the flow width, and its motion past a fixed point of observation causes records of velocity fluctuations to show alternating periods of turbulent and non-turbulent fluctuations, in other words an intermittently turbulent signal. Within the intermittency surface, the turbulence is roughly homogeneous in scale and turbulent intensity, unlike flows confined by rigid boundaries that restrict the lateral extent of eddy motions. Lastly, a free turbulent flow nearly always spreads into the surrounding fluid and it is necessarily inhomogeneous in the stream direction as well as in a transverse direction.

6.2   Equations of motion: the boundary-layer approximation

For most free turbulent flows, the mean value equations may be simplified by using a boundary-layer approximation, similar to that introduced by Prandtl for laminar boundary-layer flow at large Reynolds numbers but needing separate justification for turbulent flows. Consider a free turbulent flow bounded by non-turbulent flow with velocity gradients small compared with those in the turbulent flow. While the exact direction of mean flow may vary somewhat from one part of the flow to another, a direction, loosely called the direction of mean flow, is easily selected in any particular flow as the co-ordinate direction Ox. It is a matter of observation that gradients of mean values in the Ox direction are considerably less than in the transverse yOz plane, i.e. the length scale for variation of mean quantities in the Ox direction, say L, is an order of magnitude greater than the scale of variation in transverse directions, say l (fig. 6.1). Defining U_1 as the flow velocity just outside the turbulent flow and $U_s = U_{max} - U_{min}$ as the total variation of mean velocity over a transverse section of constant x, it follows from the continuity equation for the mean flow,

$$\frac{\partial U}{\partial x} + \frac{\partial V}{\partial y} + \frac{\partial W}{\partial z} = 0,$$

that the transverse mean velocities are of order $(U_s + L\, dU_1/dx)l/L$. Unless the free stream velocity, U_1, changes by a large factor in the distance L, transverse velocities are smaller than longitudinal velocities by, at least, a factor of order l/L.

Introducing q_0 as the scale velocity for the fluctuations, it is possible to estimate the relative magnitudes of the terms in the equations for the transverse components of mean velocity. The equation for the y component is

$$U \frac{\partial V}{\partial x} \quad + V \frac{\partial V}{\partial y} + W \frac{\partial V}{\partial z} + \frac{\partial \overline{uv}}{\partial x} + \frac{\partial \overline{v^2}}{\partial y} + \frac{\partial \overline{vw}}{\partial z}$$

$$\frac{l}{L^2}(U_1 - U_s)\left(U_s + L^2 \frac{d^2 U_1}{dx^2}\right) \quad \frac{l}{L^2}\left(U_s + L \frac{dU_1}{dx}\right)^2 \quad \frac{q_0^2}{L} \quad \frac{q_0^2}{l}$$

$$= -\frac{\partial P}{\partial y} + \quad \nu \nabla^2 V$$

$$\frac{\nu\left(U_s + L \frac{dU_1}{dx}\right)}{lL}$$

(6.2.1)

(the orders of magnitude are written below the terms). For turbulent flow, the flow Reynolds number $U_s l/\nu$ is necessarily large and, unless $(U_s + L^2\, d^2 U_1/dx^2)(U_s + U_1)$ is comparable with $q_0^2 L^2/l^2$ (which does not occur†), the equation is approximately

$$\frac{\partial \overline{v^2}}{\partial y} + \frac{\partial \overline{vw}}{\partial z} = -\frac{\partial P}{\partial y}.$$

(6.2.2)

Then, $\partial P/\partial y = O(q_0^2/l)$ and

$$P - P_1 = O(q_0^2),$$

(6.2.3)

where P_1 is the pressure just outside the flow at the particular section.

† Except in distorted wakes.

For most flows of interest, the ambient flow is irrotational with constant total head, and then

$$\frac{dP_1}{dx} + U_1 \frac{dU_1}{dx} = 0. \tag{6.2.4}$$

Making use of (6.2.4), the equation for the Ox component of mean velocity may be expressed as

$$U\frac{\partial(U-U_1)}{\partial x} + V\frac{\partial(U-U_1)}{\partial y} + W\frac{\partial(U-U_1)}{\partial z} + \frac{\partial\overline{u^2}}{\partial x} + \frac{\partial\overline{uv}}{\partial y} + \frac{\partial\overline{uw}}{\partial z}$$

$$\underbrace{\qquad}_{\frac{U_s(U_1+U_s)}{L}} \qquad \underbrace{\qquad}_{\frac{U_s(U_s+L\,dU_1/dx)}{L}} \qquad \qquad \underbrace{}_{\frac{q_0^2}{L}} \quad \underbrace{}_{\frac{q_0^2}{l}}$$

$$= (U_1-U)\frac{dU_1}{dx} - \frac{\partial}{\partial x}(P-P_1) + v\nabla^2(U-U_1) \tag{6.2.5}$$

$$\underset{U_s\frac{dU_1}{dx}}{\qquad} \qquad \underset{\frac{q_0^2}{L}}{\qquad} \qquad\qquad \underset{\frac{vU_s}{l^2}}{\qquad}$$

listing orders of magnitude as before. Omitting terms of order l/L compared with those retained, the equation appears in the 'boundary-layer' form,

$$U\frac{\partial U}{\partial x} + V\frac{\partial U}{\partial y} + W\frac{\partial U}{\partial z} + \frac{\partial\overline{uv}}{\partial y} + \frac{\partial\overline{uw}}{\partial z}$$

$$= U_1\frac{dU_1}{dx} + v\left(\frac{\partial^2 U}{\partial y^2} + \frac{\partial^2 U}{\partial z^2}\right). \tag{6.2.6}$$

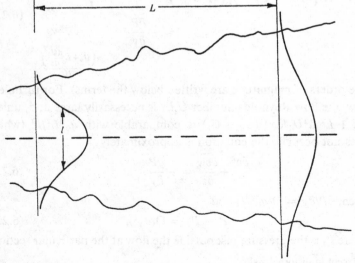

(a)

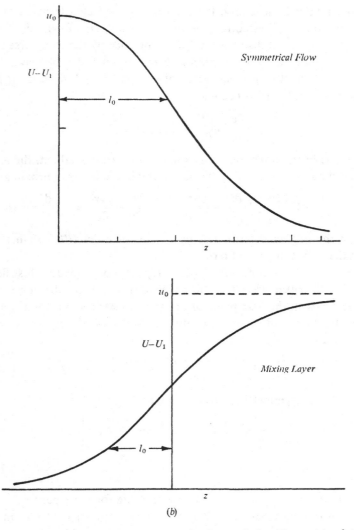

Fig. 6.1. Lateral and longitudinal scales for free turbulent flows: (*a*) relative magnitudes; (*b*) definition of lateral scales.

The ratio of the longitudinal to the lateral scale, L/l, depends on the kind of flow and is least for jets issuing into still fluid and for boundary layers in strong adverse pressure gradients. For jets, the ratio is about eight, and it would appear that (6.2.6) is not a particularly good approximation with the neglected terms over 10% of those retained. In reality, the total of the neglected terms is much smaller than this, essentially because the transverse Reynolds stress, $-\overline{vw}$, arises from distortion of the turbulence by the transverse rate of deformation, $\partial V/\partial z + \partial W/\partial y$, which is an order of magnitude less than the longitudinal rates of distortions, $\partial U/\partial y$ and $\partial U/\partial z$. Equation (6.2.1) then becomes

$$\frac{\partial \overline{v^2}}{\partial y} + \frac{\partial P}{\partial y} = O(q_0^2/L). \tag{6.2.7}$$

If Oy is in the direction of maximum transverse gradient, the error is least and $P + \overline{v^2} = P_1$†. Then, we obtain the better approximation

$$U\frac{\partial U}{\partial x} + V\frac{\partial U}{\partial y} + W\frac{\partial U}{\partial z} + \frac{\partial(\overline{u^2 - v^2})}{\partial x} + \frac{\partial \overline{uv}}{\partial y} + \frac{\partial \overline{uw}}{\partial z} = U_1\frac{dU_1}{\partial x} \tag{6.2.8}$$

with the neglected terms less by a ratio of order $(l/L)^2$ than those retained, but the added term, $(\partial/\partial x)(\overline{u^2 - v^2})$, is usually small.

So far, no assumptions of symmetry have been made. Most flows to be described are either homogeneous in the Oy direction with mean flow in the xOz plane or are axisymmetric about the Ox axis. For the first kind of flow, two-dimensional flow, the equation for U becomes

$$U\frac{\partial U}{\partial x} + W\frac{\partial U}{\partial z} + \frac{\partial(\overline{u^2 - w^2})}{\partial x} + \frac{\partial \overline{uw}}{\partial z} = U_1\frac{dU_1}{\partial x} \tag{6.2.9}$$

and, for axisymmetric flow, it becomes

$$U\frac{\partial U}{\partial x} + W\frac{\partial U}{\partial r} + \frac{\partial}{\partial x}\left[\overline{u^2} - \overline{w^2} + \int_r^\infty \frac{\overline{v^2 - w^2}}{r}\,dr\right]$$
$$+ \frac{1}{r}\frac{\partial \overline{uw}r}{\partial r} = U_1\frac{dU_1}{dx}. \tag{6.2.10}$$

In these forms they are sufficiently accurate for most purposes.

The leading terms of (6.2.5) for a flow with zero velocity in the free stream are

$$U\frac{\partial U}{\partial x} + V\frac{\partial U}{\partial y} + W\frac{\partial U}{\partial z}$$

† If the gradient of $\overline{w^2}$ in the Oz direction is not small compared with that of $\overline{v^2}$ in the Oy direction, (6.2.7) is consistent with the similar equation for W only if $\overline{v^2} = \overline{w^2}$ everywhere in the flow. For most flows, intensities of the two transverse components are nearly the same.

with magnitude $U_s{}^2/L$ and

$$\frac{\partial \overline{uv}}{\partial y} + \frac{\partial \overline{uw}}{\partial z}$$

with magnitude $q_0{}^2/l$. It follows that

$$\frac{q_0{}^2}{U_s{}^2} \approx \frac{l}{L} \tag{6.2.11}$$

and so the observation that turbulent flows spread at small angles has as a direct consequence that the relative turbulent intensity, $q_0{}^2/U_s{}^2$, is considerably less than one.

6.3 Integral constraints on free turbulence flows

The requirements of overall conservation of momentum and energy can be expressed as relations between integrals of mean values over planes parallel to yOz. The most useful is the equation for the integrated flux of momentum. If the free stream velocity is the same on all sides of the flow at the section, integration of the approximate equation (6.2.8) over the whole section leads to

$$\frac{\mathrm{d}}{\mathrm{d}x} \int \int U(U - U_1)\, \mathrm{d}y\, \mathrm{d}z + \frac{\mathrm{d}U_1}{\mathrm{d}x} \int \int (U - U_1)\, \mathrm{d}y\, \mathrm{d}z$$

$$+ \frac{\mathrm{d}}{\mathrm{d}x} \int \int (\overline{u^2} - \overline{v^2})\, \mathrm{d}y\, \mathrm{d}z = 0 \tag{6.3.1}$$

after some use of the continuity condition. Compared with the first term, the third term is of order $l(\overline{u^2} - \overline{v^2})/(Lq_0{}^2)$ and so small that it is usually omitted. For two-dimensional mean flow, the equation becomes

$$\frac{\mathrm{d}}{\mathrm{d}x} \int_{-\infty}^{\infty} U(U - U_1)\, \mathrm{d}z + \frac{\mathrm{d}U_1}{\mathrm{d}x} \int_{-\infty}^{\infty} (U - U_1)\, \mathrm{d}z$$

$$+ \frac{\mathrm{d}}{\mathrm{d}x} \int_{-\infty}^{\infty} (\overline{u^2} - \overline{w^2})\, \mathrm{d}z = 0 \tag{6.3.2}$$

and, for axisymmetric mean flow,

$$\frac{\mathrm{d}}{\mathrm{d}x} \int_0^{\infty} U(U - U_1)r\, \mathrm{d}r + \frac{\mathrm{d}U_1}{\mathrm{d}x} \int_0^{\infty} (U - U_1)r\, \mathrm{d}r$$

$$+ \frac{\mathrm{d}}{\mathrm{d}x} \int_0^{\infty} (\overline{u^2} - \tfrac{1}{2}(\overline{v^2} + \overline{w^2}))r\, \mathrm{d}r = 0. \tag{6.3.3}$$

Using the boundary-layer approximation, it may be shown that the equation for the turbulent kinetic energy takes the form,

$$\left(U\frac{\partial}{\partial x}+V\frac{\partial}{\partial y}+W\frac{\partial}{\partial z}\right)\tfrac{1}{2}\overline{q^2}+\overline{uv}\,\frac{\partial U}{\partial y}+\overline{uw}\,\frac{\partial U}{\partial z}+\overline{u^2}\,\frac{\partial U}{\partial x}+\overline{v^2}\,\frac{\partial V}{\partial y}$$

$$+\overline{w^2}\,\frac{\partial W}{\partial z}+\frac{\partial}{\partial y}(\overline{pv}+\tfrac{1}{2}\overline{q^2v})+\frac{\partial}{\partial z}(\overline{pw}+\tfrac{1}{2}\overline{q^2w})+\varepsilon=0.\quad(6.3.4)$$

By integrating over a section of the flow, we obtain an equation for the total flux of turbulent kinetic energy,

$$\frac{\mathrm{d}}{\mathrm{d}x}\iint\tfrac{1}{2}U\overline{q^2}\,\mathrm{d}y\,\mathrm{d}z+\iint\left(\overline{uv}\,\frac{\partial U}{\partial y}+\overline{uw}\,\frac{\partial U}{\partial z}\right)\mathrm{d}y\,\mathrm{d}z+\iint\varepsilon\,\mathrm{d}y\,\mathrm{d}z$$

$$+\iint\left(\overline{u^2}\,\frac{\partial U}{\partial x}+\overline{v^2}\,\frac{\partial V}{\partial y}+\overline{w^2}\,\frac{\partial U}{\partial z}\right)\mathrm{d}y\,\mathrm{d}z=0\quad(6.3.5)$$

and the last term is usually small enough to be neglected. For two-dimensional mean flow, it becomes

$$\frac{\mathrm{d}}{\mathrm{d}x}\int_{-\infty}^{\infty}\tfrac{1}{2}U\overline{q^2}\,\mathrm{d}z+\int_{-\infty}^{\infty}\overline{uw}\,\frac{\partial U}{\partial z}\,\mathrm{d}z+\int_{-\infty}^{\infty}\varepsilon\,\mathrm{d}z$$

$$+\int_{-\infty}^{\infty}\frac{\partial U}{\partial x}(\overline{u^2}-\overline{w^2})\,\mathrm{d}z=0\quad(6.3.6)$$

and, for axisymmetric mean flow,

$$\frac{\mathrm{d}}{\mathrm{d}x}\int_{0}^{\infty}\tfrac{1}{2}U\overline{q^2}r\,\mathrm{d}r+\int_{0}^{\infty}\overline{uw}\,\frac{\partial U}{\partial r}r\,\mathrm{d}r+\int_{0}^{\infty}\varepsilon r\,\mathrm{d}r$$

$$+\int_{0}^{\infty}\frac{\partial U}{\partial x}(\overline{u^2}-\tfrac{1}{2}(\overline{v^2}+\overline{w^2}))r\,\mathrm{d}r=0.\quad(6.3.7)$$

Equations for the total kinetic energy of the flow are more useful. For a flow bounded in all directions, overall conservation of energy requires that

$$\frac{\mathrm{d}}{\mathrm{d}x}\iint\tfrac{1}{2}U[(U-U_1)^2+\overline{q^2}]\,\mathrm{d}y\,\mathrm{d}z+\frac{\mathrm{d}U_1}{\mathrm{d}x}\iint(U-U_1)^2\,\mathrm{d}y\,\mathrm{d}z$$

$$+\iint\varepsilon\,\mathrm{d}y\,\mathrm{d}z=0,\quad(6.3.8)$$

for two-dimensional flow, that

$$\frac{d}{dx} \int_{-\infty}^{\infty} \tfrac{1}{2}U[(U-U_1)^2 + \overline{q^2}]\, dz + \frac{dU_1}{dx} \int_{-\infty}^{\infty} (U-U_1)^2\, dz$$

$$+ \int_{-\infty}^{\infty} \varepsilon\, dz = 0, \quad (6.3.9)$$

and, for axisymmetric flow, that

$$\frac{d}{dx} \int_{0}^{\infty} \tfrac{1}{2}U[(U-U_1)^2 + \overline{q^2}]r\, dr + \frac{dU_1}{dx} \int_{0}^{\infty} (U-U_1)^2 r\, dr$$

$$+ \int_{0}^{\infty} \varepsilon r\, dr = 0. \quad (6.3.10)$$

To reduce complexity, the small terms arising from inequality of the normal Reynolds stresses have been omitted.

Equations expressing overall conservation of momentum and energy can be found for flows which do not have the same free stream velocity on all sides or which require a different application of the boundary-layer approximation.

6.4 Self-preserving development of free turbulence flows

In a developing flow, the transverse distributions of mean velocity and other mean quantities change with distance downstream, but it is often assumed that the distributions retain the same functional forms, merely changing their transverse length-scale and the scales of the mean value quantities. It must be borne in mind that there may be features about any particular flow that prevent it developing in a self-preserving way, and that, even if such a development is possible, it may exist only as an asymptotic condition not closely approached over the range of observation. For a flow to be self-preserving in form, the variation of a mean value quantity M must be of the form,

$$M = M_1 + m_0 \mathrm{fn}(y/l_0, z/l_0) \quad (6.4.1)$$

where l_0 is a scale of length, and m_0 is a scale of the quantity. Both l_0 and m_0 are functions of x alone, as may be M_1, the reference level of the quantity in the free stream.

Apart from being a convenient mathematical hypothesis, the assumption embodies the principle of moving equilibrium. In any developing flow, turbulent fluid is carried along by the mean flow and the instantaneous conditions at any section of the flow are determined to a considerable extent by the conditions at some earlier time when the volume of fluid now at the point of observation was some distance upstream. In this sense, the flow at any section is determined by the flow at positions upstream, and the existing eddy structures have developed from earlier ones of different but proportional scales and intensities. The principle, as distinct from the assumption, of self-preservation asserts that a moving equilibrium is set up in which the conditions at the initiation of the flow are largely irrelevant, and so the flow depends on one or two simple parameters and is geometrically similar at all sections.

Almost without exception, self-preserving flows must be either axisymmetric or homogeneous in one transverse direction. The reason is that the rate of spreading in (say) the Oy direction depends on transport of stream momentum in that direction, i.e. on the general magnitude of the Reynolds stress $-\overline{uv}$ which appears as a response to distortion of the turbulence by the mean velocity gradients. If gradients of U in the Oy direction are larger than those in the Oz direction, the distortion leads to larger values of $\overline{uv}$ than of $\overline{uw}$, and so the flow tends to spread more rapidly in the direction for which its width is least†. To preserve constant ratios of flow widths in different directions, it is necessary either that the distributions are axisymmetric or that the width in one direction is effectively infinite.

The mean values that are the subjects of an assumption of self-preserving development are those describing the mean flow and the energy-containing components of the turbulent motion, none of which are affected directly by the viscous stresses in the fluid. The conditions under which self-preserving flow is possible may be established by substituting the self-preserving distributions in the

† Two qualifying remarks should be made. First, Reynolds stresses take time to be generated and time to decay, and existing rates of spread are determined by conditions some way upstream. Local axisymmetry of the mean velocity distributions need not imply axisymmetry of the stresses and equality of rates of spread. Secondly, the nature of flow in the free stream may be such that the flow is compressed along one direction preferentially. Then self-preserving development may be possible without axisymmetry.

equations of momentum and turbulent energy and examining the resulting equations for consistency. For two-dimensional mean flow, we use the distributions,

$$
\left.
\begin{aligned}
U &= U_1 + u_0 f(z/l_0), \\
\overline{uw} &= q_0^2 g_{13}(z/l_0), \\
\overline{q^2} &= q_0^2 g(z/l_0), \\
\overline{u^2} \equiv \overline{u_1^2} &= q_0^2 g_1(z/l_0) \text{ etc.,} \\
\overline{pw} + \tfrac{1}{2}\overline{q^2 w} &= q_0^3 k(z/l_0), \\
\varepsilon &= q_0^3/l_0 h(z/l_0),
\end{aligned}
\right\}
\tag{6.4.2}
$$

where U_1 is the velocity in the free stream, u_0 is the scale of mean velocity variation, q_0 is the scale of the turbulent velocities, and l_0 is the length scale of the flow. All the scales are functions of x alone, and the functions are independent of position and are characteristic of the whole flow. Substituting in equation (6.2.9) for the stream component of velocity, we find

$$
u_0 \frac{dU_1}{dx}(f - \eta f') + U_1\left(\frac{du_0}{dx} f - \frac{u_0}{l_0}\frac{dl_0}{dx} \eta f'\right)
$$

$$
+ u_0 \frac{du_0}{dx}(f)^2 - \frac{u_0}{l_0}\frac{d(u_0 l_0)}{dx} f' \int_0^{} f \, d\eta_1 + \frac{q_0^2}{l_0} g_{13}'
$$

$$
+ (g_1 - g_2)\frac{dq_0^2}{dx} - \frac{q_0^2}{l_0}\frac{dl_0}{dx} \eta(g_1' - g_2') = 0,
\tag{6.4.3}
$$

where $\eta = z/l_0$, dashes signify differentiation with respect to η, and the origin of z has been chosen so that $W(0) = 0$. Substituting in equation (6.3.4) for the turbulent kinetic energy, we find that

$$
\tfrac{1}{2}U_1\left(\frac{dq_0^2}{dx} g - \frac{q_0^2}{l_0}\frac{dl_0}{dx} \eta g'\right) - \frac{1}{2}\frac{dU_1}{dx} q_0^2(\eta g' - g_1 + g_2)
$$

$$
+ \tfrac{1}{2}u_0 \frac{dq_0^2}{dx} fg - \frac{1}{2}\frac{q_0^2}{l_0}\frac{d(u_0 l_0)}{dx} g' \int_0^{} f(\eta') \, d\eta' + \frac{u_0 q_0^2}{l_0} f' g_{13}
$$

$$
+ q_0^2(g_1 - g_2)\left(\frac{du_0}{dx} f - \frac{u_0}{l_0}\frac{dl_0}{dx} \eta f'\right) + \frac{q_0^3}{l_0} k' + \frac{q_0^3}{l_0} h = 0.
\tag{6.4.4}
$$

If the flow is self-preserving, the two equations relate the universal functions and must be identical in meaning for all values of x and of η. It is necessary that the coefficients of the various terms in each equation should either be zero or be proportional to each other. In a turbulent shear flow, the Reynolds stresses and velocity gradients are certainly not zero, and dividing the equations by the coefficients of the Reynolds stress terms shows that the non-dimensional coefficients,

$$\frac{u_0 l_0}{q_0^2}\frac{dU_1}{dx}, \qquad \frac{U_1 l_0}{q_0^2}\frac{du_0}{dx}, \qquad \frac{U_1 u_0}{q_0^2}\frac{dl_0}{dx}, \qquad \frac{u_0 l_0}{q_0^2}\frac{du_0}{dx},$$

$$\frac{u_0^2}{q_0^2}\frac{dl_0}{dx}, \qquad \frac{l_0}{q_0^2}\frac{dq_0^2}{dx}, \qquad \frac{dl_0}{dx}, \qquad \frac{U_1 l_0}{u_0 q_0^2}\frac{dq_0^2}{dx},$$

$$\frac{U_1}{u_0}\frac{dl_0}{dx}, \qquad \frac{l_0}{u_0}\frac{dU_1}{dx}, \qquad \frac{q_0}{u_0},$$

must all either be zero or be independent of x. The last coefficient refers to the energy dissipation term and is not zero. It follows that q_0 is proportional to u_0 and may be replaced by it. Then the necessary conditions for self-preserving development are that the quantities,

$$\frac{U_1 l_0}{u_0^2}\frac{du_0}{dx}, \qquad \frac{U_1}{u_0}\frac{dl_0}{dx}, \qquad \frac{l_0}{u_0}\frac{du_0}{dx}, \qquad \frac{dl_0}{dx} \qquad (6.4.5)$$

should be constant or zero.

The first way of satisfying the conditions for self-preserving development is if

$$\left.\begin{array}{l} u_0 = U_1 = \text{constant,} \\ dl_0/dx = \text{constant.} \end{array}\right\} \qquad (6.4.6)$$

Mixing layers between two uniform flows of different velocity may be self-preserving in this way. A second possibility is that

$$\left.\begin{array}{l} u_0 \propto U_1 \propto (x - x_0)^a, \\ dl_0/dx = \text{constant.} \end{array}\right\} \qquad (6.4.7)$$

Development of this kind occurs for wakes and jets in suitable external pressure gradients defined by the first condition. The equation for the momentum integral (6.3.2) becomes after substitution of the self-preserving distribution functions,

$$(2a+1)\frac{u_0 U_1 l_0}{x - x_0}\left(I_1 + \frac{u_0}{U_1}I_2\right) + a\frac{u_0 U_1 l_0}{x - x_0}I_1 = 0, \qquad (6.4.8)$$

where $I_n = \displaystyle\int_{-\infty}^{\infty} (f(\eta))^n \, d\eta$. Choosing u_0 to be the extreme value of $(U - U_1)$ (occurring at $z = 0$), and recasting (6.4.8) into the form,

$$a = -\frac{I_1 + (u_0/U_1)I_1}{3I_1 + (u_0/U_1)I_2} \tag{6.4.9}$$

we find that the exponent a must lie between

$$a = -\tfrac{1}{2} \quad \text{for } u_0/U_1 \text{ very large,}$$

i.e. a jet in still fluid, and

$$a = -(I_1 - I_2)/(3I_1 - 2I_2) \quad \text{for } u_0/U_1 = -1,$$

the practical limit of velocity defect (*N.B.* since $f(\eta) \leqslant 1$, $I_1 > I_2$).

Besides the flows which may satisfy exactly the conditions for self-preserving development, flows in which u_0/U_1 varies slowly may be approximately self-preserving. The most important are wakes with small velocity defect, i.e. with $|u_0/U_1| \ll 1$. Then the third and fourth coefficients of (6.4.5) are small compared with the first two, and the conditions are that

$$\frac{U_1 l_0}{u_0{}^2}\frac{du_0}{dx} = \text{constant}, \qquad \frac{U_1}{u_0}\frac{dl_0}{dx} = \text{constant}. \tag{6.4.10}$$

The equation for the momentum integral becomes

$$\frac{d}{dx}\left[U_1 u_0 l_0 \int_{-\infty}^{\infty} f(\eta)\,d\eta\right] + \frac{dU_1}{dx} u_0 l_0 \int_{-\infty}^{\infty} f(\eta)\,d\eta = 0 \tag{6.4.11}$$

showing that

$$I_1 U_1{}^2 u_0 l_0 = \text{constant} = U_1{}^2 \int_{-\infty}^{\infty} (U - U_1)\,dz. \tag{6.4.12}$$

For $U_1 \propto (x - x_0)^a$, solutions of (6.4.10, 6.4.12) are

$$u_0/U_1 \propto (x - x_0)^{-1/2(1 + 3a)}, \quad l_0 \propto (x - x_0)^{1/2(1 - 3a)}. \tag{6.4.13}$$

If u_0/U_1 is to *remain* small, the exponent $-a$ must be less than $\tfrac{1}{3}$. The most interesting of the family is the wake in a uniform stream ($a = 0$) for which

$$u_0 \propto (x - x_0)^{-1/2}, \quad l_0 \propto (x - x_0)^{1/2}. \tag{6.4.14}$$

This analysis of the dynamical possibility of self-preserving flow involves only two of the mean value equations, and the conditions found are necessary and not sufficient. It is possible that other dynamical conditions may prevent the establishment of the moving equilibrium, in particular the inability of the turbulent motion to develop a rate of energy dissipation sufficiently large to absorb both the energy transferred from the mean flow and the net gain of turbulent energy by mean flow advection. (It is probable that axisymmetric wakes are not self-preserving for this reason.) The available evidence for flows satisfying the conditions (6.4.6) or (6.4.7), i.e. mixing layers and jets, is that self-preserving development is possible and approached as an asymptotic state not far from the flow origin.

The treatment of self-preserving flow is readily extended to axisymmetric flows. Mixing layers are not possible, and the only exactly self-preserving flows satisfy the conditions,

$$u_0 \propto U_1 \propto (x - x_0)^a, \qquad l_0 \propto (x - x_0). \qquad (6.4.15)$$

It may be shown that

$$a = -\frac{2[I_1 + (u_0/U_1)I_2]}{3I_1 + 2I_2 u_0/U_1} \qquad (6.4.16)$$

and the exponent a must lie between -1 and $-2(I_1 - I_2)/(3I_1 - 2I_2)$, where

$$I_n = \int_0^\infty [f(\eta)]^n \eta \, d\eta \quad (\eta = r/l_0).$$

Axisymmetric wakes of small velocity defect may be self-preserving with a free-stream velocity varying as $(x - x_0)^a$ and

$$\frac{u_0}{U_1} \propto (x - x_0)^{-(a+2/3)}, \qquad l_0 \propto (x - x_0)^{1/3 - a}. \qquad (6.4.17)$$

The velocity defect or excess remains small compared with the stream velocity only if $a > -\frac{2}{3}$. For the special case of a wake in a uniform stream, the variation of scales is

$$u_0 \propto (x - x_0)^{-2/3}, \qquad l_0 \propto (x - x_0)^{1/3}. \qquad (6.4.18)$$

To sum up, by testing the consistency of an assumption of self-preserving development with the mean value equations for momentum and turbulent energy, it is possible to ascertain the possibility of self-preservation in several families of reference flows and to predict the form of the variations of the velocity and length scales with distance downstream. The forms of the distribution functions and the constants of proportionality cannot be found without more detailed assumptions about the nature of the flows.

6.5 The distributions of mean velocity and Reynolds stress

For self-preserving free turbulent flows, the mean flow problem may be split into two parts, first the form of the distribution functions and secondly the magnitude of the parameters describing the rate of development. Two specific assumptions about the distribution functions will be used. The first is that the velocity profiles in flows possessing the same symmetry are similar, and are nearly of the forms,

$$U - U_1 = u_0 \exp(-\tfrac{1}{2}z^2/l_0{}^2) \qquad (6.5.1)$$

for flows with a plane of symmetry, and

$$U - U_1 = u_0(2\pi)^{-1/2} \int_{z/l_0}^{\infty} \exp(-\tfrac{1}{2}x^2)\,\mathrm{d}x \qquad (6.5.2)$$

for the mixing-layer family of asymmetrical flows. In these profiles, the length scale has the standard values that will be used from here on, the variance of the velocity distribution in symmetrical flows or the variance of the distribution of velocity gradient for asymmetrical flows, i.e.

$$l_0{}^2 = \int (U-U_1)z^2\,\mathrm{d}z \Big/ \int (U-U_1)\,\mathrm{d}z \qquad (6.5.3)$$

or

$$l_0{}^2 = \int \frac{\partial U}{\partial z} z^2\,\mathrm{d}z \Big/ \int \frac{\partial U}{\partial z}\,\mathrm{d}z \qquad (6.5.4)$$

(fig. 6.1).

The second assumption is of wider application and it supposes that the eddy viscosity, defined as

$$v_T = -\frac{\overline{uw}}{\partial U/\partial z} \qquad (6.5.5)$$

is constant over any section of the flow. In any self-preserving flow, it is necessary that

$$v_T = R_s^{-1}|u_0|l_0, \qquad (6.5.6)$$

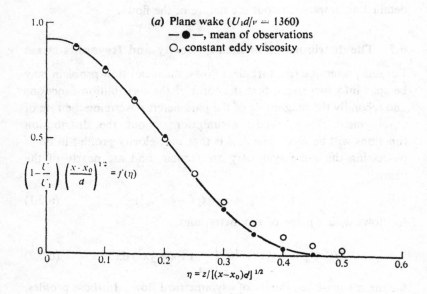

$$\left(1-\frac{U}{U_1}\right)\left(\frac{x-x_0}{d}\right)^{1/2}=f(\eta)$$

(a) Plane wake ($U_1 d/\nu = 1360$)
— ● —, mean of observations
○, constant eddy viscosity

$\eta = z/[(x-x_0)d]^{1/2}$

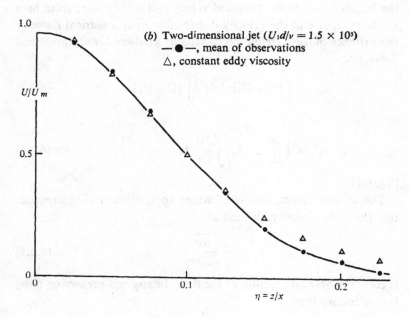

U/U_m

(b) Two-dimensional jet ($U_1 d/\nu = 1.5 \times 10^5$)
— ● —, mean of observations
△, constant eddy viscosity

$\eta = z/x$

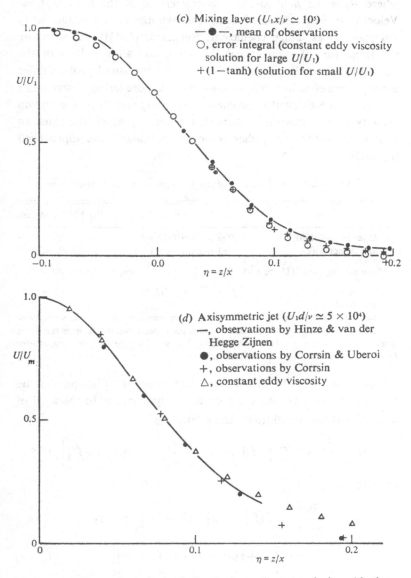

Fig. 6.2. Comparison of observed distributions of mean velocity with those calculated assuming a constant eddy viscosity.

where R_s is the *flow constant*, characteristic of the kind of flow. Velocity profiles calculated using the hypothesis of constant eddy viscosity (or exchange coefficient, Reichardt (1941)) are not quite the same for all symmetrical flows nor for all members of the mixing-layer family (see table 6.1), but the measured profiles differ among themselves in much the same way. A more serious discrepancy is that all the calculated profiles show an approach to the stream velocity that is less rapid than that observed, which indicates an eddy viscosity that diminishes as the flow boundaries are approached (fig. 6.2).

TABLE 6.1   *Velocity profiles for constant eddy viscosity*

Flow	Profile	Flow constant
2-D jet in still fluid	$f(\eta) = \operatorname{sech}^2(\eta(2/\pi)^{1/2})$	$R_s = 2/\beta$
Small defect wake	$f(\eta) = \exp(-\tfrac{1}{2}\eta^2)$	$R_s = \beta^{-1}$
Plane mixing layer ($U_1 \simeq U_2$)	$f(\eta) = (2\pi)^{-1/2}\displaystyle\int_{-\infty}^{\eta} \exp^{-1/2s^2}\,\mathrm{d}s$	$R_s = \beta^{-1}$
Axisymmetric jet in still fluid	$f(\eta) = (1 + \eta^2/2)^{-2}$	$R_s = 2/\beta$

Note: The length scales of the distributions have been defined to give the same total flow as an error law distribution, i.e. so that the integral parameter $I_1 = (\pi/2)^{1/2}$ (or 1 for an axisymmetric flow).

Given the velocity distribution, the stress distribution can be found using (6.4.3) with the coefficients appropriate to the kind of flow. For symmetrical flows, the relation is

$$g_{13}' = \frac{\mathrm{d}l_0}{\mathrm{d}x}\left\{(a+1)f'\int_0^{} f\,\mathrm{d}\eta_1 - af^2 - \frac{U_1}{u_0}[2af-(a+1)\eta f']\right\} \quad (6.5.7)$$

or, after integration,

$$g_{13} = \frac{\mathrm{d}l_0}{\mathrm{d}x}\left\{(a+1)f\int_0^{} f\,\mathrm{d}\eta_1 - (2a+1)\int_0^{} f^2\,\mathrm{d}\eta_1 \right.$$
$$\left. + \frac{U_1}{u_0}\left[(a+1)\eta f - (3a+1)\int_0^{} f\,\mathrm{d}\eta\right]\right\} \quad (6.5.8)$$

if $|u_0|/U_1$ is not small, or

$$g_{13}' = \frac{1-a}{1-3a}\frac{U_1}{u_0}\frac{\mathrm{d}l_0}{\mathrm{d}x}[f+\eta f'] \quad (6.5.9)$$

and

$$g_{13} = \frac{1-a}{1-3a} \frac{U_1}{u_0} \frac{dl_0}{dx} [\eta f]$$ (6.5.10)

if $|u_0|/U_1$ is small. Since the scales of velocity and length are related by the integral condition for conservation of momentum (see (6.4.9) and (6.4.12)), the remainder of the mean flow problem is finding the magnitude of the coefficient outside the square brackets in the preceding equations. A standard form for the coefficient is the *entrainment parameter* defined as

$$\beta = \frac{U_1 + \frac{1}{2}u_0}{|u_0|} \frac{dl_0}{dx}$$ (6.5.11)

which is a constant for any one kind of self-preserving flow but varies considerably from one kind to another.

The flow constant R_s determines the development of the flow and is closely related to the entrainment parameter. On the centre-line of a two-dimensional flow,

$$U = U_1 + u_0$$

and the momentum equation becomes

$$\frac{d}{dx}[u_0(U_1 + \frac{1}{2}u_0)] = v_T \frac{u_0}{l_0^2} f''(0)$$ (6.5.12)

assuming constant eddy viscosity near the centre. For the error law profile (6.5.1), $f''(0) = -1$ and varies little from flow to flow. It follows that

$$R_s^{-1} = (a + 1)\beta$$ (6.5.13)

if $|u_0|/U_1$ is not small, and that

$$R_s^{-1} = \beta(1 - a)/(1 - 3a)$$ (6.5.14)

if $|u_0|/U_1$ is small.

6.6 The balance of turbulent kinetic energy

To explain the magnitudes of the flow constants, we must consider the nature and mechanics of the turbulence, and a good starting-point is the conservation equation for the turbulent kinetic energy,

$$U \frac{\partial(\frac{1}{2}\overline{q^2})}{\partial x} + W \frac{\partial(\frac{1}{2}\overline{q^2})}{\partial z} + \overline{uw} \frac{\partial U}{\partial z} + \frac{\partial}{\partial z}(\overline{pw} + \frac{1}{2}\overline{q^2 w}) + \varepsilon = 0$$ (6.6.1)

Advection − Generation + Diffusion + Dissipation = 0.

Using techniques of hot-wire anemometry, it is possible to measure all the terms in the equation except the pressure–velocity product $\overline{pw}$, which may be obtained by difference. Measurements are available for the simple wake ($|u_0| \ll U_1$, $a = 0$), for the two-dimensional jet ($U_0 = 0$, $a = -\frac{1}{2}$), for the axisymmetric jet ($U_1 = 0$, $a = -1$), for a mixing layer ($U_0 = 0$), and for several kinds of boundary layer, and lateral distributions of the four terms are shown in fig. 6.3. Important features of the distributions are:

(1) generation of turbulent energy is greatest near the position of maximum shear and is zero along an axis of symmetry (if one exists),

(2) viscous dissipation is distributed over the whole flow and it is not especially large in regions of strong generation,

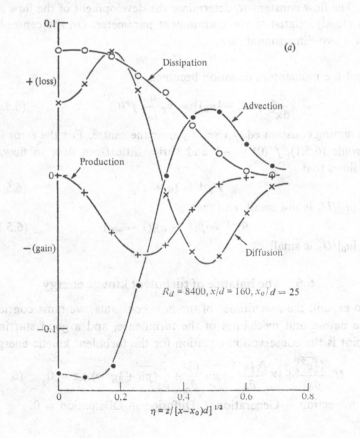

$R_d = 8400$, $x/d = 160$, $x_0/d = 25$

$\eta = z / [x - x_0)d]^{1/2}$

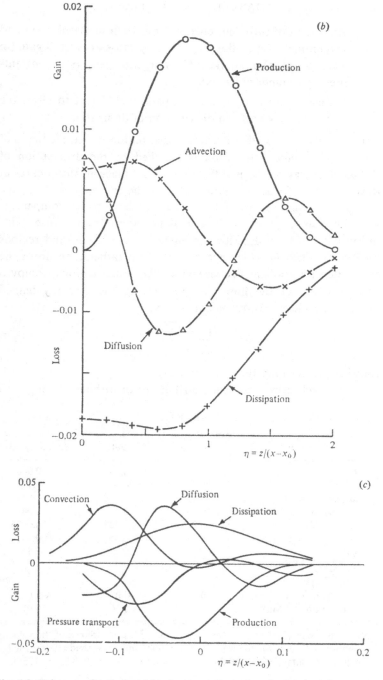

Fig. 6.3. Balances of turbulent kinetic energy for wakes, jets and mixing layers. (b) From Bradbury (1965), (c) from Wygnanski & Fiedler 1970.

(3) Advection of turbulent energy varies both in distribution and magnitude between flows, and it may cause an overall gain (in jets and wakes) or an overall loss (in mixing layers or constant-pressure boundary layers),

(4) Diffusive transport occurs in quantity sufficient to allow the comparatively uniform distribution of dissipation.

The nearly uniform distribution of dissipation within the limits of the mean velocity distribution is matched by the distribution of turbulent energy which is diffused from its place of production at least as efficiently as the momentum of the mean flow. Since the energy-containing eddies of the turbulence have scales comparable with the scale-width of the flow, it is not appropriate to use a diffusion coefficient to describe the energy flux in detail and regions can be identified in which energy flux is up rather than down the local intensity gradient (Townsend 1949a). Such regions occupy a small fraction of the flow and, in general, the formally-defined diffusion coefficient for turbulent energy,

$$K_e = -\frac{\overline{(pw + \tfrac{1}{2}q^2 w)}}{(\partial/\partial z)(\tfrac{1}{2}\overline{q^2})} \qquad (6.6.2)$$

is rather greater than the eddy viscosity.

Similar and nearly uniform distributions of turbulent energy and

TABLE 6.2

Quantity	Wake	Jet	Boundary layer
β	0.40	0.18	0.048
a/η_0	0.38	0.22	0.17
$(\overline{q^2}/U_m^2)_{y=0}$	0.21	0.13	0.08
$(\overline{q^2}/U_m^2)_{max}$	0.24	0.15	0.08
$\|\overline{uv}\|_{max}/U_m^2$	0.061	0.025	0.0105
L_u/η_0	0.5	0.55	0.7
η_0/l_0	2.1	2.0	2.15
$\left[\dfrac{\overline{(\partial\eta/\partial x)^2}}{(\eta - \eta_0)^2}\right]^{\frac{1}{2}}\eta_0$	2	3.1	2.6
Lateral rate $\begin{cases} y = 0 \\ \text{Max} \end{cases}$ of strain	0 \\ 0	0.09 \\ -0.10	-0.01 \\ $+0.01$

Note: L_u is the integral scale of the turbulence, l_0^2 is the variance of the velocity distribution. Sources: for the wake, Grant (1958) and Townsend (1956); for the plane jet, Bradbury (1965); for the boundary layer Corrsin & Kistler (1955).

dissipation rate imply a nearly uniform distribution of the dissipation length parameter, L_{ϵ}, and approximate homogeneity of scale across the flows. Measurements confirm the inference, showing the longitudinal integral scale L_{11} to be roughly equal to the width scale l_0, and about one-third of the dissipation scale (see table 6.2).

6.7 The bounding surface of free turbulence flows

All free turbulent flows exhibit the phenomenon of intermittency in their outer regions, that is to say, the signal from an anemometer is observed to alternate between sharply defined intervals of intense, rapid and characteristically 'turbulent' fluctuations and intervals of

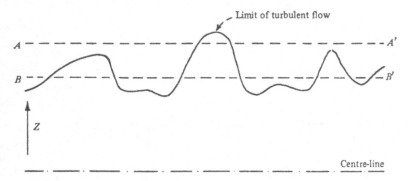

Limit of turbulent flow

Centre-line

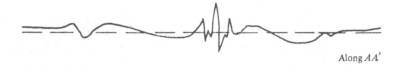

Along AA'

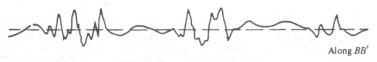

Along BB'

(a)

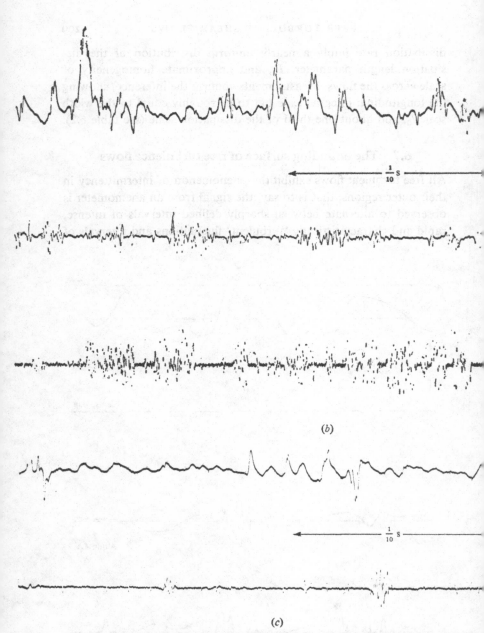

Fig. 6.4. Intermittency of anemometer signals. (*a*) Relation to the form of the boundary between turbulent and irrotational flow. (*b*) Records of u, $\partial u/\partial t$ and $\partial^2 u/\partial t^2$ in a wake near position of maximum shear. (*c*) Records far from the region of appreciable variation of mean velocity.

much weaker, slow fluctuations. The distinction is especially clear if the anemometer responds to velocity gradients or to vorticity rather than simply velocity (fig. 6.4). The reason for the phenomenon is that entrainment of non-turbulent ambient fluid (usually in irrotational flow) occurs across a well-defined bounding surface whose shape and position changes continuously. Within the surface, the flow is fully turbulent with intense, fine-scale fluctuations of velocity gradient and vorticity and the position of the surface can be defined to within a 'diameter' of the smallest eddies of the turbulence. Outside the surface, the velocity fluctuations arise from irrotational flow fields induced by the turbulent eddies, a motion that is not 'turbulence' in the usual sense. So an anemometer at a fixed position may be on either side of the bounding surface depending on its current configuration and be measuring either fully turbulent flow or fluctuations in the potential flow.

Let us suppose that the bounding surface may be defined by $\zeta(x, y, t)$, its displacement in the Oz direction from the xOy plane and a single-valued function (the validity of the assumption is examined later). Then the proportion of time that a detector situated at (x, y, z) is within the turbulent fluid is

$$\int_z^\infty P(\zeta, x) \, d\zeta = \gamma(x, z), \qquad (6.7.1)$$

where $P(\zeta, x)$ is the probability density function for a displacement ζ at a position (x, y) (homogeneity in the Oy direction is assumed). Measurements of the intermittency factor γ are usually made by constructing an intermittency signal $\delta(x, t)$, defined to be one if the anemometer is within the fully turbulent flow and zero if it is not. Clearly,

$$\overline{\delta(\mathbf{x}, t)} = \gamma(x, z) \qquad (6.7.2)$$

and

$$P(\zeta, x) = -\frac{\partial}{\partial z} \gamma(x, z). \qquad (6.7.3)$$

Corrsin & Kistler (1955) have shown that the inferred distributions of surface displacement follow closely Gaussian error-law distributions,

$$P(\zeta) = (2\pi\sigma^2)^{-1/2} \exp{-\tfrac{1}{2}[(\zeta - \zeta_0)^2/\sigma^2]}, \qquad (6.7.4)$$

8

where σ is the variance of the displacement about its mean value ζ_0 (note that $\gamma(\zeta_0) = \frac{1}{2}$). Reference to the measurements of table 6.2 shows that ζ_0 is nearly the same fraction of the width scale l_0 in all the flows, but that the ratio σ/ζ_0 varies considerably, being small in boundary layers and large in wakes. It measures the relative depth of the corrugations of the bounding surface, which might be regarded as an index of the activity of entrainment at the interface.

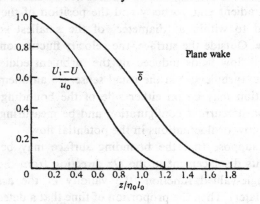

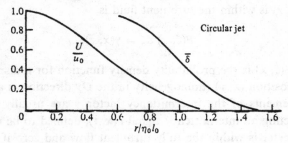

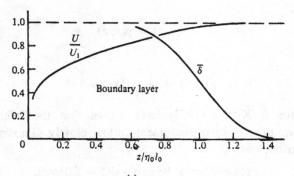

(a)

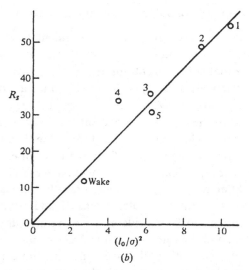

Fig. 6.5. (a) Distributions of mean velocity and intermittency factor for a plane wake, a circular jet and a boundary layer in zero pressure gradient; (b) Correlation between the variance of the intermittency factor and the flow constant (from Gartshore 1966). 1, wall jet, $V = 0$; 2, wall jet, $V = 0.52$; 3, wall jet, $V = 1.10$; 4, wall jet, $V = 3.04$; 5, free jet, $V = 0$. ($V = U_1/u_0$)

Indeed, Gartshore (1966) has shown that the ratio σ/ζ_0 correlates very closely with the magnitude of the flow constant R_s in the direction indicated by this view (see table 6.2 and fig. 6.5).

The intermittency signal may be used to 'gate' signals from an anemometer and so to obtain zone averages that are restricted to the turbulent fluid within the bounding surface. Thus the average value of a flow variable $m(\mathbf{x}, t)$ within the turbulent fluid is

$$\overline{\overline{m}} = \overline{m(\mathbf{x}, t)\delta(\mathbf{x}, t)}/\overline{\delta(\mathbf{x}, t)} \qquad (6.7.5)$$

or, for quantities that are known to be nearly zero in the nonturbulent fluid,

$$\overline{\overline{m}} = \overline{m}/\gamma. \qquad (6.7.6)$$

Zone averages of turbulent intensity and dissipation do not vary greatly within the region of appreciable variation of mean velocity, indicating very effective diffusion of turbulent energy within the turbulent fluid, and the distributions of these quantities are nearly similar in all free turbulent flows.

6.8 Distributions of turbulent intensity and Reynolds stress

Knowing that the observed structure of turbulent flow resembles closely that arising from rapid finite shearing of isotropic turbulence, we shall suppose that the basic entrainment process at the bounding surface produces parcels of fluid, each initially in a state of chaotic, roughly isotropic motion. With time each parcel becomes more deeply embedded in the region of turbulent flow, undergoing shear by the mean velocity field, and the total shear is the time integral of the local rate of shear along the path from the place of entrainment. Different parcels at a particular position will have been formed (i.e. entrained) at different places, will have travelled along different paths, and so will have different total strains. In addition, the parcel must be large enough to be considered to contain a sample of the turbulent motion, and the rate of strain will not be uniform over its volume. Qualitatively, both path variations and finite parcel size affect the average total strain in much the same way as diffusion, and it is plausible that the effective total strain, α, satisfies an equation,

$$U \frac{\partial \alpha}{\partial x} + W \frac{\partial \alpha}{\partial z} = \frac{\partial U}{\partial z} + \frac{\partial}{\partial z}\left(D_T \frac{\partial \alpha}{\partial z}\right), \tag{6.8.1}$$

where D_T is an eddy diffusivity for effective strain.

In a self-preserving flow for which

$$\left.\begin{array}{l} U = U_1 + u_0 f(z/l_0), \\ \alpha = k(z/l_0), \end{array}\right\} \tag{6.8.2}$$

the strain equation (6.8.1) takes a self-preserving form,

$$-\frac{U_1}{l_0}\frac{\mathrm{d}l_0}{\mathrm{d}x}\eta k' - \frac{1}{l_0}\frac{\mathrm{d}(u_0 l_0)}{\mathrm{d}x} k' \int_0^\eta f \,\mathrm{d}\eta = \frac{u_0}{l_0} f' + \frac{D_T}{l_0^2} k'' \tag{6.8.3}$$

(for constant eddy diffusivity), if the necessary conditions are satisfied by the coefficients in the momentum equation (6.4.3). Solutions of the equations for specific flows have been found, and results for the case of equal eddy viscosity and diffusivity are given in table 6.3 (Townsend 1970).

From the table, it appears that the effective strain has a maximum value of about one-fifth of the flow constant, a result that would be

expected if the distributions of mean velocity and effective strain
are similar in different flows. At the position of maximum effective
strain, (6.8.1) reduces to

$$0 = \frac{\partial U}{\partial z} + D_T \frac{\partial^2 \alpha}{\partial z^2} \qquad (6.8.4)$$

and, in terms of the distribution functions, f and k,

$$\alpha_m = \frac{u_0 l_0}{D_T} f'(\eta_m)/k''(\eta_m) \propto R_s. \qquad (6.8.5)$$

Naturally, the argument only works for similar flows and the small
change in the ratio R_s/α_m between the wakes and jets, and the
mixing layers is, to some extent, fortuitous.

TABLE 6.3

Flow	Effective strain	Flow constant R_s
Wake	2.6	12.5
Jet	6.1	28
Mixing layer	7.5	30
Boundary layer (constant-pressure)	10–15	55

The calculations of maximum effective strain for free turbulent,
self-preserving flows confirm that the strains are not large, but they
show that there are considerable variations between flows. Accepting
the relevance of the 'rapid-distortion' calculations to the structure
of the turbulence, inspection of fig. 3.15 shows that the shear coefficient, $\tau/\overline{q^2}$, reaches a maximum value for total strains near 2.5
and decreases as the strain increases further. Neither the position of
the maximum nor the shape of the curve is changed greatly if a
viscous term is used to model the effects of energy transfer to the
smaller eddies of the dissipation chain. Beyond the maximum, the
variation of stress–intensity ratio is slow, and, accepting the calculated values of the maximum effective shear, we expect the ratio
near the position of maximum stress to decrease somewhat as the
flow constant R_s increases. The expectation is borne out by the
rather erratic measurements in wakes, jets and boundary layers.

Within observational uncertainty, intensities of the three velocity components are distributed similarly in all the symmetrical flows (fig. 6.6), the differences being almost entirely a result of the changes in the stress–intensity ratio at the position of maximum shear. A simple application of the hypothesis that the effective strain determines the local turbulent structure leads to a distribution of total intensity,

$$\overline{q^2}/\tau_m = \frac{g_{13}(\eta)}{g_{13}(\eta_m)F(k(\eta))} \tag{6.8.6}$$

where $F(\alpha)$ is the calculated stress–intensity ratio for a total strain of α, and η_m is the position of maximum Reynolds stress. The distribution has a deep minimum at the flow centre, not found in real flows since such a rapid spatial variation is inconsistent with the considerable size of the energy-containing eddies. From the standpoint of the turbulent energy equation, large intensity at the flow centre depends on efficient lateral transfer of energy from the energy-generating regions of strong shear, an effect that would arise from the presence of eddies of size sufficient to reach between the centre and the regions of shear.

Accepting that diffusion (or finite size) reduces the variations of total intensity to near uniformity within the turbulent fluid, the average value of the intensity–stress ratio should be near the average value from (6.8.6). A fair approximation to the calculated values of $F(\alpha)$ is

$$F(\alpha) = \frac{2}{15}\frac{\alpha}{1+0.16\alpha^2} \tag{6.8.7}$$

and using the form $C\eta \exp(-\tfrac{1}{2}\eta^2)$ for the distribution functions $g_{13}(\eta)$ and $k(\eta)$, the average stress–intensity ratio for a flow is given by

$$a_1 = \frac{\tau_m}{q_0^2} = \frac{2}{15}\frac{\alpha_m}{1+0.096\alpha_m^2}. \tag{6.8.8}$$

The ratios of the intensities of the three components show considerable variation over the flow, and this is consistent with eddies generated by finite effective strains. Near the position of maximum shear, the effective strain is larger than elsewhere and the calculations of § 3.12 indicate that the differences in intensity of the components should be much greater there than either at the centre or near the edge of the flow. Since the total intensity is expected to and does reach a broad minimum at the flow centre with a maximum near the position of maximum shear, the distributions of $\overline{u^2}$ and $\overline{v^2}$ may be expected to have obvious maxima at the region of strong shear while $\overline{w^2}$ has a maximum only at the flow centre (fig. 6.6d).

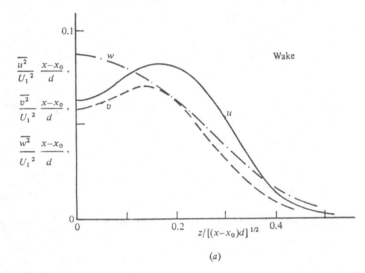

(a)

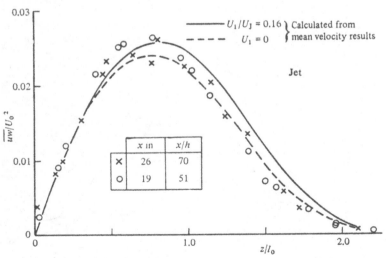

(b)

edge of the flow. Since the total intensity is approximately and does tend to be symmetric about the flow centre with a maximum near the position of maximum shear, the distributions of $\overline{u^2}$, etc., may be expected to be at levels... within of the region distinct on span physics... at the origin of the zero-intensity...

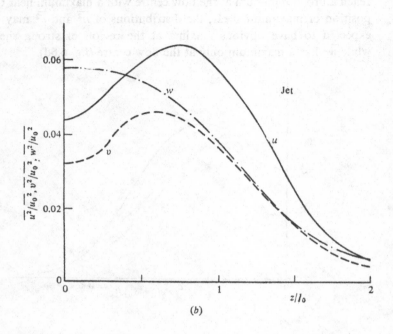

(b)

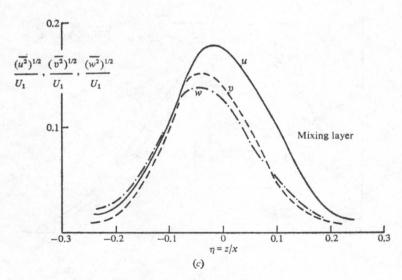

(c)

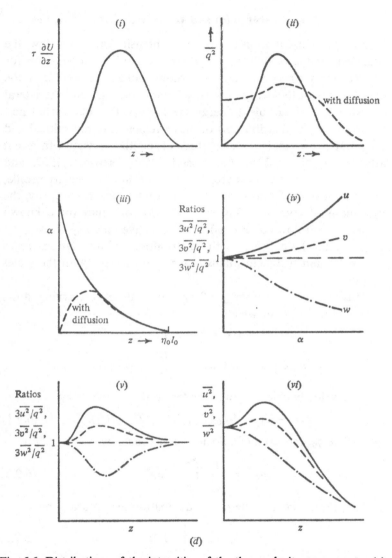

Fig. 6.6. Distributions of the intensities of the three velocity components; (a) Plane wake; (b) Plane jet into still fluid (from Bradbury 1965); (c) Plane mixing layer; (d) Distributions from rapid-distortion model, (i) Rate of energy production, (ii) Total intensity expected with constant dissipation length, with and without diffusion, (iii) Total strain with and without diffusion, (iv) Calculated intensity ratios as a function of total strain, (v) Expected intensity ratios, (vi) Expected distributions of intensity.

6.9 Flow constants for self-preserving jets and wakes

Knowing the distribution of the stress–intensity ratio in a flow, the equation for the turbulent energy may be used as an equation for the Reynolds stress and the mean-flow problem is soluble if the distributions of the dissipation length parameter, L_ε, and the lateral diffusive flux of turbulent energy are known. For free turbulence, the forms of the distributions of mean velocity are not critical and the point of interest is the magnitude of the flow constant. In recent attacks on this problem (Townsend 1966b, Newman 1967, and others), the basic elements are: (a) assumption of a velocity profile, either a universal form or one derived from application of, say, the hypothesis of constant eddy viscosity, (b) constancy (or a known variation with flow conditions) of the effective intensity–stress ratio q_0^2/τ_m, and (c) similarity of the distributions of dissipation length parameter and turbulent intensity when normalised with the scales q_0 and l_0.

Consider the 'jet' family of symmetrical self-preserving flows described in § 6.4, for which

$$U_1 \propto (x-x_0)^a,$$

$$u_0/U_1 = \text{constant}, \qquad \frac{dl_0}{dx} = \text{constant},$$

and the velocity defect ratio is related to the exponent by

$$a = -(I_1 + I_2 u_0/U_1)/(3I_1 + 2I_2 u_0/U_1). \qquad (6.9.1)$$

We define q_0, the scale of velocity fluctuations, by

$$q_0^2 = \frac{1}{\eta_0 l_0} \int_0^\infty \overline{q^2}\, dz \qquad (6.9.2)$$

and L_ε, the effective value of the dissipation length scale, by

$$\eta_0 l_0 q_0^3/L_\varepsilon = \int_0^\infty \varepsilon\, dz, \qquad (6.9.3)$$

where $\eta_0 l_0$ is the mean distance of the bounding surface from the centre-line ($\eta_0 \approx 2.0$). Then, approximating the integral of the turbulent energy flux,

$$\tfrac{1}{2}\int_0^\infty \overline{q^2}(U - U_1)\, dz$$

by $\frac{1}{2}q_0^2 u_0 l_0 I_1$ (i.e. by putting $\overline{q^2} = q_0^2$ everywhere), we can express equation (6.3.9) for conservation of total kinetic energy in the form,

$$\frac{U_1}{u_0}\frac{dl_0}{dx}\left[\frac{5a+1}{2}I_2 + \frac{3a+1}{2}I_3\frac{u_0}{U_1} + \frac{3a+1}{2}\frac{q_0^2}{u_0^2}\left(\eta_0 + \frac{u_0}{U_1}I_1\right)\right]$$
$$= -\eta_0\frac{l_0}{L_\varepsilon}\frac{q_0^3}{u_0^3}. \qquad (6.9.4)$$

Assuming knowledge of the velocity profile, (6.9.4) gives the rate of spread in terms of the exponent a, of the length ratio $\eta_0 l_0/L_\varepsilon$, and the relative turbulent intensity, $q_0^2/u_0^2 = q_0^2/\tau_m \times \tau_m/u_0^2$.

Being a length scale of the main turbulent motion, L_ε is determined by the width of the turbulent flow, i.e. by $\eta_0 l_0$, and we may expect the length ratio to have nearly the same value in all symmetrical flows. The value of the relative turbulent intensity is the product of the value of q_0^2/τ_m, which may depend on the maximum effective strain, and on the value of τ_m/u_0^2, which can be found from the momentum equation. For the moment, we assume q_0^2/τ_m to be invariant.

The ratio, τ_m/u_0^2, depends on the rate of spread and could be found directly from the momentum equation using an assumed velocity profile. When considering the whole family of flows, it is more convenient to use approximate methods based assumptions of similarity in shape of the stress profiles. Three relations have been used. At the centre-line, the momentum equation becomes

$$(U_1 + u_0)\frac{d(U_1 + u_0)}{dx} - U_1\frac{dU_1}{dx} = \left(\frac{\partial\tau}{\partial z}\right)_{z=0} \qquad (6.9.5)$$

and so

$$\frac{l_0}{u_0^2}\left(\frac{\partial\tau}{\partial z}\right)_{z=0} = 2a\beta, \qquad (6.9.6)$$

where β is the entrainment parameter defined by equation (6.5.11). Taking the first moment of the momentum equation over one half of the flow leads to

$$\frac{1}{l_0}\int_0^\infty \frac{\tau}{u_0^2}\,dz = -\frac{U_1}{u_0}\frac{dl_0}{dx}\Bigg[(4a+2)I_{1a}$$
$$+ \frac{u_0}{U_1}\left((2a+1)I_{2a} + \frac{a+1}{2}I_1^2\right)\Bigg], \qquad (6.9.7)$$

where

$$I_{1a} = \int_0^\infty x f(x)\, dx, \qquad I_{2a} = \int_0^\infty x[f(x)]^2\, dx.$$

Lastly, integrating the momentum equation from the centre-line to the 'half-velocity' point where $U = U_1 + \tfrac{1}{2}u_0$ leads to

$$\frac{\tau(\eta_{0.5})}{u_0^2} = \frac{U_1}{u_0}\frac{dl_0}{dx}\left\{ (3a+1)I_{1\frac{1}{2}} - \frac{a+1}{2}\eta_{0.5} \right.$$
$$\left. + \frac{u_0}{U_1}\left[(2a+1)I_{2\frac{1}{2}} - \frac{a+1}{2}I_{1\frac{1}{2}} \right] \right\}, \tag{6.9.8}$$

where

$$f(\eta_{0.5}) = \tfrac{1}{2}, \qquad I_{1\frac{1}{2}} = \int_0^{\eta_{0.5}} f(x)\, dx, \qquad I_{2\frac{1}{2}} = \int_0^{\eta_{0.5}} (f(x))^2\, dx.$$

Since maximum stress occurs near the half-velocity point, Newman (1967) uses (6.9.8) to provide a good approximation to τ_m/u_0^2. Another way is to suppose that the stress profile does not change its shape though its length scale may not be a constant fraction of l_0, and to use (6.9.6) and (6.9.7) to calculate

$$\tau_e = \left[\left(\frac{\partial\tau}{\partial z}\right)_{z=0} \int_0^\infty \tau\, dz \right]^{1/2}$$

as a measure of τ_m. The simplest is to assume similarity of the stress distributions in all flows, i.e. invariance of the function $g_{13}(\eta)$, in which case

$$\frac{\tau_m}{u_0^2} = 2|a|\beta C_1 \tag{6.9.9}$$

where $C_1 \approx 0.6$.

Substituting (6.9.9) in (6.9.4), we obtain an equation for the entrainment parameter,

$$\frac{5a+1}{2}I_2 + \frac{3a+1}{2}\frac{u_0}{U_1}I_3 + \frac{3a+1}{2}\left(C_1\frac{q_0^2}{\tau_m}\right)2|a|\left(\eta_0 + \frac{u_0}{U_1}I_1\right)\beta$$
$$+ \eta_0\frac{l_0}{L_\varepsilon}\left(C_1\frac{q_0^2}{\tau_m}\right)^{3/2}2^{3/2}|a|^{3/2}\left(1 + \frac{1}{2}\frac{u_0}{U_1}\right)\beta^{1/2} = 0 \tag{6.9.10}$$

which involves only the distribution integrals, the length ratios η_0 and l_0/L_ε, and the shear coefficient in the combination $C_1 q_0^2/\tau_m$. Although the general magnitudes of these non-dimensional factors are fixed by the implications of previous results, their exact magnitudes are best found by comparison with a particular flow. For a jet issuing into still surroundings, the ratio of turbulent to mean flow kinetic energy at any section is

$$\int_0^\infty \overline{q^2}\, dz \Big/ \int_0^\infty (U - U_1)^2\, dz = C_1 \frac{q_0^2}{\tau_m} \frac{\beta}{I_2}. \tag{6.9.11}$$

Using measurements of Bradbury (1965) and assuming a velocity distribution, $f(\eta) = \exp(-\tfrac{1}{2}\eta^2)$, we find

$$C_1 q_0^2/\tau_m = 2.2 \quad \text{or, say,} \quad q_0^2/\tau_m = 3.6. \tag{6.9.12}$$

It is interesting that (6.9.11) is valid for the small-defect wake, and the measurements there lead to nearly the same value.

The length ratio, l_0/L_ε, occurs as the combination, $(C_0 q_0^2/\tau_m)^{3/2} \times \eta_0 l_0/L_\varepsilon$, in the dissipation term of (6.9.10). Substituting in that equation, the values for a simple jet,

$$a = -\tfrac{1}{2}, \qquad \beta = 0.044, \qquad \eta_0 = 2.0$$

and

$$A_1 = C_1 q_0^2/\tau_m = 2.2,$$

we find that

$$A_2 = (C_1 q_0^2/\tau_m)^{3/2} \eta_0 l_0/L_\varepsilon = 2.02,$$

equivalent to $l_0/L_\varepsilon = 0.31$. In grid turbulence, the integral scale is about one-third of the dissipation scale, and the value of A_2 implies an integral length scale of nearly l_0, and energy-containing eddies that fit comfortably in one-half of the flow.

Assuming invariance of the various coefficients, the variation of the entrainment parameter among the flows of the jet family is easily calculated with results given in table 6.4. Comparison with experimental results by Fekete and by Gartshore (quoted by Newman 1967) shows the calculated values to be about 20 % too large for values of a near $-\tfrac{1}{3}$ or small $|u_0/U_1|$. A possible reason is the change in the value of

$$C_1 = \tau_m \Big/ \left[l_0 \left(\frac{\partial \tau}{\partial z} \right)_0 \right]$$

caused by changes in shape in the stress profile. For example, assumption of constant eddy viscosity leads to the velocity and stress profiles,

$$f(\eta) = \operatorname{sech}^2 \alpha\eta, \qquad g_{13}(\eta) = -2R_s^{-1}\alpha \tanh \alpha\eta \operatorname{sech}^2 \alpha\eta \tag{6.9.13}$$

for a simple jet $(a = -\frac{1}{2})$, and

$$f(\eta) = \exp(-\tfrac{1}{2}\eta^2), \qquad g_{13}(\eta) = -R_s^{-1}\eta\, e^{-1/2\eta^2} \qquad (6.9.14)$$

for jets with small defects $(a = -\frac{1}{3})$. The respective values of C_1 are 0.61 and 0.52 after choosing $\alpha = 0.75$ to make the two profiles intersect at the half-velocity point, a difference more than sufficient to account for the discrepancy. In the simple form, errors of 10–20 % may occur but the predicted entrainment parameters should provide a good description of the general behaviour.

TABLE 6.4   *Flow parameters for free turbulent flows*

Flow	β	R_s	R_0	Notes
Jet into still fluid $(a = -\frac{1}{2})$	0.044	22.8	28.5	
Jet in moving stream $(a = -\frac{1}{3})$	0.070	21.4	26.8	Observed
	0.060	25.0	31.3	Calculated
Small-defect wake $(a = 0)$	0.080	12.5	15.5	
Mixing layer $(U_2 = 0)$	0.028	35.7	28.5	

Similar methods may be used with the 'wake' family of self-preserving flows with small defects, i.e. with

$$U_1 \propto (x - x_0)^a, \qquad u_0 \propto (x - x_0)^{-1/2(1+a)}$$

$$(6.9.15)$$

$$l_0 \propto (x - x_0)^{1/2(1-3a)}.$$

The energy equation becomes

$$\frac{U_1}{u_0}\frac{dl_0}{dx}\left[\tfrac{1}{2}(1+a)I_2 + \tfrac{1}{2}(1+3a)\frac{q_0{}^2}{u_0{}^2}\eta_0\right] = \eta_0\frac{l_0}{L_e}\frac{q_0{}^3}{u_0{}^3}(1-3a). \quad (6.9.16)$$

At the flow centre,

$$\left(\frac{\partial\tau}{\partial z}\right)_0 = u_0\frac{dU_1}{dx} + U_1\frac{du_0}{dx}$$

$$= \frac{a-1}{1-3a}\frac{u_0 U_1}{l_0}\frac{dl_0}{dx} \qquad (6.9.17)$$

and, assuming similarity of the stress distributions,

$$\frac{\tau_m}{u_0^2} = C_1 \left| \frac{1-a}{1-3a} \right| \beta, \tag{6.9.18}$$

since $\beta = (U_1/u_0)(dl_0/dx)$. The energy equation then becomes

$$(1+a)I_2 + (1+3a)C_1 \frac{q_0^2}{\tau_m} \eta_0 \beta' = 2\eta_0 \frac{l_0}{L_\varepsilon} \left(C_1 \frac{q_0^2}{\tau_m} \right)^{3/2} (1-a)\beta' \tag{6.9.19}$$

where $\beta' = \beta(1-a)/(1-3a)$.

Only one flow of the family has been studied in any detail, the simple plane wake with $a = 0$, and the observations of a cylinder wake are consistent with $A_2 = 2.17$ significantly different from the value of 2.02 found for the simple jet. Again the difference may be traced to the small change in shape of the stress profile, and, indeed, the change in A_2 would nearly resolve the discrepancy between the observed and calculated values of β for small-defect jets.

It is an interesting property of equation (6.9.19) that it has no real roots with the constants, $A_1 = 2.2$, $A_2 = 2.17$, $\eta_0 = 2.0$, if $a > 0.038$, implying that then self-preserving development is not possible although the normal conditions are satisfied.

Axisymmetric flows may be treated on similar lines. Defining

$$q_0^2 = \frac{2}{\eta_0^2 l_0^2} \int_0^\infty \overline{q^2} r \, dr, \qquad \frac{q_0^3}{L_\varepsilon} = \frac{2}{\eta_0^2 l_0^2} \int_0^\infty \varepsilon r \, dr \tag{6.9.20}$$

and the distribution integrals by

$$I_n = \int_0^\infty [f(\eta)]^n \, \eta \, d\eta$$

we find for the axisymmetric 'jet' family, with

$$a = -2(I_1 + I_2 u_0/U_1)/(3I_1 + 2I_2 u_0/U_1),$$

$$u_0/U_1 = \text{constant}, \qquad \frac{dl_0}{dx} = \text{constant},$$

an energy equation,

$$\frac{U_1}{u_0} \frac{dl_0}{dx} \left\{ \frac{5a+2}{2} I_2 + \frac{3a+2}{2} I_3 \frac{u_0}{U_1} + \frac{3a+2}{2} \frac{q_0^2}{u_0^2} \left[\tfrac{1}{2} \eta_0^2 + I_1 u_0/U_1 \right] \right\}$$

$$+ \tfrac{1}{2} \eta_0^2 \frac{l_0}{L_\varepsilon} \frac{q_0^3}{u_0^3} = 0. \tag{6.9.21}$$

On the centre-line of the flow,

$$\left(\frac{\partial \tau}{\partial r}+\frac{\tau}{r}\right)_{r=0} = (U_1+u_0)\frac{d(U_1+u_0)}{dx} - U_1\frac{dU_1}{dx} \qquad (6.9.22)$$

and if the stress profiles are similar

$$\frac{\tau_m}{u_0^2} = C_1|a|\beta. \qquad (6.9.23)$$

Then, the energy equation is

$$\frac{5a+2}{2}I_2 + \frac{3a+2}{2}I_3\frac{u_0}{U_1} + \frac{3a+2}{2}|a|C_1\frac{q_0^2}{\tau_m}\left(\tfrac{1}{2}\eta_0^2 + \frac{u_0}{U_1}I_1\right)\beta$$

$$+ \tfrac{1}{2}\eta_0^2\frac{l_0}{L_\varepsilon}\left(C_1\frac{q_0^2}{\tau_m}\right)^{3/2}|a|^{3/2}\left(1 + \frac{1}{2}\frac{u_0}{U_1}\right)\beta^{1/2} = 0. \qquad (6.9.24)$$

From observations of simple axisymmetric jets issuing into still air $(a = -1)$, the constants are

$$B_1 = C_1 q_0^2/\tau_m = 2.2, \qquad B_2 = \tfrac{1}{2}\eta_0^2\frac{l_0}{L_\varepsilon}\left(\frac{q_0^2}{\tau_m}\right)^{3/2} = 2.02$$

$$\eta_0 = 2.0,$$

almost identical with the constants for plane jets.

The 'wake' family of small-defect axisymmetric flows with

$$u_0 \propto (x - x_0)^{-2/3}, \qquad l_0 \propto (x - x_0)^{1/3-a}$$

has an energy equation,

$$\frac{1}{(\tfrac{1}{3}-a)}\frac{dl_0}{dx}\left[-\tfrac{2}{3}I_2 - (a+\tfrac{2}{3})\frac{q_0^2}{u_0^2}\tfrac{1}{2}\eta_0^2\right] + \tfrac{1}{2}\eta_0^2\frac{l_0}{L_\varepsilon}\frac{q_0^3}{u_0^2U_1} = 0. \qquad (6.9.25)$$

The usual argument gives the stress ratio as

$$\frac{\tau_m}{u_0^2} = \frac{1}{2}\left|\frac{a-\tfrac{2}{3}}{a-\tfrac{1}{3}}\right|C_1\beta \qquad (6.9.26)$$

and the energy equation becomes

$$\tfrac{2}{3}I_2 + (a+\tfrac{2}{3})\tfrac{1}{4}\eta_0^2C_1\frac{q_0^2}{\tau_m}\beta' = \tfrac{1}{2}\eta_0^2\frac{l_0}{L_\varepsilon}\left(C_1\frac{q_0^2}{\tau_m}\right)^{3/2}\frac{(\tfrac{2}{3}-a)}{2^{3/2}}\beta'^{1/2}, \qquad (6.9.27)$$

where $\beta' = \beta|(a-\tfrac{2}{3})/(a-\tfrac{1}{3})|$. There are real roots only if $a < -0.41$, and it would appear that the simple axisymmetric wake in a uniform stream $(a = 0)$ is not a self-preserving flow. The nature of the failure to achieve self-preserving development will be discussed in a later section.

6.10 The flow constants of plane mixing layers

Consider the plane mixing layer formed by the junction of two free streams of different uniform velocities, coming into contact along the Oy axis. For large positive values of z, the uniform stream velocity is (U_1, V_1, W_1) and, for large negative values, (U_2, V_2, W_2), where U_1, V_1, U_2, V_2 and $(W_1 + W_2)$ are set by the nature of the free streams. Then dimensional considerations show that

$$\left.\begin{array}{l} U = U_2 + (U_1 - U_2)F_1(z/x, \, U_1 x/v), \\ V = V_2 + (V_1 - V_2)F_2(z/x, \, U_1 x/v), \end{array}\right\} \tag{6.10.1}$$

where the functions depend on the velocity ratios of the free streams. If the Reynolds number of the flow is very large, the influence of viscosity is small, and the flow is self-preserving with constant velocity scale and length-scale proportional to x.

Although skewed mixing layers are of some importance, little is known of their behaviour and we shall consider only plane mixing layers with mean flow parallel to the xOz plane. It is convenient to choose axes so that $U_1 > U_2 \geqslant 0$ and $W = 0$ for $z = 0$. Using the velocity scale, $u_0 = U_1 - U_2$, the distributions of mean velocity and Reynolds stress have the self-preserving forms,

$$\left.\begin{array}{l} U = U_2 + u_0 f(z/l_0), \\ -\overline{uw} = -u_0{}^2 g_{13}(z/l_0) \end{array}\right\} \tag{6.10.2}$$

and the momentum equation takes the form,

$$g_{13}' = \left[\frac{U_2}{u_0}\eta f' + f' \int_0^\cdot f \, d\eta\right]\frac{dl_0}{dx}. \tag{6.10.3}$$

Notice that the Reynolds stress has its maximum value at $\eta = 0$, since the axes have been chosen so that $W(0) = 0$. Overall conservation of momentum imposes the conditions,

$$\begin{aligned} \frac{\tau_m}{u_0{}^2(dl_0/dx)} &= \frac{U_2}{u_0} \int_{-\infty}^0 f(\eta) \, d\eta + \int_{-\infty}^0 [f(\eta)]^2 \, d\eta \\ &= \frac{U_2}{u_1} \int_0^\infty (1 - f(\eta)) \, d\eta \\ &\quad + \int_0^\infty f(\eta)[(1 - f(\eta))] \, d\eta \end{aligned} \tag{6.10.4}$$

and it follows that the velocity profile can be antisymmetric about

$\eta = 0$ only if u_0/U_2 is very small. A fair approximation to observed profiles is provided by the error integral,

$$f(\eta) = (2\pi)^{-1/2} \int_{-\infty}^{\eta + \eta_0} e^{-\frac{1}{2}x^2} dx \qquad (6.10.5)$$

which now defines the length scale l_0. Using the condition (6.10.4), we find that η_0 must satisfy

$$\eta_0 \left(\frac{U_2}{u_0} + \frac{1}{\sqrt{(2\pi)}} \int_{-\infty}^{\eta_0} e^{-\frac{1}{2}x^2} dx \right) = \frac{\sqrt{2}-1}{\sqrt{(2\pi)}}. \qquad (6.10.6)$$

If η_0 is not too large, approximations lead to

$$\eta_0 = \frac{\sqrt{2}-1}{\sqrt{(2\pi)}} \left(\frac{U_2}{u_0} + \frac{1}{2} \right)^{-1}. \qquad (6.10.7)$$

The maximum value of η_0 is 0.330 for $U_2/u_0 = 0$.

For a mixing layer, the equation for the conservation of total kinetic energy is

$$\frac{1}{2} \int_{-\infty}^{\infty} \frac{\partial}{\partial x} [U(U-U_1)^2 + \overline{q^2}] dz$$

$$+ \frac{1}{2} W_1 (U_1 - U_2)^2 + \int_{-\infty}^{\infty} \varepsilon \, dz = 0. \qquad (6.10.8)$$

Defining

$$q_0^2 = \frac{1}{(\eta_1 - \eta_2)l_0} \int_{-\infty}^{\infty} \overline{q^2} \, dz,$$

$$\frac{q_0^3}{L_\varepsilon} = \frac{1}{(\eta_1 - \eta_2)l_0} \int_{-\infty}^{\infty} \varepsilon \, dz, \qquad (6.10.9)$$

where η_1, η_2 define the mean position of the bounding surfaces on either side of the layer, and

$$I_1 = \int_{-\infty}^{0} f(\eta) \, d\eta, \qquad I_2 = \int_{0}^{\infty} (1 - f(\eta)) \, d\eta,$$

$$I_a = \int_{-\infty}^{\infty} \eta f^2 f' \, d\eta, \qquad I_b = \int_{-\infty}^{\infty} \eta f f' \, d\eta,$$

$$I_{12} = \int_{-\infty}^{0} f^2 \, d\eta,$$

and making use of (6.10.4), the energy equation appears in the form,

$$\frac{U_2}{u_0}I_b + \tfrac{3}{2}I_a - \tfrac{1}{2}I_2 - \frac{1}{2}\frac{q_0^2}{u_0^2}\left[\frac{U_2}{u_0}(\eta_1 - \eta_2) + \int_{\eta_2}^{\eta_1} f\,d\eta\right]$$
$$= (\eta_1 - \eta_2)\frac{l_0}{L_\varepsilon}\frac{q_0^3}{u_0^3}\left(\frac{dl_0}{dx}\right)^{-1}. \tag{6.10.10}$$

For error integral profiles, the profile integrals are

$$I_1 = (2\pi)^{-1/2} + \tfrac{1}{2}\eta_0, \qquad I_2 = (2\pi)^{-1/2} - \tfrac{1}{2}\eta_0,$$
$$I_a = (4\pi)^{-1/2} + \tfrac{1}{3}\eta_0, \qquad I_b = (4\pi)^{-1/2} - \tfrac{1}{2}\eta_0,$$
$$I_{12} = \frac{2^{1/2}-1}{2\pi^{1/2}} + \tfrac{1}{4}\eta_0$$

for moderate values of η_0, and the energy equation reduces to

$$2^{-1/2} - \frac{1}{2}\frac{q_0^2}{\tau_m}\frac{dl_0}{dx}\left[\frac{U_2}{u_0}(\eta_1 - \eta_2) + \int_{\eta_2}^{\eta_1} f\,d\eta\right]$$
$$= (2\pi)^{-1/4}(\eta_1 - \eta_2)\frac{l_0}{L_\varepsilon}\left(\frac{U_1}{u_0} + \frac{1}{2}\right)^{1/2}\left(\frac{q_0^2}{\tau_m}\right)^{3/2}\left(\frac{dl_0}{dx}\right)^{1/2}. \tag{6.10.11}$$

From the measurements of Watt (1967), it may be estimated that $\eta_2 = -1.0$, $\eta_1 = +1.4$ for $U_2/u_0 = 0$ and, very nearly,

$$\int_{\eta_2}^{\eta_1} f\,d\eta = \tfrac{1}{4}(\eta_1 - \eta_2).$$

Then the entrainment parameter is given by the equation

$$1 - 2^{-1/2}(\eta_1 - \eta_2)\frac{q_0^2}{\tau_m}\beta$$
$$= (2\pi)^{-1/4}(\eta_1 - \eta_2)\frac{l_0}{L_\varepsilon}\left(\frac{q_0^2}{\tau_m}\right)^{3/2}2^{1/2}\beta^{1/2} \tag{6.10.12}$$

in which the velocity ratio does not appear explicitly.

Using the approximate forms for the profile integrals, I_1 and I_2, (6.10.4) shows that

$$\frac{\tau_m}{u_0^2} = (2\pi)^{-1/2}R_s^{-1} = (2\pi)^{-1/2}\beta. \tag{6.10.13}$$

Since α_m is proportional to R_s (6.8.5), the effective value of q_0^2/τ_m depends only on β and the equation for the entrainment parameter indicates that it is nearly independent of the velocity ratio U_2/u_0.

Observations by Sabin (1963) confirm the conclusion and indicate that

$$2\beta = \frac{U_1 + U_2}{U_1 - U_2}\frac{dl_0}{dx} = 0.0280 \qquad (6.10.14)$$

over a wide range of velocity ratios. Using the same value of $q_0{}^2/\tau_m$ as before, substitution in (6.10.12) leads to

$$A_2 = (\eta_1 - \eta_2)\frac{l_0}{L_\varepsilon}\left(\frac{q_0{}^2}{\tau_m}\right)^{3/2} = 5.5, \quad \text{or} \quad (\eta_1 - \eta_2)l_0/L_\varepsilon = 0.77.$$

It is likely that the value of $q_0{}^2/\tau_m = 3.7$ is low, and so that $(\eta_1 - \eta_2)l_0/L$, the ratio of the total width of the turbulent flow to the dissipation length-scale, is not significantly different from $\eta_0 l_0/L$ in symmetrical flows. It will be recalled that the scale of turbulent motion between parallel walls is twice as large in plane Couette flow as in pressure flow with stationary walls and the same as in free-surface flow of the same depth (§ 5.12). All these measurements are consistent with the hypothesis that the dissipation length-scale is nearly a constant fraction of the distance between the flow 'boundaries', which may be either (1) rigid surfaces, or (2) bounding surfaces to the turbulent motion, not fixed in position, or (3) planes of zero Reynolds shear stress and zero gradient of mean velocity.

6.11 The entrainment of ambient fluid

In the previous sections, it is argued that all flows with similar symmetry have geometrically similar structures of their turbulent motions, and it follows that their rates of spread are determined within narrow limits by the basic requirements of overall con-servation of momentum and energy. On the other hand, the actual entrainment of the ambient, non-turbulent fluid takes place through sharply-defined bounding surfaces by processes which must act at rates sufficient to produce the set rate of spread. In some respects an analogy may be drawn with the closed control loops of servo-mechanisms and feedback circuits in which all elements of the loop are essential for its operation but the important properties are controlled by one particular element and are nearly independent of the remainder. If the rate of entrainment through the bounding surface is determined by structural similarity of the whole flow, the actual mechanism of entrainment must be flexible enough to adjust to a wide variation of rates without major changes in the whole flow.

The rate at which non-turbulent fluid flows through the bounding surface and becomes turbulent fluid can be found by using the intermittency function $\delta(\mathbf{x}, t)$ (defined in § 6.7) to form $(\mathbf{U} + \mathbf{u})\delta$, the volume flux of turbulent fluid. Then in a two-dimensional flow, the mean rate of formation of turbulent fluid per unit area projected on the xOy plane is defined as the *entrainment velocity*,

$$u_e = \frac{\mathrm{d}}{\mathrm{d}x} \int_0^\infty [\overline{U\delta} + \overline{u\delta}] \, \mathrm{d}z \approx \frac{\mathrm{d}}{\mathrm{d}x} \int_0^\infty \gamma U \, \mathrm{d}z \qquad (6.11.1)$$

to the boundary-layer approximation $(\gamma = \bar{\delta})$. For self-preserving flows of the 'jet' family,

$$\frac{u_e}{u_0} = (a+1)(\eta_0 + I_\gamma u_0/U_1)/(1 + \tfrac{1}{2}u_0/U_1)\beta \qquad (6.11.2)\dagger$$

and, for flows of the 'wake' family,

$$\frac{u_e}{u_0} = \eta_0 \beta (1-a)/(1-3a). \qquad (6.11.3)$$

Table 6.5 lists some values of the entrainment ratios, u_e/u_0 and $u_e/\tau_m^{1/2}$ and it is obvious that (1) the ratios vary over a very wide range, and that (2) neither the mean velocity variation u_0 nor the measure of turbulent velocity fluctuations $\tau_m^{1/2}$ determines the average rate of advance of the bounding surface.

TABLE 6.5   *Ratios of entrainment velocities*

Flow	β	u_e/u_0	$(\overline{q^2})^{1/2}/u_0$	$u_e/(\overline{q^2})^{1/2}$
Plane wake	0.080	0.21	0.46–0.49	0.46–0.43
Plane jet	0.045	0.054	0.36–0.39	0.15–0.14
Boundary layer ($a = 0$)	0.012	0.024	0.28	0.085

Note: The values of $(\overline{q^2})^{1/2}/u_0$ and $u_c/(\overline{q^2})^{1/2}$ for the jet and the wake refer to the central and maximum values of the turbulent intensity.

The entrainment velocity has been defined as the volume rate of production of turbulent fluid per unit projected area and the actual

$\dagger$ $I_\gamma = \int_0^\infty \gamma f(\eta)\mathrm{d}(\eta).$

entrainment velocity at a section of the bounding surface may be much less if the surface is folded so that its superficial area is much greater than its projected area. In that case, the variations of entrainment ratio might be explained by assuming a constant rate of entrainment per unit area of the bounding surface expressed as a fraction of the 'friction velocity' $\tau_m{}^{1/2}$ but different degrees of folding of the surface in the different flows. That there is a correlation between the depths of the indentations of the surface and the entrainment ratios has been shown by Gartshore (1966) (fig. 6.5), but so far clear evidence of variation in the degree of folding does not exist. However, the form of the folding is quite different in wakes and in boundary layers and mixing layers, and the observations may be reconciled by postulating an entrainment process made up from both (1) a basic entrainment carried out by the ordinary eddies of the turbulent motion, and (2) additional folding carried out by a distinct group of eddies, the entrainment eddies, which develop in intensity sufficient to produce the large entrainment ratios in wakes and jets.

6.12 Basic entrainment processes

An essential characteristic of turbulent fluid is the presence of vorticity in large amounts, and the development of vorticity in previously irrotational fluid depends in the first place on viscous diffusion of vorticity across the bounding surface. Since the rate of entrainment is not dependent on the magnitude of the fluid viscosity, the slow process of diffusion into the ambient fluid must be accelerated by interaction with the velocity fields of eddies of all sizes, from the viscous eddies to the energy-containing eddies so that the overall rate of entrainment is set by large-scale parameters of the flow.

A qualitative description of the transfer of vorticity to the irrotational fluid may be given by considering the influence on the viscous diffusion of eddies in groups of increasing size. The smallest eddies would produce ripples on an initially sharp interface with wavelengths comparable with the Kolmogorov length, $l_s = (v_3/\varepsilon)^{1/4}$, and induce rates of strain of order $\beta = (\varepsilon/v)^{1/2}$ which compress the diffused layer at the wave crests and stretch it in the troughs (fig. 6.7). After a time t, the average advance of vorticity into the ambient fluid will be roughly

$$\left[\frac{\nu}{\beta}(e^{2\beta t}-1)\right]^{1/2} \approx l_s\, e^{\beta t}. \qquad (6.12.1)$$

The flow pattern will persist for a finite time, of order β^{-1}, and so the average velocity of advance is of order $Cl_s\beta$, in which the factor C will be larger or smaller as the persistence time is long or short compared with β^{-1}, but always considerably more than the unaided viscous rate. Eddies one order of magnitude larger produce ripples on an interface that is being diffused by viscosity in combination with the viscous eddies, and they generate irrotational flows that extend outside the surface to a distance comparable with their diameter. Again, the effect of the eddies is to stretch the diffused layer in the troughs to a thickness comparable with the eddy diameter and so to the depth of the ripples. In short, just as the interaction between viscous diffusion and the smallest eddies fills in with vorticity the troughs in the small scale distortion of the initial interface, so does interaction between the convected diffused layer with the larger eddies fill in the troughs formed by these eddies, and extension of the argument up the whole sequence of eddy sizes is possible. The process might be described as a cascade in reverse.

An interesting point about the model of the vorticity transfer is that it presents the process as one of filling in troughs and, by compression, keeping strong gradients of vorticity over the crests, and so it gives some explanation to the comparatively smooth and flat form that is observed. On the other hand, the interaction that makes the rate of entrainment by all eddies up to a particular size depend only on the velocities of the largest is essentially one between eddies of not too dissimilar sizes, and the absence of eddies in a particular size range would interrupt the cascade and halt the entrainment at the value before the missing step in the cascade. In that case, eddies larger than the missing ones will distort but not 'diffuse' the bounding surface, and the result may be 'pockets' of non-turbulent fluid almost surrounded by turbulent fluid. Judging from the similar phenomenon of dissipation 'intermittency' in flows of large Reynolds number (Batchelor & Townsend 1949, Sandborn 1959, Stewart, Wilson & Burling 1970), abnormally large fluctuations of entrainment rate over the bounding surface may occur purely as a consequence of normal statistical fluctuations.

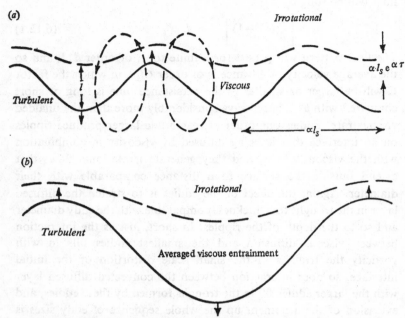

Fig. 6.7. Basic entrainment by interaction between viscous diffusion and straining by induced irrotational flow. (a) First stage (viscous eddies)—scales $u_s = (\nu\alpha)^{1/2}$, $l_s = (\nu/\alpha)^{1/2}$. Note: Viscous spread at troughs, l, given by

$$l^2 = \frac{\nu}{\alpha}(e^{2\alpha t} - 1)$$

so entrainment velocity $u_e \approx u_s\, e^{\alpha\tau_s}$ where $\tau_s = O(\alpha^{-1})$ is duration of flow. (b) Second (and later) stages—scales U, L. Note: Near peaks and troughs, large-scale rates of strain, U/L, add to smaller-scale rates but persist much longer. Hence, average entrainment rate is

$$u_e = u_s \exp(\alpha\tau_s + U\tau_L/L)$$

i.e. an increase by a factor of order e for the stage.

Observations concerning the nature of basic entrainment begin with those of Mobbs (1968) on the development of a turbulent wake with no significant variation of mean velocity. He found sharply defined bounding surfaces, but measurements of the intermittency factor showed that the entrainment velocity, as defined by (6.11.1), was negative, i.e. the total volume of turbulent fluid was decreasing during development of the flow. Strictly, once fluid acquires vorticity it can never lose it entirely and return to an irrotational motion, and his observations must mean (1) that entrainment in the absence of mean velocity gradients is very slow, and (2) that the dissipation rate in some parts of the turbulent fluid becomes so small that they do not register as turbulent fluid on the detecting instruments. Mobbs also found that normal positive rates of entrainment were recovered if the wake were allowed to enter a distorting duct that imposed mean velocity gradients on the flow. One effect of the mean velocity gradient is to align eddies with their axes parallel to the direction of extension and to the mean position of the bounding surface, and it is possible that organised eddies of this kind are more effective agents for entrainment than the randomly orientated eddies of the zero-momentum wake. Whatever the explanation, the evidence is that positive entrainment velocities depend on the presence of a mean velocity gradient in the turbulent fluid.

In table 6.2, the two flows with the lowest entrainment ratios are the plane mixing layer and the boundary layer in an unaccelerated stream, and detailed studies have been made of entrainment in both these flows (Wygnanski & Fiedler 1970, Kovasznay, Kibens & Blackwelder 1970). Conditional sampling techniques were used in which measurements are made only if the boundary is near a reference point in the flow, and it is possible to obtain an average form of the velocity distribution in the neighbourhood of the bounding surface. The measurements show this average velocity to be the free stream velocity if the measuring point is outside the turbulent fluid and to vary linearly with distance inside the turbulent fluid. No sign of a velocity jump or even large velocity gradients near the surface were found.

Other measurements, particularly of the intermittency signal, provide statistical information about the indentations of the bounding surface but, to appreciate their meanings, it is necessary to keep in mind the actual shape of the surface. Fiedler & Head (1966) and Imaki (1968) have obtained by different techniques what are in effect maps of the intermittency function over planes normal to the wall, and their results show that the folding of the surface is intense. Some tracings of smoke photographs by Fiedler & Head are shown in fig. 6.8, both for longitudinal and transverse sections, and, as often as not, the surface has folded so that ambient fluid is trapped between layers of turbulent fluid. The use of smoke to distinguish turbulent from non-turbulent fluid rests on the assumption that it is rapidly diffused through the entire turbulent fluid, but the agreement with the results of Imaki, who used hot-wire anemometers to discriminate, shows that the folding is real.† Consequently, any analysis that assumes the z co-ordinate of a point on the surface to be a single-valued function of x and y must be used with great caution.

The simplest measurements with an intermittency signal are of

(a)

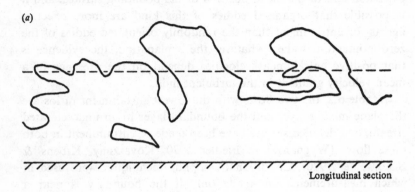

Longitudinal section

(b)

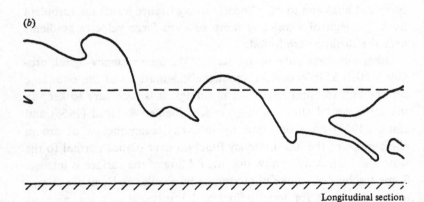

Longitudinal section

† Most photographs of smoke-filled flows or Schlieren photographs of hot flows (e.g. bullet wakes) do not show the folding. The reason is that turbulent flow is indicated if there is *any* turbulent fluid along the line of sight, and so pockets of non-turbulent fluid are not seen.

(c)

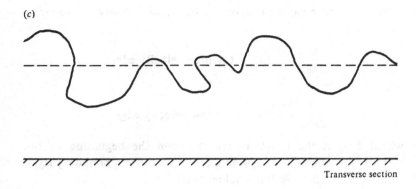

Transverse section

(d)

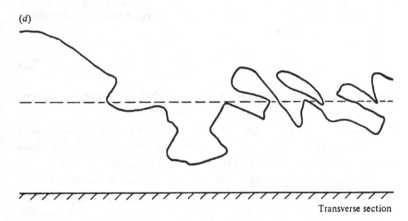

Transverse section

Fig. 6.8. Sections of a free turbulent flow (a boundary layer) showing the shape of the interface, made visible by injection of smoke (from cine film by Fiedler & Head 1966).

the distributions of its mean value, the intermittency factor, and of the frequency with which it changes between its two values. The first leads to a standard deviation that is closely related to the average departure of the surface from its mean position, and the second measures the mean time interval between successive patches of turbulent fluid. Knowing the convection velocity of the intermittency pattern, the ratios of mean indentation depth to mean interval in space can be calculated. For a mixing layer, Wygnanski & Fiedler find

$$\frac{\sigma}{2\pi\lambda_y} = 0.156 \text{ at the high-velocity edge,}$$

and

$$\frac{\sigma}{2\pi\lambda_y} = 0.048 \text{ at the low-velocity edge,}$$

where $2\pi\lambda_y$ is the mean interval between the beginning of one turbulent patch and the beginning of the next. For a boundary layer, Kovasznay, Kibens & Blackwelder find

$$\frac{\sigma}{2\pi\lambda_y} = 0.11.$$

If the form of the surface could be described by a displacement ζ that is a single-valued function of (x, y), the root-mean-square slope of the surface would be

$$\left[\overline{\left(\frac{\partial\zeta}{\partial x}\right)^2}\right]^{1/2} = \frac{\sigma}{\lambda_y} \qquad (6.12.2)$$

and the measurements seem to indicate that the slopes are comparatively small, at least over most of the surface.

The average shape of the indentations may be described in part by covariances (correlation functions) between values of the intermittency signal or the surface displacement at separate points in the flow. Some of the photographic records of Fiedler & Head have been used to produce the displacement correlations of fig. 6.9, and they show that, while the longitudinal correlation $R_\zeta(r_1, 0)$ remains positive for displacements up to the thickness of the whole layer, the transverse correlation $R_\zeta(0, r_2)$ becomes negative for displacements more than about one-third of the layer thickness. To some extent, the intermittency signal is an indicator of surface displacement, an 'on' value meaning most of the time that the point of observation is closer to the wall than the bounding surface, and the intermittency correlation,

$$R_\delta(r_1, r_2, r_3) = \overline{\delta(\mathbf{x})\delta(\mathbf{x} + \mathbf{r})}$$

is expected to have, and does have, a similar behaviour. Fig. 6.9 shows the correlation calculated from the same photographs as fig. 6.8 and the correlation in the transverse direction falls appreciably below the 'random' value of $(\bar{\delta})^2$. Measurements by Kovasznay *et al.* of correlation contours in the xOy plane show considerable elongation in the stream direction but no negative values. On the other hand, Imaki does observe them.

A quantity of great interest is the convection velocity of the indentations. In the boundary layer, the convection velocity for the

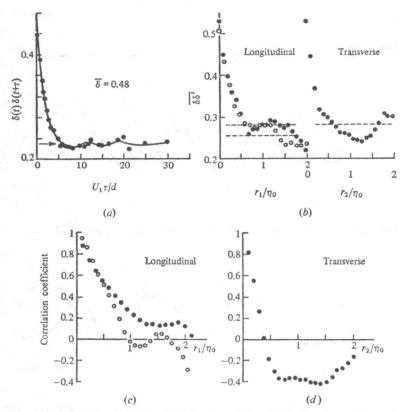

Fig. 6.9. Correlation functions for the intermittency signal and the interface distance from the reference plane. (*a*) Time delay (effectively longitudinal), correlation function of an intermittency signal in a plane wake ($U_1d/\nu = 6000$, $x/d = 150$, $\gamma = 0.48$). (*b*) Longitudinal and transverse correlation functions of intermittency signals in a constant-pressure boundary layer. (*c*), (*d*) Longitudinal and transverse correlation coefficients for displacement of the bounding surface in a constant-pressure boundary layer. (Wake data from Townsend (1970), boundary layer data from cine film of Fiedler & Head (1966)).

fronts of the bulges is about 0.96 of the stream velocity while the convection velocity of the backs is 0.90 of the velocity. The difference is due to the filling-in of the gaps, and the pattern as a whole moves with a velocity of about 0.93 U_1.

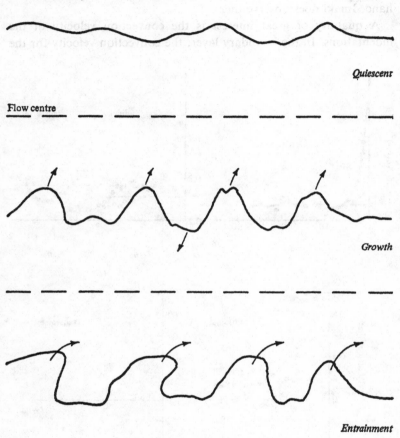

Fig. 6.10. Growth-decay cycle of the entrainment eddies in a plane wake.

On the whole, the observations of entrainment in mixing layers and boundary layers point to the main eddies of the turbulent motion as the agents of distortion of the bounding surface. In particular, the behaviour of the displacement and intermittency correlations is what might be expected if roller eddies with axes aligned with the flow performed the distortion.

6.13 Entrainment eddies in plane wakes

Of the flows listed in table 6.2, plane wakes in a uniform stream have by far the largest entrainment ratios, larger by an order of magnitude than the ratios for boundary layers and mixing layers, and, not surprisingly, studies of entrainment in wakes show a different process from that observed in the flows with small ratios. The basic difference is that, in boundary layers, entrainment is a process as nearly continuous as can be expected in a turbulent flow and indentations of the bounding surface are caused by eddies of the main turbulent motion, but entrainment in wakes depends on a cyclic growth and decay of large *entrainment eddies* which are distinct in form from the eddies of the main motion.

The growth–decay cycle was observed by Grant (1958) in the wake of a cylinder towed through a tank of still water, an arrangement that allows prolonged observation of the development of the bounding surface. The typical behaviour is a progression through four stages:

(1) A quiescent stage of slow growth and small indentations of the bounding surface.

(2) Appearance and growth of indentations in groups resembling wave packets, each with three or four crests nearly at right angles to the flow direction. The indentations move with a phase velocity intermediate between the central velocity and the free stream velocity.

(3) Overturning of the wave-like indentations, causing engulfment and enclosure of non-turbulent fluid.

(4) Rapid growth to fill in the intervals between the indentations. Fig. 6.10 is a pictorial representation of the cycle.

At the end of one cycle, the local thickness of the turbulent flow has increased by a factor of about two, and the indentations of the next cycle are similar in shape but have a wavelength increased by the same factor and develop at a slower rate. In other words, the cycle is part of the self-preserving development and the quasi-periodic groups of eddies are generated by the flow and are not vestiges of the eddies shed by the wake-producing cylinder.

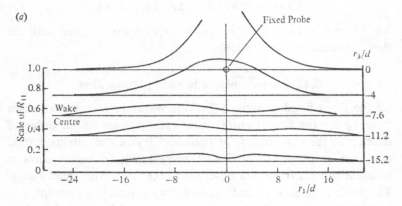

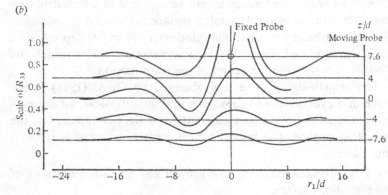

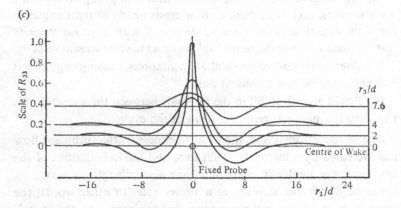

Fig. 6.11. Correlation functions in a wake, showing evidence of quasiperiodic eddies (most noticeable in the $R_{33}(r_1, 0, r_3)$ correlation with fixed r_3. From Grant 1958.) (a) $R_{11}(r_1, 0, r_3)$, fixed probe at $z/d = 7.6$; (b) $R_{33}(r_1, 0, r_3)$, fixed probe at $z/d = 7.6$; (c) $R_{33}(r_1, 0, r_3)$, fixed probe at the centre of the wake.

During growth of the indentations, the large-scale motion near the bounding surface is nearly that of a short train of equally spaced simple eddies with circulation in the xOz plane, perhaps with velocity distributions similar to eddies of type C in § 1.5. Then the correlation functions for not too small separations should show quasi-periodicity in the Ox direction extending over distances comparable with the extent of the groups. Grant (1958) has made extensive measurements of the correlations, $R_{11}(x; r)$ and $R_{33}(x; r)$, which give quantitative evidence of the existence and form of the entrainment eddies (fig. 6.11).

Other evidence comes from the observations by Keffer (1965) of the development of a plane wake subjected to irrotational extension in the Oy direction, i.e. parallel to the axis of the cylinder producing the wake. Eddies of the form inferred from the other observations have circulation in the xOz plane and extension in the Oy direction will increase their energy. Keffer used smoke to make visible the turbulent fluid and observed that the periodic groups became larger and more distinct as the flow entered the distorting section.

Correlations of the intermittency function also show that the crests of the indentations lie at right angles to the flow direction, i.e. $R_\delta(r)$ is positive for separations in the Oy direction but takes negative values for separations in the stream direction. It is possible to detect periodicity in the stream direction from intermittency correlations, but the nature of the sampling process tends to obscure the presence of periodic groups.

6.14 Mechanism of the entrainment eddies

Grant's description of the entrainment eddies shows them to resemble the eddies that arise from the Helmholtz instability of a vortex sheet, and suggests that they derive their energy by direct interaction with the mean flow. If the effect of the main turbulent motion on the entrainment eddies can be approximated by a turbulent fluid of appropriate properties, a perturbation of the mean flow will be opposed by (Reynolds) stresses induced in the turbulent fluid by its motion and it is likely that perturbations will not grow if the turbulent intensity exceeds some critical value. The existence of the growth–decay cycle, with alternation between stability and instability to the growth of the eddies, suggests that the flow is always

9

near a condition of neutral stability for the growth of the eddies, the eddies growing at the end of an entrainment period as the intensity decays to below the critical value and decaying as the turbulent intensity increases sharply as a direct consequence of the entrainment. This *equilibrium hypothesis* provides a reason for the growth–decay cycle, but it needs elaboration before it can explain the different scales of the entrainment eddies in different flows.

During a quiescent period, the rate of entrainment is comparatively small and uniform, and the bounding surface is nearly a material surface in the flow for large-scale motions. Since the flow is irrotational in the free stream, any distortion of the surface generates a pressure distribution on it which may be calculated from the equations of potential flow, and stability of the perturbation causing the distortion depends on the relative magnitudes of the irrotational pressures and the pressures generated by flow in the turbulent fluid. The latter depend on the perturbation, on the mean velocity distribution and on the Reynolds stresses arising from the perturbation. It is likely that the initial Reynolds stresses are elastic in character, i.e. determined by the total additional strain rather than by the additional rate of strain, but the stability problem is an unusual and difficult one, even using an isotropic elastic modulus to represent the response of the turbulent fluid. The only simple solution is for a plane vortex sheet with uniform, irrotational and inviscid flow on one side and an elastic jelly moving with uniform velocity on the other, corresponding to a flow with a 'top-hat' velocity profile (Townsend 1966b). There, the interface is unstable if the velocity difference across the vortex sheet exceeds $1.79\, n^{1/2}$ (where n is the 'kinematic' shear modulus of the jelly), and the phase velocity of the neutrally stable disturbances differs from the free-stream velocity by $1.16\, n^{1/2}$.

The relation of the effective shear modulus to the turbulent intensity is known only for initially isotropic turbulence, for which the rapid-distortion theory gives a value of $\frac{2}{15}\overline{q^2}$. Calculations of the response of sheared turbulence to additional small distortion show it to be markedly anisotropic, but the stresses induced by any distortion other than further plane shear are much larger than for isotropic turbulence of the same intensity. It is possible that the effective value of the shear modulus is nearly $\frac{1}{3}\overline{q^2}$.

To use the 'jelly' results in a flow with a continuous distribution of velocity in the turbulent fluid, we suppose that surface distortions of wave number k have their origins and effective centres at a depth of order k^{-1}, so that their phase velocity is the mean velocity at that depth. Our reason is that the initial perturbations are caused by eddies of the turbulent motion with 'centres' moving at definite velocities. If they are too close to the surface, the difference of velocity from the free stream is small and the pressures induced by the irrotational flow are also small. If they are deeper than k^{-1}, they will not distort the surface to any serious extent. Then the condition for neutral stability of disturbances of this wave number may be guessed from the 'jelly' calculation to be

$$U_c{}^2 = [U_1 - U(\eta_0 l_0 - \alpha k^{-1})]^2 \simeq Cn(k) \qquad (6.14.1)$$

where $n(k)$ is the effective modulus for motion of wave number k, and C is a constant about one.

Since eddies of size larger than the wavelength of the perturbation are unlikely to contribute to the incremental Reynolds stresses, the value of $n(k)$ should be determined by the intensity of eddies of sizes less than k^{-1}. Using the Kolmogorov structure function to measure this intensity,

$$\begin{aligned} n(k) &= A'B_{11}(k^{-1}, 0, 0) \\ &\simeq A\overline{q^2}(kL_e)^{-2/3} \end{aligned} \qquad (6.14.2)$$

and instability occurs if

$$U_c(k) = |U(\eta_0 l_0 - \alpha k^{-1}) - U_1| > A^{1/2}(\overline{q^2})^{1/2}(kL_e)^{-1/3}. \qquad (6.14.3)$$

Inspection of fig. 6.12, which displays the relative magnitudes of the two sides of the inequality for several values of $q_0{}^2/u_0{}^2$, shows that the bounding surface is unstable for disturbances of wave numbers less than a critical value, k_c. The growth rate of a disturbance is zero for the critical wave number and it is small for very small wave numbers. So maximum growth rate occurs for a wave number which is comparable with but less than k_c, say k_m, and the dominant components produced by amplification of an initial disturbance of broad spectral distribution will all have wave numbers close to k_m. Then the indentations at the end of the growth period are expected to take the form of a wave group of wave number k_m, and to resemble the calculated form of fig. 6.13.

The growth phase ends when the slopes of the indentations are sufficiently large to cause overturning and engulfment, and each growth phase causes a general advance of the surface by a distance

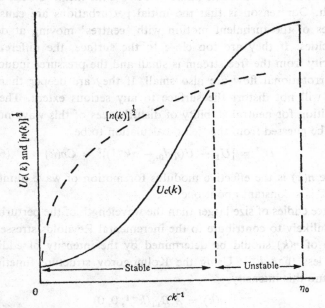

Fig. 6.12. Stability diagram for deformations of the interface, assuming elastic response of turbulent fluid (from Townsend 1966b).

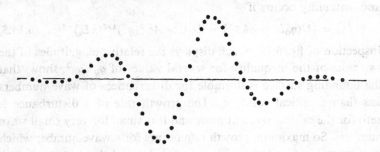

Fig. 6.13. Form of a wave group produced by finite growth from initial, aperiodic disturbances (from Townsend 1966b).

comparable with the wavelength of the indentations, i.e. with k_m^{-1}. The duration of the growth phase depends on the growth rate and on the amplification necessary to produce large indentations from the initial disturbances. The maximum growth rate is expected to be of order $k_m U_c(k_m)$ and the duration of order $[k_m U_c(k_m)]^{-1} \log \alpha_0$, where α_0 is the amplification ratio, so, supposing the growth period to occupy a fixed proportion of the whole cycle, the entrainment velocity is expected to be

$$u_c = A_c' U_c(k_m)/\log \alpha_0. \qquad (6.14.4)$$

The entrainment ratio may be written as

$$\frac{u_e}{q_0} = A_e(k_m L_e)^{-1/3}\overline{(q^2)}^{1/2}/q_0, \qquad (6.14.5)$$

where A_e is a 'constant' comparable in magnitude with $(\log \alpha_0)^{-1}$, and the intensity $\overline{q^2}$ should be a value taken near the centre of the disturbance. Finally, the depth of the indentations is of order k_m^{-1} and the relative depth of them should be related to the flow constant by

$$\frac{\sigma}{l_0} = \text{const}(k_m l_0)^{-1} = fn\left(\frac{q_0}{u_0} \text{ or } R_s\right). \qquad (6.14.6)$$

Even though the cyclic entrainment process is distinct only in flows of large entrainment ratios, the variation of σ/l_0 calculated from (6.14.3) and (6.14.6) is in quite good agreement with observations (Townsend 1966b, Newman 1967), and the variation of entrainment ratios between different flows implied by (6.14.3) and (6.14.5) is qualitatively consistent with observation and the dependence of the ratio on q_0^2/u_0^2 deduced from the similarity arguments.

6.15 Control of the entrainment rate

From the discussion of the mechanism of entrainment it would appear that the entrainment rate is determined by the relative turbulent intensity in the flow, but the similarity arguments of § 6.9 predict both the entrainment rates and the relative intensities. Both approaches predict that the flow constant R_s should be a function of q_0/u_0, but not the same functions. The inconsistency is not serious since (1) the predicted dependences are roughly the same and (2) quite small changes in the distributions of mean velocity and turbulent intensity, too small to affect appreciably the similarity calculations, could change the entrainment rates from solution of equations (6.14.3) and (6.14.5) by moderately large amounts.

Adjustment of the entrainment mechanism to produce a rate set by overall balances of energy and momentum might occur by the following process. Consider a flow with 'normal' distributions of mean velocity and turbulent intensity in which the entrainment eddies are too small to produce the entrainment rate dictated by the overall balances. Within the bounding surface, momentum transfer goes on at nearly the normal rate but the slowness of entrainment means that the momentum flux near the bounding surface is relatively small. Consequently, turbulent transfer of momentum tends to equalise mean velocity inside the turbulent flow and to produce velocity gradients near the surface that are stronger than normal. The modified velocity distribution is more unstable than the normal one so that entrainment eddies will grow faster and the quiescent stage of the entrainment cycle is likely to be shorter. The changes envisaged concern the gradients near the edge of the flow and are unlikely to change the values of the distribution integrals that occur in the energy balance equations of § 6.9.

Entrainment eddies are much less prominent in flows of small entrainment ratios, but it is possible that they exist and play a part in entrainment. The calculations indicate that their wavelengths in these flows would be fairly small, and their effect may be to produce small indentations on the larger-scale bulges induced by the main turbulent motion (fig. 6.8). In that case, it would be difficult to detect them, especially with sampling measurements, since they would have sizes comparable with eddies of main motion.

6.16 Fluctuations outside the turbulent flow: sound radiation

Although the flow outside the bounding surface is very nearly ir-rotational, it contains random fluctuations of velocity induced in it by the turbulence. It is convenient to consider the induced motion in two parts: (1) progressive sound waves radiated by the flow which dominate the motion at distances large compared with the wavelengths for frequencies characteristic of the flow, and (2) a 'near field' with fluctuations determined essentially by the instan-taneous turbulent velocities and confined to distances small com-pared with the wavelengths.

If u_0 and l_0 are characteristic scales of velocity and length, the wavelengths concerned are of order $a_0 l_0 / u_0$ which is large compared with the flow width if the Mach number of the velocity fluctuations u_0/a_0 is small. Sound radiation from turbulent flows has received considerable attention because of its importance for technology and environment, and the literature is considerable. The following brief account is based on work of Lighthill (1952a, b, 1962).

Consider a region of turbulent flow imbedded in an unbounded fluid that is at rest at infinity. From the equations for momentum and mass flux in a compressible fluid, it may be shown that the density satisfies the equation,

$$-a_0^2 \frac{\partial^2 \rho}{\partial x_i^2} - \frac{\partial^2 \rho}{\partial t^2} = \frac{\partial^2 T_{ij}}{\partial x_i \, \partial x_j} \qquad (6.16.1)$$

where

$$T_{ij} = \rho u_i u_j - p_{ij} + (p' - a_0^2 \rho') \delta_{ij}. \qquad (6.16.2)$$

Here, p' and ρ' are the differences of the pressure and density from their values at infinity, p_0 and ρ_0, p_{ij} is the viscous stress tensor, and a_0 is the velocity of sound. For unbounded fluid, the solution of (6.16.1) is

$$\rho'(r, t) = \frac{1}{4\pi a_0^2} \int \frac{[\partial^2 T_{ij}/\partial x_i \, \partial x_j]}{|r - x|} \, dV(x), \qquad (6.16.3)$$

where the square brackets mean that the quantity is to be evaluated at position x and the 'retarded' time $t - |r - x|/a_0$, and it describes the density fluctuation as the sum of contributions from sound waves radiated by quadrupole sources of strength per unit volume T_{ij}. Far from the flow, the small fluctuations of pressure and density are related by

$$p'/\rho' = a_0^2 = \gamma p_0/\rho_0 \qquad (6.16.4)$$

and $T_{ij} = \rho u_i u_j$ is negligible. It follows that the integration need be made only over the volume occupied by the turbulent flow and in its immediate neighbourhood.

Sound radiation from low speed flows is certainly weak, and it is reasonable to regard the integral in equation (6.16.3) as one over a specified velocity field. Using an origin near the centre of the flow, the value of $|r - x|$ may be approximated by

$$|r - x| = |r| + r.x/|r|$$

and the density fluctuation at large distances is

$$\rho'(r, t) = \frac{1}{4\pi a_0^2 r}(a_0 - V_l r_l/r)^{-2} \int \frac{\partial^2 T_{ij}}{\partial t^2} \frac{r_i r_j}{r^2} \, dV(\mathbf{x}) \qquad (6.16.5)$$

where the time derivative is measured in a frame moving with velocity $\mathbf{V}$ relative to the ambient fluid. The reason for introducing a translation velocity is that velocity patterns of turbulence usually move with well-defined convection velocities, and it is the time variation of T_{ij} in the convection frame that determines the sound radiation. If most of the sound is produced by eddies of size comparable with the width of the flow l_0 which change at rates of order u_0/l_0, the radiated sound comes from contributions from volumes of size about $l_0^3/(1 - M_c \cos \theta)$ with coherent, correlated motion. It follows that the mean square pressure fluctuation is

$$\overline{p'^2} = K' \frac{\rho_0^2 u_0^8 V_0}{a_0^4 l_0 r^2} (1 - M_c \cos \theta)^{-5} f(\theta), \qquad (6.16.6)$$

where θ is the angle between the convection velocity and the direction $\mathbf{r}$, $M_c = V/a_0$, and V_0 is the total volume of the turbulent flow. The angular distribution function $f(\theta)$ depends on the spatial distribution of the quadrupole sources, i.e. on the geometrical form of the turbulence.

Most of the sound from a high-speed jet comes from the cylindrical mixing layer that extends to several diameters from the exit, and it is found that the variation of power radiated is well described by

$$Q_s = K\rho_0 U_0^8 d^2/a_0^5, \qquad (6.16.7)$$

where U_0 and d are the exit velocity and diameter of the jet. Since the radial flux of sound energy is $\overline{p^2}/a_0$, this result is to be expected from the theory if the convection Mach number M_c is small. In fact, the power varies as U_0^8 up to exit velocities for which the effects of eddy convection should increase the radiation. It seems that the effects of Mach number described by equation (6.16.6) are balanced nearly by reductions in the relative turbulent intensities.

The measurements of sound power show that the energy loss in this way is very small compared with the loss by the normal processes of dissipation. In the circular jet, the volume of the turbulent flow responsible for most of the sound is roughly $3d^3$ and so the energy loss per unit volume has an average value of

$$\rho_0 \varepsilon_s = \tfrac{1}{3} K \rho_0 U_0^3 M^5/d.$$

In the whole region of the cylindrical mixing layer, the total dissipation is roughly one-third of the energy flux through the jet and an average value for the dissipation rate is

$$\rho_0 \varepsilon = (\pi/24)\rho_0 U_0^3/d.$$

Then the ratio of loss by sound radiation to the total is

$$\varepsilon_s/\varepsilon = 24KM^5/\pi. \tag{6.16.8}$$

Lighthill (1962) quotes values of the constant K from 3.10^{-5} in a low turbulence jet to 10^{-4} for a jet issuing from a long pipe, so that the ratio is of order $10^{-3} M^5$ ($M = U_0/a_0$), a very small quantity for any subsonic jet. So it is unlikely that the observed decreases of relative turbulent intensities and rates of spread as the flow Mach number increases are caused by sound radiation.

6.17 Irrotational fluctuations in the near field

In the near field, propagation delays are negligible and the flow is determined by the instantaneous pattern of the turbulent motion. Consider a two-dimensional mean flow nearly homogeneous in the Ox_1 and Ox_2 directions. Outside the turbulent flow, the motion is irrotational and may be described by a velocity potential, i.e.

$$\mathbf{u} = \operatorname{grad} \phi.$$

If the inducing motion in the turbulent flow travels with a convection velocity V_c that is large compared with the fluctuations of velocity, the potential satisfies the equation

$$\frac{\partial^2 \phi}{\partial x_1^2}(1 - M_c^2) + \frac{\partial^2 \phi}{\partial x_2^2} + \frac{\partial^2 \phi}{\partial x_3^2} = 0. \tag{6.17.1}$$

Solutions periodic in planes of homogeneity are

$$\phi = \exp(-\lambda x_3)\exp i(lx_1 + mx_3), \tag{6.17.2}$$

where

$$\lambda^2 = l^2(1 - M_c^2) + m^2 \tag{6.17.3}$$

and the irrotational motion for $x_3 > Z$ may be described by the spectrum function for ϕ in the plane $x_3 = Z$, say $\Psi(l, m)$ normalised so that

$$\iint_{-\infty}^{\infty} \Psi(l, m)\, dl\, dm = (\overline{\phi^2})_{x_3 = Z}.$$

It may then be shown that

$$
\begin{aligned}
\overline{u_1{}^2} &= (1 - M_c{}^2)^{-3/2} \int_0^\infty \int_0^{2\pi} \lambda^2 \cos^2\theta \, \Psi(l', m) \\
&\quad \times \exp[-2\lambda(x_3 - Z)] \, \lambda \, d\lambda \, d\theta, \\
\overline{u_2{}^2} &= (1 - M_c{}^2)^{-1/2} \int_0^\infty \int_0^{2\pi} \lambda^2 \sin^2\theta \, \Psi(l', m) \\
&\quad \times \exp[-2\lambda(x_3 - Z)] \, \lambda \, d\lambda \, d\theta, \\
\overline{u_3{}^2} &= (1 - M_c{}^2)^{-1/2} \int_0^\infty \int_0^{2\pi} \lambda^2 \Psi(l', m) \\
&\quad \times \exp[-2\lambda(x_3 - Z)] \, \lambda \, d\lambda \, d\theta,
\end{aligned}
\tag{6.17.4}
$$

where $l' = \lambda \cos\theta/(1 - M_c{}^2)$, $m = \lambda \sin\theta$. In particular, the intensities are related by

$$
\overline{u_1{}^2}(1 - M_c{}^2) + \overline{u_2{}^2} = \overline{u_3{}^2}.
\tag{6.17.5}
$$

For large values of $x_3 - Z$, much larger than the scale of the motion, $\Psi(l', m)$ may be approximated by its value for $l' = m = 0$, and then

$$
\begin{aligned}
\overline{u_1{}^2} &= \frac{\pi}{48}\Psi(0, 0)(1 - M_c{}^2)^{-3/2}(x_3 - Z)^{-4}, \\
\overline{u_2{}^2} &= \frac{\pi}{48}\Psi(0, 0)(1 - M_c{}^2)^{-1/2}(x_3 - Z)^{-4}, \\
\overline{u_3{}^2} &= \frac{\pi}{24}\Psi(0, 0)(1 - M_c{}^2)^{-1/2}(x_3 - Z)^{-4}.
\end{aligned}
\tag{6.17.6}
$$

Since the original work by Phillips (1955), the predictions (6.17.5) and (6.17.6) have been tested against observations in a number of different flows, with surprisingly good agreement even when no effort is made to separate the contributions of the turbulent flow (for example, in a boundary layer (Bradshaw 1967a), in a plane mixing-layer (Wygnanski & Fiedler 1970).

6.18 Development of nearly self-preserving flows

Self-preserving development is an asymptotic state attained only after considerable development, and any real flow deviates from it to some extent. There are three main reasons for deviation:

(1) The external flow conditions do not permit self-preserving flow.

(2) There has been insufficient development to erase memory of the initial conditions.

(3) The mean value equations cannot be satisfied *exactly* by self-preserving distributions. Supposing that the departures from self-preserving development are small, the deviation of the entrainment parameter from the asymptotic self-preserving value can be expressed by an equation of the form,

$$\frac{d\beta}{dX} + \frac{1}{L}\beta = F_e + F_u, \qquad (6.18.1)$$

where X is a suitably-defined function of x_1, L is an adjustment 'length', F_e is a forcing function dependent on the deviation of the external conditions from the ideal, and F_u is a forcing function dependent on the departure from exact satisfaction of the mean value equations (e.g. for the 'wake' family, the magnitude of u_0/U_1). Forms of the forcing functions can be found from the conservation equations and the behaviour of the asymptotic flow, and the remaining element is the adjustment length which describes the rate at which initial conditions are forgotten in ideal circumstances.

As an example, consider a two-dimensional jet issuing into still fluid. Assuming the distribution functions to be unchanged by the small deviation from the ideal, conservation of momentum requires that

$$u_0^2 l_0 = M \text{ (a constant)}, \qquad (6.18.2)$$

flow similarity and conservation of energy that

$$\frac{1}{2}\frac{d}{dx}\left[u_0^3 l_0\left(I_3 + \frac{q_0^2}{u_0^2}I_1\right)\right] = -\frac{\eta_0 l_0}{L_\varepsilon}q_0^3, \qquad (6.18.3)$$

and, from the acceleration at the flow centre,

$$\frac{\tau_m}{u_0^2} = -C_1\frac{l_0}{u_0}\frac{du_0}{dx}. \qquad (6.18.4)$$

Defining an entrainment parameter

$$\beta = -\frac{l_0}{u_0}\frac{du_0}{dx}, \quad \text{and} \quad X = \int\frac{dx}{l_0},$$

the energy equation becomes an equation for the entrainment parameter,

$$I_3 + A_1 I_1 \beta - A_1 I_1 \frac{1}{\beta} \frac{d\beta}{dX} = 2A_2 \beta^{1/2}. \tag{6.18.5}$$

For small deviations from the asymptotic solution, $\beta = \beta_0$,

$$\frac{d\beta}{dX} + \beta_0(\alpha - 1)(\beta - \beta_0) = 0 \tag{6.18.6}$$

with solutions,

$$\beta - \beta_0 = C \exp(-(\alpha - 1)\beta_0 X), \tag{6.18.7}$$

where

$$\alpha = \frac{A_2}{A_1 I_1 \beta_0^{1/2}}. \tag{6.18.8}$$

In terms of distance from some flow origin,

$$\beta - \beta_0 \propto x^{-(\alpha - 1)} \tag{6.18.9}$$

to the same approximation.

Next, consider a two-dimensional wake developing in a nearly uniform stream with small but not negligible velocity defect. Again assuming similarity of the distributions and using equations like (6.18.2, 6.18.3, 6.18.4), it is possible to show that the variation of entrainment parameter is described by equation (6.18.1) with

$$X = \int \frac{u_0 \, dx}{U_1 l_0}, \tag{6.18.10}$$

$$\beta = -\frac{l_0}{u_0} \frac{d}{dx}[u_0(U_1 + \tfrac{1}{2}u_0)], \tag{6.18.11}$$

$$L^{-1} = \alpha - 1 = \frac{A_2}{A_1 \eta_0 \beta_0} - 1, \tag{6.18.12}$$

$$F_e = (\alpha + 1)\frac{1}{U_1}\frac{dU_1}{dX} = (\alpha + 1)\frac{l_0}{u_0}\frac{dU_1}{dx},$$

$$F_u = \frac{u_0}{U_1}\frac{2\beta_0}{A_1 \eta_0}[I_3 + A_1 I_1 \beta_0 - (1 + I_2/I_1)A_2 \eta_0^{1/2}].$$

For small deviations from the asymptotic entrainment parameter, the solution is

$$\beta - \beta_0 = \text{const } x^{-\frac{1}{4}(\alpha-1)} - (\alpha+1) x^{-\frac{1}{4}(\alpha-1)}$$

$$\times \int^x x'^{\frac{1}{4}(\alpha-1)} \frac{1}{U_1} \frac{dU_1}{dx'} \, dx' + F_u/(\alpha-2) \qquad (6.18.13)$$

expressed in terms of real distance which is related to X by

$$x = e^{2\beta_0 X}. \qquad (6.18.14)$$

Substitution of observed values of the entrainment parameters shows that the rate of approach to asymptotic self-preserving flow is much slower in wakes than in jets, essentially because of the relatively larger turbulent intensity. In a jet, $\alpha - 1 = 3$ and an initial departure will be reduced to 1% of the original value while the jet width increases by a factor of three. In a wake, $\alpha - 1 = 0.75$ and the same reduction would require an increase by a factor of several hundred. It is interesting that the effect of initial conditions decays as $x^{-0.38}$ while the effect of finite u_0/U_1 falls off as $x^{-1/2}$, so that observations of the plane wake may not be representative of the asymptotic flow even after long development.

6.19 Development of a jet in a moving stream of constant velocity

If a jet is injected into an ambient flow of constant velocity, the mean velocity profiles are observed to remain very nearly similar in shape and their development may be described by the variation of scales of velocity and length with distance downstream. Unless the velocity excess at the jet centre is either very small or very large compared with the stream velocity, the mean value equations for momentum and turbulent energy cannot be satisfied even approximately by similarity distributions with the mean velocity scales. Leaving aside for the moment the question of the turbulent motion, let us assume similarity of the mean velocity profiles, i.e. that

$$U = U_1 + u_0 f(z/l_0).$$

The scales of velocity and length are then related by the condition for overall conservation of momentum,

$$\int_{-\infty}^{\infty} U(U - U_1) \, dz = u_0 U_1 l_0 [I_1 + I_2 u_0/U_1]$$

$$= U_1^2 D, \qquad (6.19.1)$$

where D is a scale of the whole flow.

A very simple approach is to assume that the entrainment 'constant',

$$\beta = \left(\frac{U_1}{u_0} + \frac{1}{2}\right)\frac{dl_0}{dx},$$

is the same at all sections, and then the momentum condition (6.19.1) leads to a differential equation for l_0,

$$\frac{1}{2}\frac{dl_0}{dx}\left[I_1\frac{l_0}{D} + \left(I_1{}^2\frac{l_0{}^2}{D^2} + 4I_2\frac{l_0}{D}\right)^{1/2}\right] = \beta, \qquad (6.19.2)$$

which can be integrated numerically. However, the development of the jet is between a state approximating to a jet in still surroundings to an asymptotic state of small velocity excess similar except in sign to the small-defect wake, and the entrainment constants for the two self-preserving flows differ by a factor of two.

A less restrictive assumption is that structural similarity of the turbulent motion is preserved during development with nearly universal profiles of turbulent intensity, Reynolds stress and dissipation rate, specified by the same length-scale but a separate velocity scale not a constant fraction of the mean velocity scale. Then the equation for the total kinetic energy of the flow may be used to derive a differential equation for the entrainment parameter,

$$\beta_1 = Mu_0{}^{-3}\,du_0/dx, \qquad (6.19.3)$$

where $M = \int_0^\infty U(U - U_1)\,dz$ is the constant momentum flux. Fig. 6.14 compares observations of the variation of (Bradbury & Riley 1967) with calculations using two alternative relations between the distributions of Reynolds stress and the turbulent velocity scales. In each case, the two constants are chosen to give the observed value of the entrainment parameter for a plane jet in still fluid. The first relation uses the central stress gradient in the same way as equation (6.9.9) to give

$$q_0{}^2 = -CU_1 l_0\,du_0/dx = Cl_0|\partial\tau/\partial z|_{z=0}. \qquad (6.19.4)$$

The second uses the stress integral in the form

$$q_0{}^2 = C'l_0{}^{-1}\int_0^\infty |\tau|\,dz = -C'\int_0^\infty\left(U\frac{\partial U}{\partial x} + W\frac{\partial U}{\partial z}\right)dz. \qquad (6.19.5)$$

It will be seen that the first assumption leads to a variation of β_1

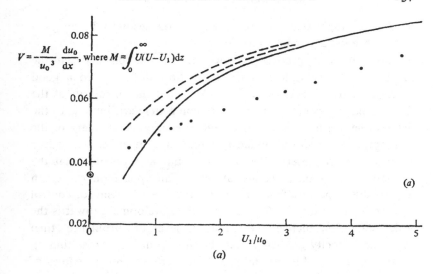

$$V = -\frac{M}{u_0^3}\frac{du_0}{dx}, \text{ where } M = \int_0^\infty U(U-U_1)dz$$

(a)

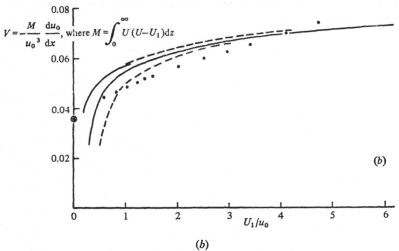

$$V = -\frac{M}{u_0^3}\frac{du_0}{dx}, \text{ where } M = \int_0^\infty U(U-U_1)dz$$

(b)

Fig. 6.14. Calculated and measured rates ($\bullet$, Bradbury & Riley 1967) of entrainment for a plane jet developing in a moving stream of uniform velocity. (a) Average turbulent intensity given by (6.19.4). (b) Average intensity given by (6.19.5), τ determined by

$$\frac{\partial \tau}{\partial z} = U\frac{\partial U}{\partial x} + W\frac{\partial U}{\partial z}$$

using $U = U_1 + u_0 \exp(-\tfrac{1}{2}z^2/l_0^2)$. (*Note:* The different theoretical curves are for different initial levels of turbulent intensity. The circled points are for jets issuing into still ambient fluid.)

much more rapid than that observed, but the second relation gives results in good agreement with observation.

Although an appropriate choice of the relation between the turbulent velocity scale and the mean flow does lead to a good description of the behaviour, it is likely that the 'memory' of the flow for past entrainment rates is not completely attributable to the delay in generating additional turbulent energy. A corollary of the assumption of structural similarity is that entrainment eddies appear as necessary to preserve the similarity, but it is possible that the specialised entrainment eddies of wakes can appear only after an appreciable departure from similarity. If the mechanism for control of entrainment rate of § 6.15 is valid, the development towards the wake state first increases the relative level of turbulence, then intensifies velocity gradients near the flow boundary, so setting up conditions favourable for the production of entrainment eddies of large scale, but these eddies only become effective at a considerable distance downstream.

CHAPTER 7

BOUNDARY LAYERS AND WALL JETS

7.1 Wall layers in general

Comparatively thin layers of turbulent flow are usually found on the surfaces of rigid bodies in moving streams, typically as boundary layers of retarded flow but also as wall jets produced by injection of high-velocity fluid. Such wall layers are bounded on one side by a nearly rigid wall and on the other by fluid in steady, and usually irrotational, motion, and naturally their properties are a mixture of the properties of free turbulence and of wall turbulence. The outermost part of the flow is expected to resemble closely free turbulence, with entrainment of non-turbulent fluid taking place across well-defined bounding surfaces and with properties of the turbulence strongly dependent on conditions far upstream of the section of observation. Close to the wall, the conditions for existence of an equilibrium layer are usually met and the flow there is similar in most important respects to that in other equilibrium layers, the mean flow being determined by the local stress distribution and the surface roughness. By its nature, an equilibrium layer can form only a small part of a complete flow and most of the flow is free turbulence comparable with that in jets and wakes.

As for channel flows, characteristic distributions of eddy viscosity across the wall layers show a rapid increase with distance from the wall, followed by nearly constant values to the outer edge of the flow. It is useful to take the region of rapid variation as the inner, equilibrium layer, and the remainder as an outer layer of free turbulence (fig. 7.1). In many flows and particularly in self-preserving flows, the observed distributions of mean velocity change almost abruptly from equilibrium-layer forms to constant eddy-viscosity forms typical of free turbulence, and the two-layer model provides a useful method for discussing the distributions of mean velocity and Reynolds stress. In most layers, the inner layer occupies between one-tenth and one-fifth of the total thickness, and its principal function is to establish at its outer edge inner boundary conditions of velocity and shear stress for the outer flow.

For simplicity, most of the following account assumes that the layers are formed on plane surfaces and are homogeneous in directions transverse to the flow and parallel to the surface. Effects of surface curvature and of lateral inhomogeneity are large in some flows, but many practical wall layers have properties similar to the simple two-dimensional flows.

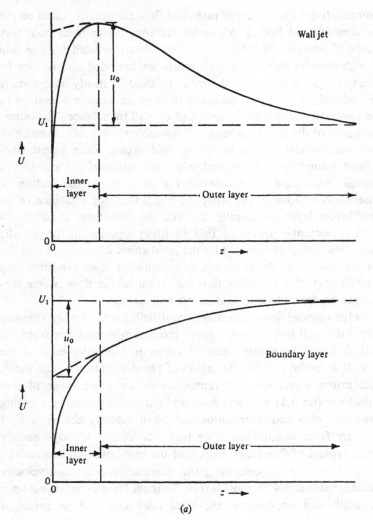

(a)

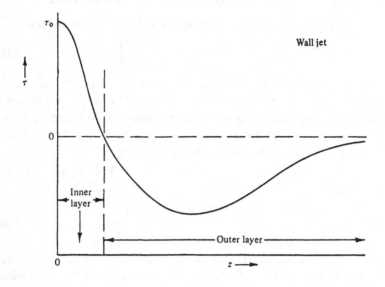

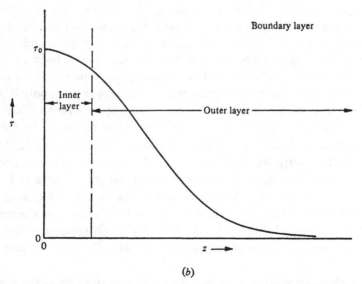

(b)

Fig. 7.1. Profiles of mean velocity and shear stress for wall jets and boundary layers.

7.2   Self-preserving development of wall layers

Self-preserving development of a wall layer with the distributions,

$$U = U_1 + u_0 f(z/l_0),$$
$$\tau = -\overline{uw} = -u_0{}^2 g_{13}(z/l_0), \text{ etc.,} \Bigg\} \qquad (7.2.1)$$

may be possible if the Reynolds equations for mean velocity and turbulent energy can be satisfied with universal distribution functions and scales dependent only on the downstream co-ordinate x. Then the arguments of § 6.4 show that necessary conditions are that the free stream velocity shall have one of the forms,

$$U_1 \propto (x - x_0)^a \; (x > x_0), \quad \text{or} \quad U_1 \propto (x_0 - x)^a \; (x < x_0), \quad (7.2.2)$$

that either

$$l_0 \propto (x - x_0) \quad \text{or} \quad l_0 \propto (x_0 - x) \Bigg\}$$
and that $\qquad\qquad\qquad\qquad\qquad\qquad\qquad\qquad (7.2.3)$
$$u_0/U_1 = \text{constant}\dagger$$

Notice that the mean velocity distributions are of two kinds, one with a limit of zero or infinite velocity upstream and the other with a limit downstream.

The conditions are derived by assuming that the viscous terms in the mean value equations are negligible and that the boundary conditions to the flow have self-preserving forms. Sufficiently close to a smooth wall, the viscous terms are not negligible and strict self-preserving development is possible in the viscous layer only if the exponent a is equal to -1, so that $u_0 l_0/\nu = \text{constant}$. If the wall is rough, the boundary conditions have the proper form only if the local scale of the roughness elements is proportional to distance from the flow origin at $x = x_0$, i.e. to the layer thickness. In either case, the restriction is apparently strong, but the part of the flow affected directly by viscosity or by flow around the roughness elements is very small if the flow Reynolds number is large and the scale of the roughness small compared with the layer thickness. If the motion in the viscous layer or around the roughness elements allows mean velocities and stresses of the self-preserving forms, self-preserving flow may be possible over the fully turbulent part of the flow.

Unless the wall stress is very small, the shear stress at the inner

---

† The ratio u_0/U_1 is never small for wall layers with flow Reynolds numbers attainable in practice, and it is unnecessary to consider small-defect flows.

boundary of the equilibrium layer, i.e. at the outer edge of the viscous or 'roughness' layer, is nearly equal to the wall stress, and the local velocity distribution has the logarithmic form,

$$U = (\tau_0^{1/2}/k) \log(z/z_0). \tag{7.2.4}$$

To form part of a self-preserving distribution, the distribution function for velocity must vary logarithmically for small values of $\eta = z/l_0$ with the form,

$$f(\eta) = \gamma(\log \eta + C), \tag{7.2.5}$$

where $\gamma = \tau_0^{1/2}/(ku_0)$ and C are constants of the flow. Then it is necessary that

$$U_1/u_0 = \gamma[\log(l_0/z_0) - C] \tag{7.2.6}$$

showing that the conditions (7.2.3) can be met only if the effective roughness length is the same fraction of the layer thickness at all sections.

As a matter of practice, the mean value equations are closely self-preserving in form with a power-law distribution of mean velocity if both dl_0/dx and U_1/u_0 change slowly during development. Since the ratio l_0/z_0 is always large (at least 10^4), large changes in the ratio change the velocity ratio by comparatively small amounts, and flows with, for example, constant roughness length can develop with distributions of velocity and stress that are very nearly those for a hypothetical layer with constant ratio of l_0/z_0 equal to the current value (see § 7.6).

7.3   General properties of self-preserving wall layers

In a typical wall layer, mean velocity gradients are considerably smaller in the outer layer than in most of the inner equilibrium layer, and it is possible without much ambiguity to extrapolate the outer velocity profile to the wall and to define a velocity scale u_0 as the difference between the extrapolated velocity and the velocity of the free stream. The length scale l_0 might be defined as the variance of the velocity distribution, but, unlike free jets and wakes, the outer profiles of wall layers do not always approximate to the error-function form. For example, the outer profile of a boundary layer in a uniform free stream resembles closely the error integral.

In the following account of self-preserving flows, the scales are defined by fitting assumed velocity distributions to the actual profiles. For wall jets, which have a velocity maximum not far from the wall, the assumed distribution is

$$U = U_1 + u_0 \exp\{-\tfrac{1}{2}[z/l_0 - \eta_1]\}^2. \tag{7.3.1}$$

The maximum velocity is $(U_1 + u_0)$ and occurs at $z = \eta_1 l_0$. The length scale is chosen to give the best fit to the outer profile. For boundary layers, the assumed distribution is

$$U = U_1 + u_0\left[\sigma\, e^{-\frac{1}{2}\eta^2} + \left(\frac{2}{\pi}\right)^{1/2}(1-\sigma)\int_\eta^\infty e^{-\frac{1}{2}t^2}\, dt\right] \tag{7.3.2}$$

and the scales and the shape parameter σ are chosen for best fit to the outer profile.†

Although the velocity change in the inner layer may be large, the layer is so thin that the values of the profile integrals,

$$I_n = u_0^{-n}\int_0^\infty (U - U_1)^n\, dz/l_0 \tag{7.3.3}$$

are only slightly different from values calculated from the outer profile and its extrapolation to the wall. Then I_1 is greater than I_2 and the choice of scales means that I_1 is not far from 1.0. In terms of the friction coefficient, $c_f = \tau_0/(U_1 + u_0)^2$, and the entrainment parameter, $\beta = |U_1/u_0 + \tfrac{1}{2}|\, dl_0/dx$, the equation for overall conservation of momentum in the layer,

$$\frac{d}{dx}\int_0^\infty U(U - U_1)\, dz + \frac{dU_1}{dx}\int_0^\infty (U - U_1)\, dz = -\tau_0, \tag{7.3.4}$$

becomes a relation between the exponent for the variation of free stream velocity (or pressure gradient) and the *velocity ratio*, $V = U_1/u_0$,

$$a = -\frac{I_1 V + I_2}{3I_1 V + 2I_2} - \frac{c_f}{\beta}\frac{|V + \tfrac{1}{2}|(V+1)^2}{3I_1 V + 2I_2}. \tag{7.3.5}$$

The friction coefficient is always small, not exceeding 0.002, and the entrainment parameter is commonly in the range 0.02 to 0.04, so that the ratio c_f/β is fairly small.

It follows that the second term of (7.3.5) is small compared with the first unless the velocity ratio V is large, and then the relation

---

† The assumed distribution is an interpolation between forms known to be good approximations for flow with negligible stress in strong adverse pressure gradients and for flow in a uniform free stream. It may not be adequate for flow in strong favourable pressure gradients.

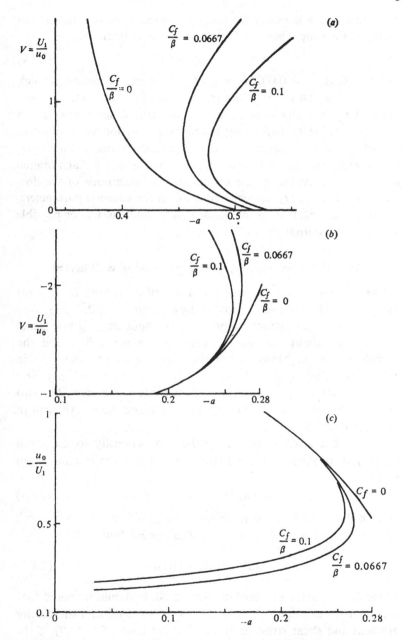

Fig. 7.2. Dependence of velocity ratio on exponent of stream velocity variation for (a) self-preserving wall jets and (b) & (c) boundary layers, assuming constant ratio of friction coefficient to entrainment constant.

between a and V is nearly the same as for free turbulent flows. For example, the simple wall jet in still air has $V = 0$, and so

$$a = -\tfrac{1}{2} - c_f/(4I_2\beta). \qquad (7.3.6)$$

For $\beta = 0.03$, $c_f = 0.002$, $I_2 = 0.89$ (for an error function profile), $a = -0.52$ which is very close to the value for a free jet, $a = -\tfrac{1}{2}$.

For large velocity ratios, the second term is large and, for a particular value of c_f/β, a expressed as a function of V has two branches, one for boundary layers with negative ratios and $a > -\tfrac{1}{3}$, the other for wall jets with positive ratios and $a < -\tfrac{1}{3}$. Each branch has an extreme value for the exponent and, since none of the flow constants, β, c_f, I_1, I_2, depend critically on the external parameters, a and $(x - x_0)/z_0$, two distinct kinds of development may be possible in the same external conditions (fig. 7.2).

7.4 Flow parameters of self-preserving wall layers

While the length and velocity scales of self-preserving flows must vary with fetch in the ways specified by equations (7.2.2, 7.2.3), their actual magnitudes cannot be predicted without making additional assumptions about the relation between the mean flow and the turbulent motion. Many assumptions have been proposed, all in effect providing a second relation between mean velocity and Reynolds stress with one or more disposable parameters, but two are specially relevant to the concepts of turbulent flow developed in previous chapters.

Mellor & Gibson (1966) relate the eddy viscosity to the mean velocity distribution. In the inner layer, it has the mixing-length form,

$$v_T = \tau/(\partial U/\partial z) = k^2 z^2 \, \partial U/\partial z \qquad (7.4.1)$$

and, in the outer layer, it is independent of distance from the wall and has the magnitude given by the Clauser relation,

$$v_T = R_0^{-1} \int_0^\infty |U - U_1| \, dz, \qquad (7.4.2)$$

where R_0 is a universal constant around 50. It should be noted that the inner distribution is simply the relation between mean velocity gradient and shear stress in an equilibrium layer of (5.14.9), if the diffusion parameter B is zero.

Bradshaw *et al.* (1967) assume (1) a constant ratio of Reynolds stress to turbulent intensity, (2) a universal distribution of the dissipation length scale in terms of the total thickness of the layer, and (3) a universal distribution for the lateral flux of turbulent energy across the layer. The assumptions allow the equation for the turbulent kinetic energy to be expressed as a relation between shear stress and mean velocity, and it becomes the necessary second relation. Like the Mellor & Gibson assumptions, the equations take the equilibrium layer form near the wall.

For self-preserving flows, both methods lead to similar results that are in good agreement with observation but the profiles must be obtained by numerical integration of the equations. To survey the whole range of wall layers, detailed calculation of the distribution functions may be avoided by assuming explicit forms, and the flow constants may then be found by requiring overall conservation of momentum, conformity with the properties of the equilibrium layer, and by making assumptions that are nearly equivalent to those of the two methods.

It will be assumed that the mean velocity distribution has an equilibrium layer form for $z < \eta_1 l_0$, where η_1 is a flow constant marking the boundary between the inner and outer layers, and an outer profile of the forms (7.3.1, 7.3.2). The boundary conditions are that mean velocity and Reynolds stress are continuous, the condition for stress continuity being found by requiring overall conservation of momentum in each layer. In the outer layer, integration of the equation of mean motion from the junction to the free stream gives

$$\frac{dl_0}{dx}\left[(3a+1)I_1'V + (2a+1)I_2' + (a+1)\left(\eta_1 V + \int_0^{\eta_1} f \, d\eta\right)f(\eta_1)\right]$$
$$= g_{13}(\eta_1) = -\tau_1/u_0^2, \qquad (7.4.3)$$

where $I_n' = \int_{\eta_1}^{\infty} [f(\eta)]^n \, d\eta$. Integration from the wall to the junction gives

$$\frac{dl_0}{dx}\left[(3a+1)V\int_0^{\eta_1} f \, d\eta + (2a+1)\int_0^{\eta_1} f^2 \, d\eta\right.$$
$$\left. -(a+1)\left(\eta_1 V + \int_0^{\eta_1} f \, d\eta\right)f(\eta_1)\right] = (\tau_1 - \tau_0)/u_0^2. \qquad (7.4.4)$$

To complete the system of equations, an assumption must be made about flow in the outer layer. One that resembles the Mellor & Gibson assumptions is to impose boundary and integral conditions that would be satisfied if the eddy viscosity were constant and given by the Clauser relation. One that is closer to the Bradshaw, Ferriss & Atwell method is to assume structural similarity of the turbulent motion and to use the equation for the total kinetic energy in the same way as for free turbulent flows. Examples of both approaches will be found in the following sections.

7.5   Development of self-preserving wall jets

In a wall jet, the position of zero Reynolds stress coincides very nearly with the maximum in the mean velocity profile, and it is possible and convenient to choose that position for the junction between the two layers. Suitable velocity profiles are

$$
\left.
\begin{aligned}
U &= U_1 + u_0 \exp[-\tfrac{1}{2}(\eta - \eta_1)^2] \quad \text{for } \eta > \eta_1, \\
U &= \frac{\tau_0^{1/2}}{k}[\log(z/z_0) - z/(\eta_1 l_0)] \quad \text{for } \eta < \eta_1.
\end{aligned}
\right\} \tag{7.5.1}
$$

For continuity of velocity at the junction,

$$
V + 1 = \gamma[\log(\eta_1 l_0/z_0) - 1] \tag{7.5.2}
$$

where $\gamma = \tau_0^{1/2}/(ku_0)$ and $V = U_1/u_0$. Reynolds stress and velocity gradient are both zero at the junction (the inner profile is one for linear variation of $\tau^{1/2}$ in an equilibrium layer), and the two equations for conservation of momentum (7.4.3, 7.4.4) become

$$
V\left[(3a+1)\int_0^{\eta_1} f\,d\eta - (a+1)\eta_1\right] + (2a+1)\int_0^{\eta_1} f^2\,d\eta
$$
$$
-(a+1)\int_0^{\eta_1} f\,d\eta = -\frac{k^2\gamma^2}{dl_0/dx} \tag{7.5.3}
$$

and

$$
V[(3a+1)I_1' + (a+1)\eta_1] + (2a+1)I_2' + (a+1)\int_0^{\eta_1} f\,d\eta = 0. \tag{7.5.4}
$$

If it is assumed that the eddy viscosity is constant in the outer layer, the maximum stress is given by

$$
\frac{\tau_m}{u_0^2} = 2|a|C\beta = \frac{\nu_T}{u_0 l_0}C, \tag{7.5.5}
$$

where C is expected to be near $e^{-1/2}$ (see § 6.9), and the energy equation for the outer layer is

$$\frac{5a+1}{2}I_2'V + \frac{3a+1}{2}I_3' + \frac{a+1}{2}\left(\eta_1 V + \int_0^{\eta_1} f\,d\eta\right)$$

$$+ A_1|a|\beta\left[(3a+1)(\eta_0 V + I_1') + (a+1)\left(V\eta_1 + \int_0^{\eta_1} f\,d\eta\right)\right]$$

$$= -A_2(2|a|)^{3/2}(V + \tfrac{1}{2})\beta^{1/2} \quad (7.5.6)$$

where A_1, A_2 and η_0 have the same meaning as for free jets. The assumption of structural similarity is in all respects similar, and equations (7.5.6) and (6.9.10) differ only by the presence of additional terms that describe the advective transport of momentum and kinetic energy across the surface of zero stress.

The set of equations (7.5.2–7.5.4, 7.5.6) can be solved numerically for the flow constants, V, β, γ and η_1, as functions of the external flow parameters, a and $z_0/(x - x_0)$, with results shown in fig. 7.3. One set of results has been calculated for the same values of A_1, A_2 and η_0 that were used for free jets, and the predicted rates of spread are considerably less than for free jets. Even so, they are nearly one-third larger than the rates measured by Patel & Newman (1961).[†] The discrepancy can be removed by increasing A_2 from 2.02 to 2.3, a change that could depend on two effects. First, the presence of a solid boundary will restrict eddies of the outer layer to sizes that are somewhat less than in a free jet of the same width, so causing a reduction in the dissipation length scale. Next, a smaller entrainment constant implies a larger effective strain of the turbulence and a somewhat smaller ratio of Reynolds stress to turbulent intensity. Both effects lead to an increase in the value of A_2 (the second also increases A_1 but the relevant term in (7.5.6) is small), possibly to the value that brings the calculations into fair agreement with the measurements.

Although the Clauser condition has not been used, the 'constant' R_0 can be calculated from equation (7.5.5) to be

$$R_0 = \frac{1}{v_T}\int_0^\infty (U - U_1)\,dz = \frac{I_1}{2|a|\beta}. \quad (7.5.7)$$

† The ratio of roughness length to fetch is not constant for these wall jets on smooth surfaces, but the effect on the entrainment constants is very small.

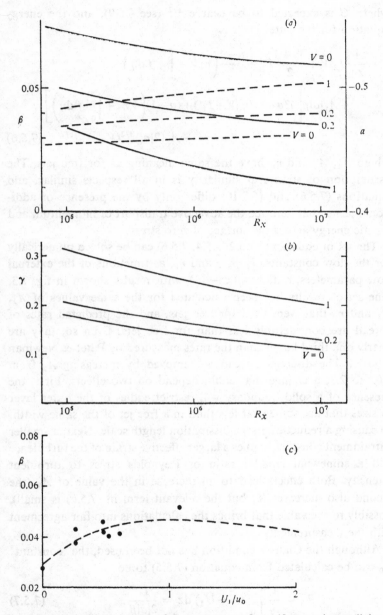

Fig. 7.3 Calculated variations of flow constants for self-preserving wall jets. (a) Entrainment constant and exponent as a function of Reynolds number for various velocity ratios. (b) Friction parameter as a function of Reynolds number for various velocity ratios. (c) Entrainment constant as a function of velocity ratio for a Reynolds number of 5×10^5 ($\bullet$, measurements from Gartshore 1966: ——, a; --- β.)

The values of R_0 so found do not change greatly over the whole range of flow conditions. For ratios of fetch to roughness length in the range of laboratory observations, R_0 lies in the range 65–75 for all possible velocity ratios, but it decreases slowly with increase of the ratio $(x - x_0)/z_0$, i.e. with increase of Reynolds number. Very similar values of the flow parameters would be found using the Clauser condition instead of the structural similarity assumption and the energy balance.

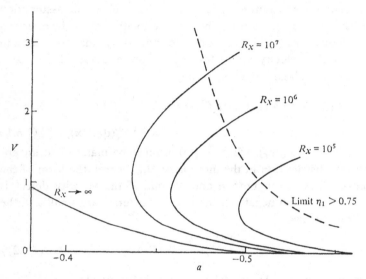

Fig. 7.4. Calculated relation between velocity ratio and exponent of velocity variation for various Reynolds numbers.

It is necessary that the inner layer should be considerably smaller in thickness than the outer layer in order that it may satisfy the conditions to be an equilibrium layer. Fig. 7.4 shows the dependence of the velocity ratio, $V = U_1/u_0$, on the exponent for fixed values of $(x - x_0)/z_0$, indicating the limits set by this requirement. It appears that two kinds of self-preserving development may be possible with the same external parameters, one with a small velocity ratio and a large angle of spread and one with a large ratio but a smaller angle of spread. In laboratory conditions, the two kinds of flow might be observable in free streams with exponents rather less negative than $-\frac{1}{2}$.

7.6 Development of self-preserving boundary layers

The velocity and stress distributions in wall jets are nearly similar in the outer flow, essentially because of the restraint imposed by the existence of maximum velocity and zero stress at the inner boundary. In boundary layers, there is no velocity maximum and the variations in shape of the distributions cannot be ignored. Clauser (1954) was the first to show that the outer velocity profiles of self-preserving flows resemble closely profiles calculated from the equation of mean flow assuming constant eddy viscosity. The result of the assumption is to change the equation for the mean velocity in its self-preserving form (equation (6.4.3) ignoring normal Reynolds stresses and equating the velocity scales for mean flow and turbulence) to a form of the Falkner–Skan equation,

$$V[2af - (a+1)\eta f'] + af^2 - (a+1)f' \int_0^\eta f \, d\eta$$
$$= -R_s^{-1} f''/(dl_0/dx), \qquad (7.6.1)$$

(where $R_s = |u_0| l_0 / \nu_T$), whose solution must be matched to an appropriate distribution in the inner layer. In general, the form of the solution depends on both a and V, but, if the velocity defect is small (V large and negative), the non-linear terms are small and the solutions are nearly

$$f(\eta) = \frac{2^{n/2}(n/2)!}{\sqrt{(\frac{1}{2}\pi)}} Hh_n(R\eta), \qquad (7.6.2)\dagger$$

where $n = 2a/(a + 1)$ and $R^2 = -(a + 1)R_s V \, dl_0/dx$.

If the velocity defect is not small, a solution of the Falkner–Skan equation (7.6.1) can be approximated by one of the small-defect solutions (7.6.2), but not usually the solution for the same value of the exponent a. Most interest is attached to boundary layers in zero or adverse pressure gradients ($a \leqslant 0$) for which the extreme small-defect solutions are (with an obvious choice of length scale)

$$f(\eta) = (2/\pi)^{1/2} Hh_0(\eta) = (2/\pi)^{1/2} \int_\eta^\infty \exp(-\tfrac{1}{2}t^2) \, dt \qquad (7.6.3)$$

for zero pressure gradient ($a = 0$ and $n = 0$), and

$$f(\eta) = Hh_{-1}(\eta) = \exp(-\tfrac{1}{2}\eta^2) \qquad (7.6.4)$$

† $Hh_n(x)$ is the solution of Hermite's differential equation, $y'' + xy' - ny = 0$, that behaves as $x^{-(n+1)} \exp(-\tfrac{1}{2}x^2)$ for large positive values of x.

for flow with zero stress at the wall ($a = -\frac{1}{3}$ and $n = -1$). An approximation to the intermediate forms is given by

$$f(\eta) = \sigma \exp(-\tfrac{1}{2}\eta^2) + (1-\sigma)(2/\pi)^{1/2} \int_\eta^\infty \exp(-\tfrac{1}{2}t^2)\,dt \qquad (7.6.5)$$

where σ lies in the range 0–1 and should be chosen to make the assumed distribution resemble as closely as possible a solution of the Falkner–Skan equation.

The magnitudes of the Reynolds stresses in the inner layer, though not necessarily their local gradients, may be described by a linear distribution,

$$\tau = \tau_0 + \alpha z \qquad (7.6.6)$$

and, if the Reynolds stress just within the fully turbulent flow is very nearly equal to the wall stress τ_0 (possible if $\tau_0 > 200\alpha z_0$), the velocity distribution is the equilibrium distribution

$$U = \frac{\tau_0^{1/2}}{k} \log\left[\frac{(\tau_0+\alpha z)^{1/2}-\tau_0^{1/2}}{(\tau_0+\alpha z)^{1/2}+\tau_0^{1/2}}\frac{4\alpha z_0}{\tau_0}\right]$$
$$+ \frac{2(1-B)}{k}[(\tau_0+\alpha z)^{1/2}-\tau_0^{1/2}]. \qquad (7.6.7)$$

Given the velocity distribution, the stress difference across the inner layer could be found from the momentum condition (7.4.4) and the average stress gradient α equated to the difference divided by the thickness. It is simpler and hardly less accurate to use

$$\alpha = (U_1 + u_0)\,d(U_1 + u_0)/dx - U_1\,dU_1/dx \qquad (7.6.8)$$

which supposes the flow to be nearly parallel to the wall with an effective average velocity of $(U_1 + u_0)$. It follows that the non-dimensional stress gradient is related to the entrainment constant by

$$\alpha l_0/u_0^2 = -2a\beta. \qquad (7.6.9)$$

The composite velocity distribution must satisfy the condition of overall conservation of momentum expressed by (7.4.3). Then,

$$(3a+1)I_1'V + (2a+1)I_2' + (a+1)f(\eta_1)\left(V\eta_1 + \int_0^{\eta_1} f\,d\eta\right)$$
$$= -\left(\frac{\tau_0}{u_0^2} + \frac{\alpha l_0}{u_0^2}\eta_1\right)\bigg/\frac{dl_0}{dx}. \qquad (7.6.10)$$

Next, the outer velocity profile must have a value of the shape parameter σ that makes it a good approximation to the profile for constant eddy viscosity. This may be done by requiring that the integral over the outer layer of the Reynolds stresses calculated from the profile and the equation of mean flow should have the same value as for an exact solution of (7.6.1), i.e. that

$$\int_{\eta_1 l_0}^{\infty} \tau \, dz = -u_0 v_T f(\eta_1). \tag{7.6.11}$$

The condition is that

$$(4a+2)I_{1a}V + (2a+1)I_{2a} + (a+1)\eta_1 f(\eta_1)\left(V\eta_1 + \int_0^{\eta_1} f \, d\eta\right) +$$

$$+ \tfrac{1}{2}(a+1)I_1'^2 = -\eta_1 \frac{\tau(\eta_1)}{u_0^2} - R_s^{-1} f(\eta_1) \tag{7.6.12}$$

where $I_{na} = \displaystyle\int_{\eta_1}^{\infty} [f(\eta)]^n \, \eta \, d\eta.$

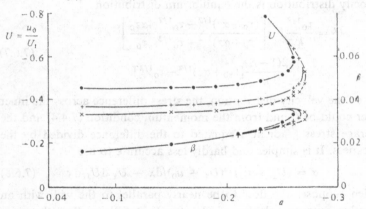

Fig. 7.5. Calculated variations of velocity ratio and entrainment constant for self-preserving boundary layers, as a function of exponent for various Reynolds numbers. x/z_0: ●, 10^5; +, 10^6; ×, 10^7.

Further conditions are (1) that mean velocity is continuous at the junction between the layers, i.e.

$$V + f(\eta_1) = \gamma\left[\log\left(\frac{X-1}{X+1}\frac{4\tau_0}{\alpha z_0}\right) + 2(1-B)(X-1)\right], \tag{7.6.13}$$

where $X^2 = 1 + \alpha l_0 \eta_1 / \tau_0$, (2) that stress at the junction calculated with the outer layer eddy viscosity (given by (7.6.12)) should agree with stress calculated from flow in the inner layer, i.e.

$$R_s^{-1} f'(\eta_1) = -\tau(\eta_1)/u_0^2 \tag{7.6.14}$$

and (3) that the junction should be smooth without change of gradient, i.e.

$$f'(\eta_1) = -\left(\frac{\tau(\eta_1)}{u_0^2}\right)^{1/2}\frac{1}{k\eta_1}\left(1 - B\frac{\alpha l_0\eta_1}{\tau(\eta_1)}\right). \qquad (7.6.15)$$

There are now five equations for the six flow parameters, β, γ, V, R_s, σ and η_1, of a flow developing in external conditions defined by a and $(x - x_0)/z_0$.

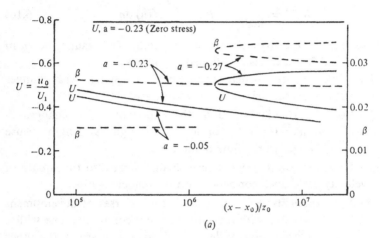

(a)

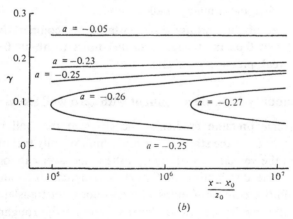

(b)

Fig. 7.6. Calculated variations of flow constants for self-preserving boundary layers. (a) Entrainment constant and velocity ratio as a function of Reynolds number for various exponents of velocity variation. (b) Friction parameter as a function of Reynolds number for various exponents.

10

Either the Clauser relation between eddy viscosity and velocity profile or the assumption of structural similarity might be used to complete the set of equations, but the relation between maximum Reynolds stress and velocity scale depends on the shape of the distribution and the method of structural similarity is difficult to apply. It is simpler to use a Clauser condition based on the outer profile and its extrapolation to the wall, putting

$$R_s^{-1} = \frac{v_T}{|u_0|l_0} = R_0^{-1} \int_0^\infty f(\eta)\,\mathrm{d}\eta, \qquad (7.6.16)$$

where $f(\eta)$ is the distribution function (7.6.5). The results shown in figs. 7.5 and 7.6 have been obtained in this way.

The main features, which are not likely to change with changes in the details of the various assumptions, are:

(1) For a particular value of the exponent, large changes in Reynolds number (or ratio of fetch to roughness length) cause only small changes in the flow parameters.

(2) Unless the wall stress is very small, its ratio to the square of the velocity scale does not depend on the velocity ratio.

(3) For exponents less than -0.25, two courses of development appear to be possible with the same external conditions, one with a comparatively large velocity defect and large wall stress, the other with a larger defect and a much smaller stress.

(4) The values of the entrainment constants are comparatively small, from about 0.03 in strongly retarded flows to about 0.015 in flows with weak pressure gradients.

7.7 Boundary-layer development with zero wall stress

In strong adverse pressure gradients, the stress at the wall may differ appreciably from the stress at the inner limit of fully turbulent flow, and then the velocity distribution in the inner layer cannot be represented by the distribution of (7.6.7). The actual distribution includes an additive constant (representing a velocity of translation) that depends on flow in the viscous layer or around the roughness elements, but it only appears in the condition for velocity continuity at the junction and the dependence of β, γ, σ on a and V is hardly affected since (except for the very small contribution of the viscous layer to the I_n) the governing equations refer only to the fully turbulent flow.

The self-preserving flow with zero wall stress has been studied by Stratford (1959a, b) and is the limiting member of the low-stress flows. With zero wall stress, the velocity distribution in an equilibrium layer over a smooth surface has the form,

$$U = \frac{2}{k_0}(\alpha z)^{1/2} + C(\alpha v)^{1/3}, \qquad (7.7.1)$$

where α is the stress gradient, and $k_0 = k/(1 - B)$ and C are constants. The second term is usually small, and an immediate consequence is that the equilibrium layer and the whole flow has a structure that is independent of Reynolds number.

Using values of the Clauser constant determined from measurements in constant-pressure boundary layers, good agreement has been obtained with the measurements of Stratford, both by using an assumed velocity profile in the outer layer of the form,

$$U = U_1 + u_0 \exp(-\tfrac{1}{2}\eta_1^2) \qquad (7.7.2)$$

matched to the inner profile by conditions of continuity of velocity and eddy viscosity (Townsend 1960) and by the method of Mellor & Gibson (1966).

The zero-stress layer is an extreme member of the boundary-layer family and it is interesting to carry through the calculation of the flow constants, using the energy equation and assuming similarity of the turbulent motion. The conditions for continuity of velocity and velocity gradient at the junction are

and

$$\left.\begin{aligned} V + \exp(-\tfrac{1}{2}\eta_1^2) &= -\frac{2}{k_0}X\eta_1^{1/2}, \\[2mm] \eta_1 \exp(-\tfrac{1}{2}\eta_1^2) &= \frac{1}{k_0}X\eta_1^{-1/2}, \end{aligned}\right\} \qquad (7.7.3)$$

where

$$X^2 = \frac{\alpha l_0}{u_0^2} = 2|a|\beta \qquad (7.7.4)$$

to satisfy momentum conservation in the inner layer. Overall conservation of momentum requires that

$$(3a+1)I_1V+(2a+1)I_2 = 0 \tag{7.7.5}$$

and the equation for the total kinetic energy becomes

$$\frac{5a+1}{2}I_2'V+\frac{3a+1}{2}I_3'+\frac{(a+1)}{3}\eta_1(V+\mathrm{e}^{-\frac{1}{2}\eta_1^2})\mathrm{e}^{-\eta_1^2}+2a(V+\tfrac{1}{2})\eta_1\mathrm{e}^{-\frac{1}{2}\eta_1^2}$$

$$+A_1\left\{\frac{3a+1}{2}[(\eta_0-\eta_1)V+I_1']+\frac{(a+1)}{3}\eta_1(V+\mathrm{e}^{-\frac{1}{2}\eta_1^2})\right\}X^2$$

$$= (A_2(1-\eta_1/\eta_0)+D\eta_1^{3/2})2a(V+\tfrac{1}{2})X. \tag{7.7.6}$$

It has been assumed that the inward flux of turbulent energy is

$$\overline{pw} + \tfrac{1}{2}\overline{q^2w} = D\tau^{3/2}, \tag{7.7.7}$$

the same assumption that was used to derive the equilibrium velocity distribution of equation (7.6.7). It is related to B by $B = \frac{3}{2}kD$, and the value used here is $D = 0.29$ corresponding to $B = 0.18$ and $k_0 = 0.50$.

Flow constants using the Clauser condition are compared with observations in table 7.1. Using the energy method, fair agreement is obtained with the observations for the same values of the constants A_1 and A_2 that were used for wall jets, values about 15% more than for free jets, and the differences are comparable with the experimental uncertainty.

TABLE 7.1   *Flow constants for self-preserving boundary-layer flow with zero wall stress*

Flow parameter	Calculated ($R_0 = 55$)	Measured (Stratford)
a	−0.225	−0.23
u_0/U_1	−0.81	−0.8
β	0.033	0.034
dI_0/dx	0.045	0.0475

Both Clauser (1956) and Bradshaw (1967b) have produced and studied self-preserving layers with exponents in the range, −0.15 to −0.255, and with considerable wall stress, and there can be no doubt that a self-preserving layer with finite wall stress could be produced in a free stream with velocity varying in the same way as for the zero-stress layer of Stratford. It is therefore possible for a boundary layer to develop in two distinct, self-preserving ways in the same external conditions of surface roughness and pressure gradient specified by an exponent near -0.23.

No examples of the alternative small-stress layers that are expected for exponents more negative than -0.23 have yet been reported. To make one, it would be necessary to produce a free-stream with an exponent in the narrow range between the zero-stress

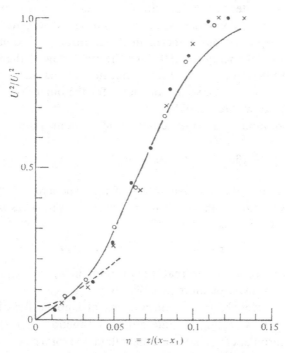

Fig. 7.7. Mean velocity distribution for self-preserving boundary layer flow with zero wall stress. (Full line is the calculated distribution for $k_0 = 0.50$ and $R_s = 74$, experimental points from Stratford, 1959b. c_p: ●, 0.682; ×, 0.624; ○, 0.489.)

value and the upper limit set by flow Reynolds number (see fig. 7.5), and to introduce into it a layer with thickness and stress distribution resembling the values for the self-preserving development. Existing work has started with initially thin layers which have developed in nearly uniform flow. It might be better to introduce a layer that is approaching separation in a strong adverse pressure gradient.

7.8 Wall layers with convergent flow

It is possible that the necessary conditions for self-preserving development may be satisfied if the rate of increase of flow width, dl_0/dx, is negative and invariant. Then

$$l_0 \propto (x_0 - x), \qquad U_1 \propto (x_0 - x)^a, \qquad u_0/U_1 = \text{constant},$$

where x is less than x_0, the position of the virtual origin ('end' is more appropriate) of the flow. Such a converging flow is clearly impossible with free boundaries, for it would be a divergent flow with the same exponent but reversed in direction and it would have a negative rate of energy transfer from the mean flow to the turbulent motion. In boundary layers though not in wall jets, the presence of wall friction provides a source of energy for the outer flow and some converging flows are possible.

The condition for overall conservation of momentum,

$$\frac{dl_0}{dx}[(3a+1)I_1 V + (2a+1)I_2] = -\tau_0/u_0{}^2 \qquad (7.8.1)$$

imposes a restriction on the values of the exponent a defining the velocity variation in the free stream. Since the wall stress is positive and the rate of spread negative,

$$a < -(I_1 V + I_2)/(3I_1 V + 2I_2) \qquad (7.8.2)$$

the opposite condition to that for diverging flow, and implying that the exponent must be more negative than $-\frac{1}{3}$.

Now consider the rate of entrainment of ambient fluid by a self-preserving flow. Using an intermittency function, $\delta(\mathbf{x}, t)$, to distinguish turbulent from non-turbulent fluid, the total flux of turbulent fluid across a section of the flow is

$$Q_t = \int_0^\infty \delta(\mathbf{x}, t)(U+u)\,dz = u_0 l_0 (V\eta_0 + C), \qquad (7.8.3)$$

where

$$\eta_0 = \int_0^\infty \delta\,dz/l_0,$$

and

$$C = \int_0^\infty f(\eta)\delta\,d\eta + u_0{}^{-1}\int_0^\infty \overline{\delta u}\,d\eta$$

are positive constants of the flow. Then the true rate of entrainment is

$$\frac{dQ_t}{dx} = (a+1)(V\eta_0 + C)u_0 \, dl_0/dx. \qquad (7.8.4)$$

Since $(V\eta_0 + C)$ and u_0 are negative in boundary layers, a converging layer has a positive rate of entrainment only if $a < -1$. While there is some evidence for the occurrence of negative entrainment, involving the conversion of turbulent fluid to the non-turbulent state, large negative rates are unlikely and self-preserving convergent flow is possible only if the exponent is near or more negative than -1.

The converging layer for $a = -1$ has the interesting property that it can satisfy the exact conditions for self-preserving development with smooth-wall boundary conditions. The layer could form on the walls of a converging wedge and the local Reynolds number of flow is everywhere the same. The equation of mean flow takes the form,

$$\frac{dl_0}{dx}(2Vf + f^2) = g_{13}' \qquad (7.8.5)$$

and, if the eddy viscosity is constant in the outer part of the layer, the velocity distribution is given by

$$f(1 + \tfrac{1}{2}f/V) = f'' \qquad (7.8.6)$$

choosing the scale-length so that

$$R_s^{-1} = (v_T/|u_0|l_0) = 2V \, dl_0/dx. \qquad (7.8.7)$$

In a strongly accelerated flow, the velocity defect is expected to be small outside the inner equilibrium layer and the solution of (7.8.7) may be approximated by

$$f(\eta) = \exp(-\eta) \qquad (7.8.8)$$

valid if $-2V \gg 1$.

A plausible composite distribution of velocity is obtained by joining the outer distribution smoothly to the inner distribution for linear stress variation in an equilibrium layer. The conditions are that

$$V + \exp(-\eta_1) = \gamma \log(\eta_1 l_0/z_0) \qquad (7.8.9)$$

and that

$$-\exp(-\eta_1) = \frac{(\tau(\eta_1 l_0))^{1/2}}{k u_0}\eta_1^{-1}. \qquad (7.8.10)$$

The condition for overall conservation of momentum is

$$k^2 \gamma^2 = (2I_1 V + I_2)\,dl_0/dx \qquad (7.8.11)$$

and the condition for conservation in the inner layer is (nearly) that

$$\frac{\tau_0 - \tau(\eta_1 l_0)}{u_0^2} = 2\beta\eta_1 \qquad (7.8.12)$$

(compare (7.6.9)). The various flow constants may now be calculated, either by using the Clauser relation for an eddy viscosity or by the equation for energy balance.

TABLE 7.2 *Flow constants for self-preserving boundary-layer flow within a wedge* $(a = -1)$ [a]

u_0/U_1	β	dl_0/dx	c_f	Rx	l_0/z_0
−0.40	8.2×10^{-3}	4.4×10^{-3}	3.09×10^{-3}	0.24×10^6	550
−0.35	8.4	3.6	2.40	0.85	1,500
−0.30	8.7	3.1	1.79	4.39	5,700
−0.25	8.9	2.5	1.26	4.06×10^7	36,700
−0.20	9.1	2.0	0.82	1.03×10^9	600,000

[a] Calculated for $R_0 = 50$.

Using the equation for the stress integral (7.6.12) to define an effective eddy viscosity for the whole outer layer,

$$\frac{dl_0}{dx}(2I_{1a}V + I_{2a}) = \eta_1\,\frac{\tau(\eta_1 l_0)}{u_0^2} + R_s^{-1}\,e^{-\eta_1} \qquad (7.8.13)$$

and the Clauser condition with the outer profile (7.8.8) is

$$R_s^{-1} = \frac{v_T}{|u_0|l_0} = R_0^{-1}. \qquad (7.8.14)$$

The energy equation is

$$\frac{dl_0}{dx}\left\{2I_2'V + I_3' + \frac{q_0^2}{\tau_0}k^2\gamma^2[(\eta_0 - \eta_1)V + I_1']\right\}$$
$$= \frac{\tau_0 - \tau(\eta_1 l_0)}{u_0^2}e^{-\eta_1} + \left(\frac{q_0^2}{\tau_0}\right)^{3/2}\frac{(\eta_0 - \eta_1)l_0}{L_\varepsilon}k^3\gamma^3 \qquad (7.8.15)$$

in the usual notation. Notice that the advection terms on the left represent a loss of energy, and that the sole source of energy is the wall stress. Calculated values of the flow constants are shown in table 7.2.

7.9 Almost self-preserving development

If the external conditions allow it at all, self-preserving development is usually an asymptotic state reached after long development and stable in the sense that deviations from the asymptotic distributions become less as the flow develops. An effect of the stability may be that, if the external parameters vary with fetch almost but not exactly in the prescribed way, the distributions are nearly those for self-preserving development with the current values of the parameters. The condition for almost self-preserving development is that the external parameters should deviate little from the standard values while the flow develops over a distance sufficient to reduce substantially a deviation from the equilibrium development.

With few exceptions, wall layers on ordinary surfaces cannot satisfy exactly the requirement of constant ratio of roughness length to fetch, but substantial variations of the ratio have a comparatively small effect on the flow. The reason is that the ratio affects only the velocity of translation of the turbulent flow and has a weak influence on it. If there is a region of fully turbulent flow with Reynolds stress nearly equal to the wall stress, consistency of the 'law of the wall' with the self-preserving form for the mean velocity requires that $\gamma = \tau_0^{1/2}/(ku_0)$ should be invariant and that

$$V + C_1 = \gamma(\log l_0/z_0 + C_2), \qquad (7.9.1)$$

where C_1, C_2 are not large constants. Since $\log l_0/z_0$ is commonly about ten, considerable change of the ratio need cause only small changes in V and γ. Substantial changes can be expected only after a fetch of order

$$\begin{aligned}
L_r &= \log l_0/z_0/[\mathrm{d}(\log l_0/z_0)/\mathrm{d}x] \\
&= \log l_0/z_0 \bigg/ \left[\frac{1}{l_0}\frac{\mathrm{d}l_0}{\mathrm{d}x} - \frac{1}{z_0}\frac{\mathrm{d}z_0}{\mathrm{d}x}\right] \\
&= \mathrm{O}[(x-x_0)\log l_0/z_0]
\end{aligned} \qquad (7.9.2)$$

much more than the current distance from the flow origin.

The fetch necessary for substantial reduction of a small deviation from the equilibrium development could be set by the time necessary for abnormal Reynolds stresses to settle down to the values appropriate to the velocity distribution, or by the time taken by the incremental stresses generated in response to attenuate a change in the velocity distribution. The first time may be measured by the ratio of the density of turbulent kinetic energy, $\frac{1}{2}\overline{q^2} = \tau/(2a_1)$, to the rate of energy production, $\tau(\partial U/\partial z)$. Supposing an effective convection velocity of $(U + \frac{1}{2}u_0)$, the adjustment fetch for stress perturbation is

$$L_t = (U + \tfrac{1}{2}u_0)l_0/(2a_1 u_0)$$

$$= \frac{\beta}{2a_1}(x - x_0) \tag{7.9.3}$$

(assuming u_0/l_0 to be a typical value of the mean velocity gradient). The second time may be measured by the ratio of u_0, the velocity variation in the outer flow, to τ_m/l_0, a typical value for the stress gradient, and the adjustment fetch for velocity perturbation is

$$L_m = \frac{(U + \frac{1}{2}u_0)u_0}{\tau_m/l_0} = \beta\frac{u_0{}^2}{\tau_m}(x - x_0). \tag{7.9.4}$$

In wall jets and strongly retarded boundary layers (with exponent near -0.25), the wall stress is much less than the maximum stress τ_m, and nearly $\tau_m/u_0{}^2 = e^{-1/2}2|a|/\beta$. Then,

$$L_m = \frac{e^{1/2}}{2|a|}(x - x_0). \tag{7.9.5}$$

In other boundary layers, the maximum stress is comparable with the wall stress, and using equation (7.3.4) in the form,

$$-\tau_0/u_0{}^2 = [(3a+1)I_1 V + (2a+1)I_2]\frac{dl_0}{dx} \tag{7.9.6}$$

the adjustment fetch is estimated to be

$$L_m = \frac{V + \frac{1}{2}}{(3a+1)I_1 V + (2a+1)I_2}(x - x_0). \tag{7.9.7}$$

If typical values of the flow constants are used in equations (7.9.2, 7.9.3, 7.9.5, 7.9.7), it appears (1) that the turbulent motion itself adjusts to a change in the velocity distribution within a fetch

L_c that is much less than $(x - x_0)$, (2) that adjustment of the mean flow to a change of stress distribution is substantial in a fetch L_m that is not more than about $3(x - x_0)$,† and (3) that both fetches are short compared with the fetch L_r that requires substantial changes in the flow parameters.

An exactly self-preserving flow which resembles most closely a flow developing in external conditions that require slow changes in the flow constants is one with external conditions the same as the current external conditions of the almost self-preserving flow. The effective origin is at

$$x_0' = x - l_0 \bigg/ \frac{\mathrm{d}l_0}{\mathrm{d}x}, \qquad (7.9.8)$$

the ratio of fetch to roughness length is $(x - x_0)/z_0$, and the effective exponent of the velocity variation in the free stream is

$$a' = \frac{x - x_0'}{U_1} \frac{\mathrm{d}U_1}{\mathrm{d}x}. \qquad (7.9.9)$$

Two examples may show how the development of an almost self-preserving layer may be calculated.

For a simple wall jet issuing into ambient fluid at rest, $V = 0$ and the momentum equation is

$$\frac{\mathrm{d}}{\mathrm{d}x}(I_2 u_0{}^2 l_0) = -\tau_0 = -k^2\gamma^2 u_0{}^2. \qquad (7.9.10)$$

The condition (7.5.2) for matching velocity distributions in the inner and outer layers is

$$\gamma^{-1} - \log \eta_1 = \log l_0/z_0 - 1 \qquad (7.9.11)$$

and the calculations of § 7.5 show that $\mathrm{d}l_0/\mathrm{d}x$ varies with change of a much less than does γ, and that $|\log \eta_1| \ll \gamma^{-1}$. Assuming the rate of spread to be constant, it follows that

$$a = -\frac{1}{2} - \frac{k^2}{4\beta I_2(\log(\eta_1 l_0/z_0) - 1)^2}. \qquad (7.9.12)$$

For values of $\log l_0/z_0$ near ten, the exponent is close to -0.52 and a hundred-fold increase in the ratio l_0/z_0 is necessary to change it to

---

† For layers with maximum stress comparable with the wall stress, $L_m \approx (x - x_0)$. In flows with small wall stress, L_m is larger but the development of the flow is only weakly dependent on the wall stress. So, while equation (7.9.2) describes the fetch necessary for a substantial change in γ, the perturbation of the whole layer is less, roughly in the ratio τ_0/τ_m.

-0.51. The assumption of a constant rate of spread is therefore consistent, and it is concluded that the velocity scale of the flow varies with the exponent given by (7.9.12) for any ordinary flow either on a smooth surface or on a uniformly rough surface.

If the wall stress of a boundary layer is comparable with the maximum stress, the calculations of § 7.6 show that the friction parameter γ depends mostly on the exponent a and that it is nearly independent of the velocity ratio V. For stream velocity varying as $(x - x_0)^a$, the momentum equation may be put into the form,

$$
\begin{aligned}
-k^2\gamma^2 = \exp(X - C_2) & \left[\frac{z_0}{x - x_0}(3aI_1V + 2aI_2) \right. \\
& + z_0\frac{dV}{dx}(-I_1 - 2I_2V^{-1}) + (I_1V + I_2) \\
& \left. \times \left(z_0\frac{dX}{dx} + \frac{dz_0}{dx} \right) \right],
\end{aligned}
\tag{7.9.13}
$$

where $X = (V + C_1)/\gamma$ and the variations of flow parameters other than V and $(x - x_0)/z_0$ have been neglected. Three special cases are of interest:

(1) Constant $(x - x_0)/z_0$, the condition that allows exactly self-preserving development. Then,

$$
-k^2\gamma^2\frac{x - x_0}{z_0} = \exp(X - C_2)[(3a + 1)I_1V + (2a + 1)I_2]. \tag{7.9.14}†
$$

(2) Constant roughness length. For large values of V/γ, an approximate solution is

$$
\begin{aligned}
-k^2\gamma^2\frac{x - x_0}{z_0} = \exp(X - C_2) \\
\times [(3a + 1)I_1V + (2a + 1)I_2 - 2\gamma I_1].
\end{aligned}
\tag{7.9.15}†
$$

(3) Aerodynamically smooth surface with roughness length $z_0 = (\nu/\tau_0^{1/2})\exp(-A)$. For large values of V/γ, an approximate solution is

$$
\begin{aligned}
-k^3\gamma^3\frac{(x - x_0)U_1}{\nu} = \exp(X - C_2 - A)[(3a + 1)I_1V^2 \\
+ (2a + 1)I_2V - (2a + 2)I_1\gamma V].
\end{aligned}
\tag{7.9.16}†
$$

† If complementary solutions of (7.9.13) are added, the ratios of the coefficients in the self-preserving equations of motion are not constant but vary as powers of $(x - x_0)$.

The equations give the dependence of the velocity ratio on fetch, and the leading terms are the same if the fetch is expressed as a fraction of the current roughness length.

The use of flow constants for exactly self-preserving flow is possible only if the calculated coefficients turn out to be nearly independent of fetch. With a power-law variation of velocity in the free stream, only two of the coefficients in the equations of motion and of energy are independent, and their calculated values are;

$$\frac{x - x_0}{l_0} \frac{dl_0}{dx} = 1 - (a+1)\gamma/V \quad \text{(smooth surface)}$$
$$= 1 - \gamma/V \quad \text{(uniform roughness)} \qquad (7.9.17)$$

and

$$\frac{x - x_0}{u_0} \frac{du_0}{dx} = a - (a+1)\gamma/V \quad \text{(smooth surface)}$$
$$= a - \gamma/V \quad \text{(uniform roughness)} \qquad (7.9.18)$$

The values differ from those of an exactly self-preserving flow by quantities of order γ/V, which are about 0.1 for weakly retarded flows at ordinary Reynolds numbers. The development equations (7.9.15, 7.9.16) show that they change slowly and by small total amounts in any likely range of development.

7.10 Layers with nearly uniform velocity in the free stream

The development equations for boundary layers on smooth surfaces have an application in the prediction of skin friction on slender bodies, for which most of the friction is nearly that under a boundary layer that has developed in conditions approximating to zero pressure gradient. If (7.9.16) is expressed as a relation between the (conventional) coefficient of local friction, $c_f^* = 2\tau_0/U_1^2$, and the Reynolds number of fetch, $R_x = U_1(x - x_0)/\nu$, it may be approximated by

$$\sqrt{2k}c_f^{*-1/2} = \log(R_x c_f^*) + \log[k|\gamma|/(6a+2)I_1]$$
$$+ A + C_2 - C_1/\gamma + O(c_f^{*1/2}). \qquad (7.10.1)$$

With zero longitudinal pressure gradient, i.e. $a = 0$, the total frictional force up to the position of observation is equal to the flux

of momentum defect, and the overall friction coefficient is related
to the velocity profile by

$$C_f = 2\int_0^\infty U(U_1 - U)\,dz/(U_1^2(x-x_0))$$

$$= -\frac{2l_0}{x-x_0}(I_1 V + I_2)V^{-2}. \qquad (7.10.2)$$

To the approximation in use, that c_f^* is small, it may be shown that

$$c_f^*/C_f = 1 - \frac{\gamma}{V}\frac{2I_1}{I_1 + I_2/V} \qquad (7.10.3)$$

and the relation between friction coefficient and Reynolds number
takes the form of (7.10.1), that is,

$$\sqrt{2k}C_f^{-1/2} = \log(R_x R_f) + \log[k|\gamma|/(2I_1)]$$
$$+ A + C_2 - C_1/\gamma - 1. \qquad (7.10.4)$$

The predicted variation of skin friction with Reynolds number
depends mostly on the value adopted for the Kármán constant k.
With the value used here ($k = 0.41$), $\sqrt{2}.k\log_e 10 = 0.252$, and
the relation,

$$0.252\,C_f^{-1/2} = \log_{10}(R_x C_f) + 0.3 \qquad (7.10.5)$$

describes well the variation of skin friction for layers developing in
zero pressure gradient. Schoenherr (1932) proposed the relation,

$$0.242\,C_f^{-1/2} = \log_{10}(R_x C_f) \qquad (7.10.6)$$

which is satisfactory for Reynolds numbers in the range 10^6 to 10^8,
but it implies a value of 0.396 for the Kármán constant and is
possibly in error at larger Reynolds numbers.

A boundary layer on an aerofoil may develop in a region of
negative pressure gradient before entering on an extensive region of
weak positive gradient. If it is supposed that the external conditions
are always near to those in zero gradient, the flow constants for
zero gradient may be assumed and the momentum equation leads to
an equation for the development,

$$\frac{d}{dx}\int_0^\infty U(U_1 - U)\,dz + \frac{dU_1}{dx}\int_0^\infty (U_1 - U)\,dz = \tau_0, \qquad (7.10.7)$$

where U_1 is a function of x. From the matching condition (7.9.1),

$$u_0 l_0 = \frac{\nu}{k\gamma} \exp\left[\frac{V+C_1}{\gamma} - A - C_2\right]. \qquad (7.10.8)$$

Defining a friction function,

$$Y = -(k\gamma)^{-3} \exp\left(\frac{V+C_1}{\gamma} - A - C_2\right)[I_1 V^2 + I_2 V - 2I_1 V\gamma] \qquad (7.10.9)$$

and omitting small terms, the development equation becomes

$$\frac{\nu}{U_1}\frac{dY}{dx} + \frac{2I_1 V + I_2}{I_1 V + I_2 - 2\gamma I_1 V}\frac{\nu}{U_1^2}\frac{dU_1}{dx}Y = 1. \qquad (7.10.10)$$

If the variation of free stream velocity is so slow that the second term is negligible, the solution is

$$Y = R(x) = \frac{1}{\nu}\int_{x_0} U_1(x)\,dx. \qquad (7.10.11)$$

Using this as a first approximation, a better approximation to the solution is

$$Y = \int_{x_0}\left(1 - 2R\frac{\nu}{U_1^2}\frac{dU_1}{dx}\right)\frac{U_1\,dx}{\nu}. \qquad (7.10.12)$$

The meaning is that the boundary layer will have very nearly the properties of a layer that has developed with a constant stream velocity equal to the current value over a distance sufficient to give the Reynolds number defined by the equations (7.10.11, 7.10.12).

7.11 Turbulent flow in self-preserving boundary layers

Comprehensive measurements have been made of turbulent intensities and Reynolds stresses, of one-dimensional spectra, of energy dissipation and of lateral energy transport in self-preserving boundary layers with exponents, $a = 0$, -0.15 and -0.255 (Klebanoff 1955, Bradshaw 1967b), and some of the measurements, including the balances of turbulent energy are shown in figs. 7.8, 7.9, 7.10.

The principal features are:

(1) In all three flows, the relative intensities of the three velocity components are much the same over most of the flow. In the outermost parts where entrainment is occurring, the intensities are more nearly equal than deeper in the layer where the intensity of the streamwise component is considerably greater.

(2) The ratio of Reynolds stress to total intensity, $a_1 = \tau/\overline{q^2}$, is roughly constant, though the ratio is least in the most strongly retarded flow. The variation may be attributed to 'inactive' fluctuations whose direction is nearly parallel to the wall and which transport energy and momentum to a negligible extent.

(3) The dissipation length parameter L_ε is nearly constant in the outer flow, and the variation near the outer edge is consistent with a nearly constant value within the turbulent fluid.

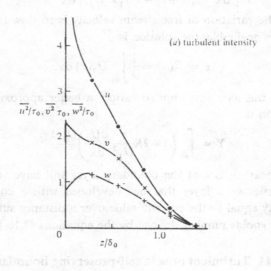

(a) turbulent intensity

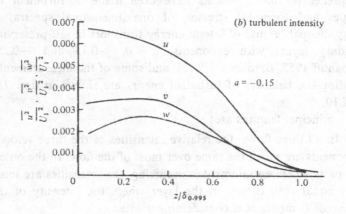

(b) turbulent intensity

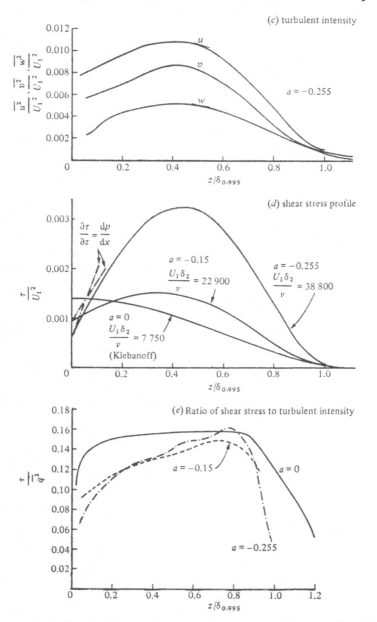

Fig. 7.8. Distributions of turbulent intensity and Reynolds stress in self-preserving boundary layers.

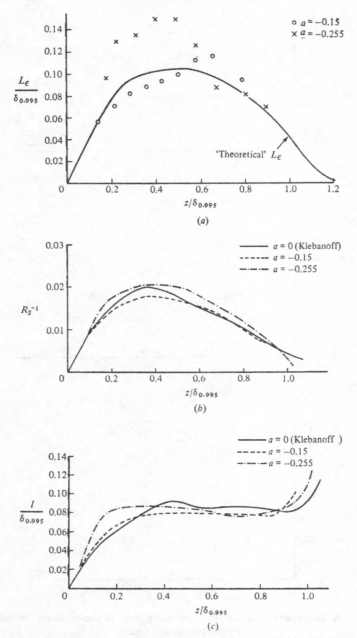

Fig. 7.9. Distributions of (a) dissipation length parameter, $L_\varepsilon = (\tau/\rho)^{3/2}/\epsilon$, (b) eddy viscosity and (c) mixing length in self-preserving boundary layers.

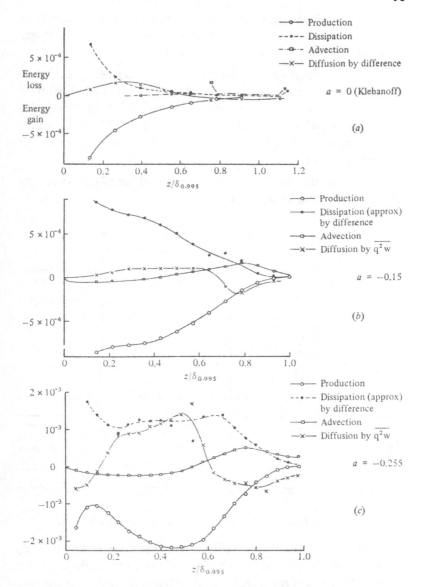

Fig. 7.10. Balances of turbulent kinetic energy for self-preserving boundary layers. Terms made dimensionless with U_1 and $\delta_{0.995}$.

(4) In the balance of turbulent energy, advection by the mean flow is significant only near the outer edge. Lateral diffusion of

energy is significant near the edge, where it balances the effects of advection, and near the position of maximum stress in strongly retarded flows.

(5) The distributions of intermittency factor are nearly similar in all three flows, with the average depth of the turbulent fluid (measured by $\eta_0 l_0$) very nearly the same fraction of the flow width $\delta_{0.995}$ (the position where $U = 0.995\ U_1$). The width of the intermittent region of flow is slightly more in the flow with the largest pressure gradient.

(6) Outside the thin inner layer, the variations of eddy viscosity and of mixing length are not large.

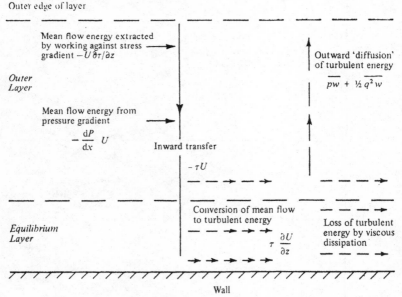

Fig. 7.11. Conversion and flow of energy in a boundary layer.

Over most of the flow, particularly if the layer is not strongly retarded, energy transfer from the mean flow to the turbulence at a rate $\tau\ \partial U/\partial z$ is approximately equal to the local rate of energy dissipation, and most of the energy is transferred in the inner layer where the shear is large. The source of most of the energy is the loss of kinetic energy of the free stream, a process that may be discussed using the equation for the mean flow kinetic energy,

$$\frac{1}{2}\left(U\frac{\partial}{\partial x}+W\frac{\partial}{\partial z}\right)(U^2+W^2) = -U\frac{dP}{dx}+\frac{\partial(\tau U)}{\partial z}-\tau\frac{\partial U}{\partial z} \qquad (7.11.1)$$

to the boundary-layer approximation. The advection terms on the left and the term $-U\,dP/dx$ (describing work done by the pressure gradient) become smaller near the wall but are distributed fairly uniformly over the layer. Near the wall where $\tau\,\partial U/\partial z$ is large, the transfer of energy to turbulence can be balanced only by $\partial(\tau U)/\partial z$, the divergence of τU, the energy flux transferred inwards by Reynolds stresses working on the mean velocity. The relations are illustrated in fig. 7.11.

An important consequence of the dependence of energy supply on Reynolds stresses further from the wall is that, if the turbulent motion at one level is destroyed (for example by buoyancy forces), energy supply to the motion nearer the wall is interrupted and all motion, both turbulent and mean flow, will decay rapidly.

7.12 Development of boundary layers in arbitrary external conditions

If the external conditions of a boundary-layer flow do not change with fetch even approximately in the ways necessary for self-preserving development, prediction of the development requires a more detailed account of the relation between Reynolds stress and velocity field than has been used for self-preserving flows. Many such relations have been proposed and used with varying degrees of success, but the one that conforms most nearly to the concepts used here was developed by Bradshaw, Ferris & Atwell (1967). They assume a basic degree of similarity between all boundary layers, specifically (1) that the stress–intensity ratio has a universal value, (2) that the dissipation length parameter has a similar distribution in all flows,

$$L_\varepsilon/\delta_{0.995} = F_1(z/\delta_{0.995}), \qquad (7.12.1)$$

and (3) that the lateral flux of turbulent energy has the universal distribution,

$$\overline{pw}+\tfrac{1}{2}\overline{q^2 w} = \frac{\tau\tau_m}{U_1}f_2(z/\delta_{0.995}). \qquad (7.12.2)$$

With these assumptions, the equation for the turbulent kinetic energy becomes a partial differential equation for the mean velocity and the Reynolds stress, and, using the equations of continuity and of mean flow, the development of a boundary layer in any external conditions of pressure distribution, surface roughness or suction can be calculated if the initial distributions of velocity and stress are known. The BFA method has been applied to calculate the development of a variety of boundary layers, some of them decidedly unusual, with considerable success in prediction, and the good performance may well be the result of the emphasis on providing a recognisable description of the turbulent motion.

In one respect, the BFA method is inconsistent with the conclusion that the translational velocity of the fully turbulent flow is not in itself relevant to the mechanics of the motion, so that velocity scales for the turbulent motion are determined by the stresses and mean velocity differences. By supposing that the scale of the diffusive flux of turbulent energy is $\tau \tau_m / U_1$, the third assumption (7.12.2) is at variance with the behaviour of self-preserving flows. For example, Bradshaw *et al.* show that the angle of spread of a layer is

$$\frac{\mathrm{d}\delta_{0.995}}{\mathrm{d}x} = \frac{W_1}{U_1} + 2a_1 \frac{\tau_m}{U_1^2} f_2(1). \tag{7.12.3}$$

For a layer in zero pressure gradient, the momentum condition is, nearly,

$$\frac{\mathrm{d}l_0}{\mathrm{d}x}(I_1 V + I_2) = -\frac{\tau_0}{u_0^2}$$

and $W_1 / U_1 = I_1 \, \mathrm{d}l_0/\mathrm{d}x \, V^{-1}$. Since the velocity ratio V becomes larger with increase of fetch Reynolds number, the conclusion is that the ratio $\delta_{0.995}/l_0$ decreases, roughly as V^{-1}, which is quite impossible.

The experimental data used in determining the universal distribution of lateral energy flux are (with the exception of mixing layers) equally well described if

$$\overline{pw} + \tfrac{1}{2}\overline{q^2 w} = \tfrac{1}{2}\overline{q^2} \mathscr{V}_0 F_2(z/\delta_{0.995}) \tag{7.12.4}$$

where $\mathscr{V}_0$ is a scale of convection velocity determined by the stress distribution in the layer. Table 7.3 compares values of the scale velocity for the BFA assumption and as given by the simple expression,

$$\mathscr{V}_0 = 2.1 \, \tau_m^{1/2} - 1.1 \, \tau_0^{1/2} \tag{7.12.5}$$

The calculations of Bradshaw *et al.* show that lateral transport of energy has little effect on the development of a layer, even close to the outer edge, and the point is mostly of theoretical interest. It is however interesting that the BFA function f_2 has a value of 33 at the edge of the flow, while the function F_2 in (7.12.4) would have a value near 1.2. In other words, the convection velocities are comparable in magnitude with the turbulent velocities, not unexpectedly since they form part of the motion, and a velocity scale defined by the Reynolds stresses is more plausible than one involving the mean velocity.

TABLE 7.3   *Comparison of scales for convection velocities*

	$a = 0$	$a = -0.15$	$a = -0.255$
$\tau_0/U_1{}^2$	0.0014	0.00094	0.00065
$\tau_m/U_1{}^2$	0.0014	0.00152	0.00325
$\tau_0{}^{\frac{1}{2}}/U_1$	0.0374	0.0306	0.0256
$\tau_m{}^{\frac{1}{2}}/U_1$	0.0374	0.0390	0.0570
$\dfrac{V_0}{U_1} \begin{cases} \text{BFA} \\ (7.12.5) \end{cases}$	0.0014 0.0374	0.00152 0.0482	0.00325 0.0915
Ratio	26.7	31.8	28.1

Note: The values refer to flow at Reynolds numbers near 10^6.

The assumption of constant stress–intensity ratio is supported both by observation and by theoretical arguments based on the rapid-distortion calculations and stability of eddy structures. The most obvious departures are in the equilibrium layers for strongly retarded flows where the intensity is strongly affected by contributions from 'inactive' fluctuations from eddies of relatively large scale. In an equilibrium layer, advection of energy is negligible and the energy equation contains the ratio only in the combination $a_1 L_\varepsilon/z$, i.e. the value of a_1 refers to eddies which dissipate energy locally and have scales comparable with distance from the wall. The ratio is larger in the entrainment region of the flow but the change is absorbed in the empirical function describing convective transfer.

The distribution of dissipation length in a flow depends on the size distribution of energy- and stress-containing eddies. In self-preserving flow, the distribution arises from an equilibrium established as the flow develops under constraints by the wall and the free edge of the flow. If the flow is disturbed, say by a change of pressure gradient, the flow expansion will distort the eddy pattern into a broadly similar form scaling with total thickness. The success of the similarity assumption (7.12.1) may mean that the distribution of eddy sizes with position is determined almost entirely by flow geometry.

The BFA calculation of boundary-layer development is based on assumptions about the nature of turbulent flow that are essentially those used to derive the properties of equilibrium layers, and so it is certain to give a good description of the flow near the wall. At the outer edge of the flow, the true entrainment rate is set by a convection velocity whose magnitude was chosen to conform with observed entrainment rates. The velocity and stress distributions over most of the layer are, in a sense, interpolated between the known limiting distributions near the wall and at the edge, and the choice of interpolation profiles is limited. Any procedure of inter-polation must lead to fairly accurate predictions, but the merit of the BFA interpolation is that it is based on assumptions of similarity of the turbulent motion that are both plausible and consistent with observation. The interpolation should be a good one, even in unusual flows.

The least satisfactory part of the scheme is the empirical relation between convection velocity at the edge and stress distribution, though the defect is shared with other methods. A more basic approach might be to derive the relation from the entrainment rates calculated for self-preserving boundary layers from the overall energy balance and assuming similarity of the distributions of dissipation length scale.

7.13 Boundary-layer development after a sudden change of external conditions

In discussing the development of boundary layers whose external conditions are nearly but not exactly those for self-preserving development, it was shown that the adjustment fetch is usually comparable with the current distance from the flow origin. It follows that the effect on the flow of a substantial change of external con-ditions may be to establish a different moving equilibrium but that the change of structure will not be complete until the layer has developed over a distance large compared with the adjustment fetch. On the other hand, the rates of energy production and dissipation are very large near the wall and the flow here is capable of responding rapidly to the change. The result is that the changes of the flow are confined initially to an internal boundary layer that spreads from the section of transition, and that the flow outside the internal layer has developed in almost the same way as in the original flow. To

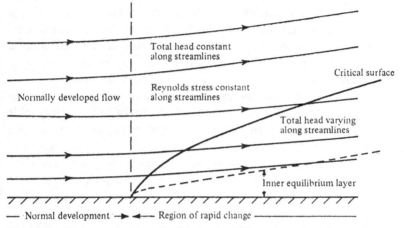

Fig. 7.12. Diagram showing development within a boundary layer after a sudden change in external conditions.

be precise, the changes of total head and Reynolds stress along streamlines of the mean flow proceed at nearly the same rates as in the original flow (fig. 7.12).

To the boundary-layer approximation, the momentum equation may be written as an equation for the total head,

$$\left[\frac{\partial}{\partial x}(P+\tfrac{1}{2}U^2)\right]_\psi = \frac{\partial \tau}{\partial z}, \tag{7.13.1}$$

where $\psi = \displaystyle\int_0^z U(z')\,dz'$ is the stream function. Unless the stress gradient has changed considerably from its original value, changes of total head are small compared with the kinetic head $\tfrac{1}{2}U^2$ over fetches small compared with $(x_1 - x_0)$, supposing the original flow to be roughly self-preserving so that $\partial \tau/\partial z = O(\tau_0/l_0)$.

Assuming a constant ratio of stress to intensity, the rate of change of Reynolds stress along streamlines is given by

$$\frac{U}{2a_1}\left(\frac{\partial \tau}{\partial x}\right)_\psi = -\frac{\partial}{\partial z}(\overline{pw}+\tfrac{1}{2}\overline{q^2 w})+\tau \frac{\partial U}{\partial z}-a_1^{-3/2}\tau^{3/2}/L_\varepsilon \qquad (7.13.2)$$

and it will be nearly unchanged unless either the mean velocity gradient or the dissipation length change substantially. In two-dimensional flow, acceleration by pressure gradients has, by itself, no effect on the velocity gradient, and the dissipation length probably remains the same fraction of the layer thickness. It follows that a change of external conditions hardly affects the way in which total head and Reynolds stress change along streamlines except in a growing region near the wall where the change has induced a diffusion of vorticity (i.e. velocity gradient) into the flow.

A rough estimate of the extent of the modified flow can be obtained from (7.13.2). Near the edge of the modified layer, Reynolds stresses and rates of shear have changed appreciably from their original values in a time of flight which must be sufficient to allow considerable replacement of turbulent energy by generation. The replacement time for the original flow is $\tfrac{1}{2}\overline{q^2}/(\tau\, \partial U/\partial z)$, and, ignoring any change of velocity on the streamline, the necessary fetch is

$$x-x_1 = \frac{1}{2a_1}U\bigg/\frac{\partial U}{\partial z}. \qquad (7.13.3)$$

If the particular streamline lay within a constant-stress layer in the original flow,

$$x-x_1 = \frac{1}{2a_1}z'\log(z'/z_0), \qquad (7.13.4)$$

where z' is the height of the streamline at $x = x_1$. Notice that the stream function for the streamline entering the modified layer at the section is

$$\psi_c = \frac{\tau_0^{1/2}}{k}z'\left(\log\frac{z'}{z_0}-1\right) = \frac{2a_1}{k}\tau_0^{1/2}(x-x_1). \qquad (7.13.5)$$

The set of equations used by Bradshaw *et al.* are naturally consistent with the development of the modified flow, but they predict a sharp boundary to it. The equations are hyperbolic in form and the characteristic directions are given by

$$\frac{\mathrm{d}\psi_c}{\mathrm{d}x} = \infty, \qquad [a_1\mathscr{V} \pm (a_1{}^2\mathscr{V}^2 + 2a_1\tau)^{1/2}], \qquad (7.13.6)$$

where ψ_c is the value of the stream function on the characteristic and $\mathscr{V} = \overline{(pw + \frac{1}{2}q^2w)}/\tau$ is the local convection velocity. Any small change has an influence that is confined within the wedge enclosed by the second and third characteristic directions, and changes originating at the wall are confined within the surface defined by

$$\frac{\mathrm{d}\psi_c}{\mathrm{d}x} = a_1\mathscr{V} + (a_1{}^2\mathscr{V}^2 + 2a_1\tau)^{1/2}. \qquad (7.13.7)$$

If the streamline entering the wedge of modified flow lay within a constant-stress layer,

$$\psi_c = (2a_1)^{1/2}\tau_0{}^{1/2}(x - x_1), \qquad (7.13.8)$$

a very similar expression to equation (7.13.5) ($a_1\mathscr{V}^2/\tau$ is small in a constant stress layer).

The problem of predicting flow in the modified layer following the change of external conditions is to interpolate between the substantially unchanged outer flow and a thin equilibrium layer near the wall. Compared with the same problem in ordinary boundary layers, the choice of interpolations is very limited and good results are possible with the simplest one, extension of the equilibrium layer profile to intersect the outer profile. The necessary conditions at the junction are that velocity and stress should be continuous.

7.14 Development in a region of strong adverse pressure gradient

It is usual for the boundary layer on a streamline body to develop first in a favourable pressure gradient with velocity and stress profiles similar to those of a self-preserving layer in zero pressure gradient, and then to enter a region of adverse pressure gradient. If the extent of the region is considerably less than the distance from the effective origin of the layer, a good account of the flow development may be obtained by assuming

(1) that total head and Reynolds stress are unchanged along streamlines of the mean flow lying outside an internal boundary layer of modified flow, and

(2) that the mean velocity distribution within the internal layer is

that for an equilibrium layer with stress varying linearly with distance from the wall. The value of this approach, first used by Stratford (1959a), is that it leads to an explicit condition for zero wall stress, a situation in which the boundary layer normally separates from the surface.

Consider a streamline of the mean flow that enters the region of modified flow at (x, z_c) and that was distant z' from the wall at the beginning of the adverse gradient. Supposing z' to be within the constant-stress layer, the conditions of negligible changes of total head and stress along the streamline are that

$$\left.\begin{array}{l} \frac{1}{2}U'^2 = \frac{1}{2}U_c^2 + \Delta P, \\ \tau_0' = \tau_c, \end{array}\right\} \tag{7.14.1}$$

where ΔP is the pressure rise. In the modified layer, the stress is assumed to vary linearly with distance from the wall, i.e.

$$\tau = \tau_0 + (\tau_0' - \tau_0)z/z_c, \tag{7.14.2}$$

and the velocity distribution to be that of an equilibrium layer,

$$U = \frac{\tau_0^{1/2}}{k} \log\left[\frac{\tau^{1/2} - \tau_0^{1/2}}{\tau^{1/2} + \tau_0^{1/2}} \frac{4\tau_0}{\alpha z_0}\right] + \frac{2}{k_0}(\tau^{1/2} - \tau_0^{1/2}). \tag{7.14.3}$$

The average stress gradient,

$$\alpha = \frac{\tau_c - \tau_0}{z_c}, \tag{7.14.4}$$

may be found by integrating the equation for the mean velocity over the modified layer, and appears as a function of dP/dx, $d\tau_0/dx$, τ_0, z_0 and z_c. Then, with the condition that (x_1, z') and (x, z_c) lie on the same streamline, the thickness of the modified layer and the wall stress can be found as functions of pressure rise and pressure gradient for given properties of the original layer and surface roughness.

The most interesting and simple result is the condition for zero wall stress. Then the velocity profile in the modified layer is (very nearly)

$$U = \frac{2}{k_0}(\alpha z)^{1/2} \tag{7.14.5}$$

and, using (7.14.1),

$$U_c = \frac{2}{k_0} \tau_0'^{1/2},$$
$$\psi_c = \frac{4}{3k_0} \tau_0'^{3/2}/\alpha.$$

(7.14.6)

At (x_1, z'), the stream function is

$$\psi_c = \frac{\tau_0'^{1/2}}{k} z' \left(\log \frac{z'}{z_0} - 1 \right).$$

(7.14.7)

The condition for conservation of total head is

$$\Delta P + \frac{2}{k_0^2} \tau_0' = \frac{1}{2} \frac{\tau_0'}{k^2} \log^2 \frac{z'}{z_0}$$

(7.14.8)

and eliminating z',

$$\frac{4k}{3k_0} \frac{\tau_0'}{\alpha z_0} = \left[\left(\frac{2\Delta P}{\tau_0'} + \frac{4}{k_0^2} \right)^{1/2} k - 1 \right] \exp \left[k \left(\frac{2\Delta P}{\tau_0'} + \frac{4}{k_0^2} \right)^{1/2} \right]$$

(7.14.9)

appears as a condition for zero stress.

To use (7.14.9) as a criterion for zero wall stress, the average stress gradient may be replaced by the pressure gradient, assuming in effect that flow accelerations in the modified layer are small compared with the pressure gradient. For pressure rises occurring in ordinary flows, the critical pressure rise is a slowly varying function of stress gradient and the assumption is not a critical one.

A comparison of the predicted pressure rises to separation with observations is made in fig. 7.13, using results of Schubauer & Klebanoff (1951) and of Bradshaw & Galea (1967). It may be noticed that the criterion is satisfied twice in each flow, the first time while the surface stress is still quite large, and it is the second intersection that predicts fairly accurately the position of separation. From a consideration of the flow as it approaches the position of zero stress, it may be shown that the critical condition must be approached from 'above', that is, pressure gradients decreasing to the critical value (Townsend 1961b). The accuracy of the basic assumptions of constant total head and stress along streamlines can be judged from fig. 7.14, showing results obtained by Bradshaw & Galea.

If the appropriate roughness length for a smooth surface, $z_0 = \nu/\tau_0^{1/2} \exp(-A)$, is substituted in (7.14.9), the variations of pressure recovery with surface stress of the initial layer and with

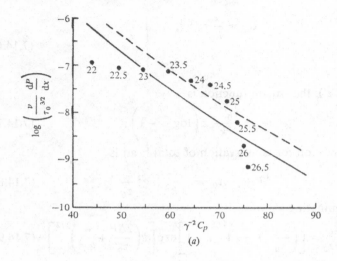

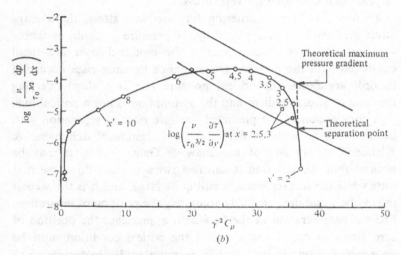

Fig. 7.13. Comparison of observed pressure rises to separation with predictions from equation (7.14.9). (a) Comparison with results of Schubauer & Klebanoff (1951) (observed separation near 25.8 ft). (b) Comparison with results of Bradshaw & Galea (1967) (observed separation near $x = 2.6$).

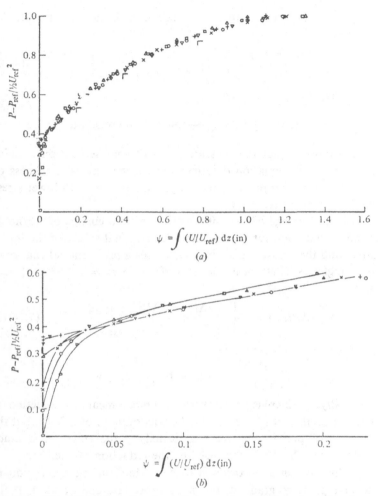

Fig. 7.14. Total head as a function of stream function in a boundary layer developing in a strong adverse pressure gradient. (a) Whole flow. (b) Near the surface showing large changes for ψ less than 0.03 (from Bradshaw & Galea 1967).

	x'	z at $\psi = 0.05$
□	∞	0.10
○	8	0.12
×	6	0.14
△	4	0.2
▽	3	0.25
+	$2\frac{1}{2}$	0.31

fluid viscosity is easily found. At constant Reynolds number, the relation is

$$\tau_0'^{1/2} \frac{d(\Delta P/\tau_0')}{d\tau_0'^{1/2}} = \frac{3}{k^2}\left[k\left(\frac{2\Delta P}{\tau_0'}+\frac{4}{k_0^2}\right)^{1/2}-1\right] \qquad (7.14.10)$$

or nearly, if $\Delta P/\tau_0'$ is large,

$$(\Delta P/\tau_0')^{1/2} = \frac{3}{2^{1/2}k}\log \tau_0'^{1/2}+\text{constant}.$$

(It is supposed that the pressure gradient near separation remains the same.) The equation describes the effect of initial thickness of boundary layer on pressure recovery, a thick layer with lower stress having a smaller recovery.

A simple change of viscosity (equivalent to a change of Reynolds number that does not alter the non-dimensional values of the initial stress and the pressure distribution) causes a proportional change in the roughness length of a smooth surface, i.e. $z_0 = v/\tau_0^{1/2} \exp(-A)$, and

$$v\frac{d}{dv}(\Delta P/\tau_0') = -\left[k\left(\frac{2\Delta P}{\tau_0'}+\frac{4}{k_0^2}\right)^{1/2}-1\right]k^{-2} \qquad (7.14.11)$$

or nearly, if $\Delta P/\tau_0'$ is large,

$$\left(\frac{\Delta P}{\tau_0}\right)^{1/2} = \text{constant}-\frac{1}{k2^{1/2}}\log v.$$

Normally, an increase of Reynolds number means a reduction of friction coefficient (described by the friction laws of § 7.10), but the increase of pressure recovery indicated by the equation is much greater than the decrease caused by the reduction of friction.

If fluid is removed through suction slots during the region of adverse pressure gradient, the same procedure can be used. If the suction is not sufficient to remove all the fluid that comes from the original constant-stress layer, the equations (7.14.6, 7.14.7) become

$$\psi_c = \frac{4}{3k_0}\frac{\tau_0'^{3/2}}{\alpha}+\psi_s = \frac{\tau_0'^{1/2}}{k}z'\left(\log\frac{z'}{z_0}-1\right), \qquad (7.14.12)$$

where ψ_s is the volume flow per unit span removed by suction. The criterion for separation of the flow is then

$$\frac{4k}{3k_0}\frac{\tau_0'}{\alpha z_0}+\frac{k\psi_s}{\tau_0'^{1/2}z_0}$$

$$= \left[k\left(\frac{2\Delta P}{\tau_0} + \frac{4}{k_0^2}\right)^{1/2} - 1 \right] \exp\left[k\left(\frac{2\Delta P}{\tau_0'} + \frac{4}{k_0^2}\right)^{1/2} \right]. \quad (7.14.13)$$

If the pressure gradient is uniform, the change of pressure recovery caused by the suction is given by

$$\frac{\tau_0'^{1/2}z_0}{k} \frac{\mathrm{d}}{\mathrm{d}\psi_s}\left(\frac{\Delta P}{\tau_0'}\right)$$

$$= k^{-2}\left[k\left(\frac{2\Delta P}{\tau_0'} + \frac{4}{k_0^2}\right)^{1/2} - 1 \right] \Bigg/ \left(\frac{k\psi_s}{\tau_0'^{1/2}z_0} + \frac{4k}{3k_0}\frac{\tau_0'}{\alpha z_0}\right). \quad (7.14.14)$$

For large values of $\Delta P/\tau_0'$, approximately

$$(\Delta P/\tau_0')^{1/2} - (\Delta P/\tau_0')^{1/2}_{\psi_s=0} = \frac{1}{2^{1/2}k}\log\left(1 + \frac{3k_0\alpha\psi_s}{\tau_0'^{3/2}}\right). \quad (7.14.15)$$

Since $\Delta P/\tau_0'$ is commonly in the range 100–200, suction in quantity $\tau_0'^{3/2}/\alpha$ will increase the pressure recovery by 20–30%. For a typical gradient of $\frac{1}{2}U'^2/(x_1 - x_0)$, the volume per unit span is approximately $\frac{1}{2}\tau_0'^{1/2}l_0'u_0'/U_1'$, or about $0.2\,\tau_0'^{1/2}l_0'$, i.e. removal of about 1% of the boundary layer.

In flows approaching separation, the layer thickens very rapidly and consequently, irrespective of the flow geometry, the pressure gradient at separation is always a small fraction of the gradient a short distance upstream. In fact, the pressure distribution near separation depends on the occurrence of separation and a really useful criterion for predicting separation should be in terms of the pressure distribution in the non-separating flow with the same geometry. It may be argued that the degree of weakening of the gradient just before separation is much the same in all separating flows, so that the criterion may be used by putting the gradient at separation equal to a fraction of the gradient in the non-separating flow (Townsend 1961b). If the argument is accepted, the results for the dependence of pressure recovery on initial wall stress (7.14.10), on Reynolds number (7.14.11), and on suction (7.14.15) are unchanged. The matter has been examined by Bradshaw & Galea (1967) in the light of measurements.

Using a suitably designed duct, Stratford (1959b) has produced a layer with continuously zero wall stress that does not separate from the surface. In that flow, the condition for zero wall stress defines a pressure distribution, and the distribution measured by

11

Stratford is in good agreement with the distribution found by integration of (7.14.9). An interesting feature of the flow is that the internal boundary layer of modified flow may be self-preserving in the special sense that changes of total head and Reynolds stress along streamlines are described by scales of velocity and stream function and by universal distribution functions, i.e.

$$P + \tfrac{1}{2}U^2 - (P' + \tfrac{1}{2}U'^2) = u_s{}^2 f(\psi/\psi_s), \left.\right\}$$
$$\tau - \tau_0' = \tau_0' g(\psi/\psi_s). \quad\quad (7.14.16)$$

Here u_s and Δ_s depend only on x, and f, g are distribution functions that approach zero in the unmodified flow (large Δ/Δ_s) and that approach forms appropriate to a zero-stress equilibrium layer near the wall (Townsend 1965c).

7.15 Layer development after a sudden change of roughness

If a boundary layer passes from one kind of surface to another of different roughness, the initial changes might be calculated assuming some plausible form for the velocity distribution in the internal boundary layer of modified flow and setting a condition for conservation of flow momentum. For a thick boundary layer in negligible pressure gradient, calculated solutions have been given by Elliott (1958) and by Panofsky & Townsend (1964) but it is more informative to look at the consistency of assuming universal forms for the velocity distribution.

With no pressure gradient, changes of velocity are negligible along streamlines outside the internal layer. If the streamline through (x, z) has been displaced by $\delta(z)$ from well upstream of the roughness change at $x = x_1$, the change of velocity along the streamline, $[U(x, z) - U_1(z - \delta)] = V$, is assumed to have the self-preserving form,

$$V = (u_0/k) f(z/l_0) \quad\quad (7.15.1)$$

($U_1(z)$ is the velocity distribution upstream of the change of roughness, u_0 and l_0 are scales of velocity and length). The changes of Reynolds stress are also assumed self-preserving, i.e.

$$\tau(x, z) - u_1{}^2 = (\tau_0 - u_1{}^2) g(z/l_0), \quad\quad (7.15.2)$$

where $u_1 = (\tau_0')^{1/2}$ is the friction velocity upstream, and it is supposed that all the streamlines in the modified layer passed through the constant-stress layer of the original flow. Necessary conditions for the development are obtained in the usual way by substituting the distributions in the mean value equations for velocity and energy, and requiring that the coefficients of significant terms should remain in constant ratios.

First consider the conditions imposed on the distributions by the boundaries. For large values of $\eta = z/l_0$ in the unmodified flow, both distribution functions become small. For small values near the wall, the conditions for an equilibrium layer are met and the distributions are

$$U = \frac{\tau_0^{1/2}}{k} \log(z/z_0), \qquad \tau = \tau_0 \qquad (7.15.3)$$

The velocity distribution upstream of the change is

$$U_1(z) = (u_1/k) \log(z/z_1) \qquad (7.15.4)$$

and it may be shown (Townsend 1965a, 1966a) that

$$\left.\begin{array}{l} f(\eta) \sim \log(\eta) + C \text{ for small } \eta, \\ \tau_0^{1/2} = u_1 + u_0\{1 + [\log(l_0/z_1) - C_0]^{-1}\}, \\ u_0/u_1 = -M/[\log(l_0/z_0) - C + 1], \end{array}\right\} \qquad (7.15.5)$$

if the distributions described by (7.15.1) and (7.15.3) are to coincide near the wall. (C_0 is a constant of order one whose value depends on the velocity distribution function, $M = \log(z_1/z_0)$.)

For large values of $\log(l_0/z_0)$ and not too large changes of roughness, the change of surface stress is small and nearly, $\tau_0 - u_1^2 = 2u_1 u_0$. Then, in the outer part of the modified flow, the changes of velocity both along a streamline and at constant height are small and the equation for the mean velocity is nearly

$$u_1 \log \frac{z}{z_1}\left[\frac{du_0}{dx} f - \frac{u_0}{l_0}\frac{dl_0}{dx} \eta f'\right]$$
$$+ \frac{u_1}{l_0}\left[u_0\frac{dl_0}{dx} f - \frac{d(u_0 l_0)}{dx}\eta^{-1}\int_0^{} f\,d\eta\right] = 2k^2 u_0 u_1 g' \qquad (7.15.6)$$

omitting terms quadratic in the velocity change. From (7.15.5), the ratio of du_0/dx to dl_0/dx is of order $(u_0/l_0)[\log(l_0/z_0)^{-1}]$ and, for the moderate values of $\log \eta$ that refer to flow outside the constant-stress layer, the equation reduces to

$$-\log\left(\frac{l_0}{z_1}\right)\frac{dl_0}{dx}\eta f' = 2k^2 g'. \qquad (7.15.7)$$

It appears that approximately self-preserving development is possible with a scale determined by

$$\log(l_0/z_1)\frac{dl_0}{dx} = 2k^2 \qquad (7.15.8)$$

if $\log(l_0/z_0)$ is large and u_0/u_1 is small, since

(1) in the outer part of the flow where flow accelerations are appreciable, the full equation for the mean velocity is nearly of self-preserving form, and

(2) in the equilibrium layer, the self-preserving distribution of velocity is consistent with the logarithmic distribution if the conditions (7.15.5) are met.

To the approximation used, the development equation (7.15.8) may be written as an equation for the stream function at height l_0,

$$\frac{d\psi(l_0)}{dx} = 2ku_1, \qquad (7.15.9)$$

very similar to equations (7.13.5, 7.13.8).

To the same approximation, the equation for the turbulent energy is satisfied by self-preserving distributions of the changes in turbulent intensity, dissipation and lateral diffusion of energy. If it is assumed (1) that the stress–intensity ratio, $a_1 = \tau/\overline{q^2}$, is constant, and (2) that the energy dissipation rate is $\varepsilon = \tau^{3/2}/(kz)$ (i.e. the dissipation length is unchanged, remaining the same fraction of distance from the wall), the energy equation becomes

$$f' - \eta^{-1}g = kD - \frac{2k^2}{a_1}\eta g', \qquad (7.15.10)$$

where $D(\eta) = l_0(\partial/\partial z)(\overline{pw} + \frac{1}{2}\overline{q^2 w})/(u_1{}^2 u_0)$ describes the lateral diffusion of energy in the modified flow. Using the scale given by equation (7.15.8) and the mean flow equation (7.15.7), it appears as an equation for the stress distribution function,

$$g + g' = \frac{2k^2}{a_1}\eta^2 g' - k\eta D. \qquad (7.15.11)$$

The stress distribution depends on the magnitude and nature of the diffusion term. If diffusion is negligible, the solution is

$$g = \left\{ \left[1 - \left(\frac{2k^2}{a_1}\right)^{1/2}\eta \right] \bigg/ \left[1 + \left(\frac{2k^2}{a_1}\right)^{1/2}\eta \right] \right\}^{(k^2/8a_1)^{1/2}} \qquad (7.15.12)$$

for $\eta < (\tfrac{1}{2}a_1)^{1/2}/k$, implying a sharp boundary to the region of modified flow. It is likely that diffusion is appreciable and one possibility is that its magnitude is described by an eddy diffusion coefficient comparable with the local eddy viscosity of the unmodified flow, that is,

$$\overline{pw}+\tfrac{1}{2}\overline{q^2 w} = -\alpha k u_1 z \frac{\partial}{\partial z}(\tfrac{1}{2}\overline{q^2}). \tag{7.15.13}$$

The equation is then

$$g+g' = \frac{2k^2}{a_1}\eta^2 g' + \frac{\alpha k^2}{a_1}(\eta g' + \eta^2 g'') \tag{7.15.14}$$

which may be integrated with boundary condition $g(0) = 1$.

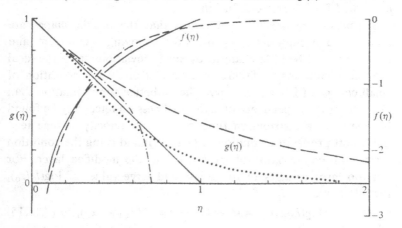

Fig. 7.15. Profiles for changes in stress and mean velocity for self-preserving development after a change of surface roughness. ——, Elliott; – – –, mixing length; . . . ., $k^2 = a_1$, $\alpha = 1$; –.–.–, $k^2 = a_1$, $\alpha = 0$.

Stress distribution functions calculated from (7.15.14) are shown in fig. 7.15 for $a_1 = k^2$ and two values of the diffusion constant, $\alpha = 0$ and 1. They may be compared with the distributions implied by (7.15.7) for the velocity profiles assumed by Elliott,

$$f(\eta) = \log(\eta), \qquad g(\eta) = 1 - \eta$$

by Panofsky & Townsend,

$$f(\eta) = \log(\tfrac{1}{2}\eta) + 1 - \tfrac{1}{2}\eta, \qquad g(\eta) = \tfrac{1}{4}(2 - \eta)^2$$

and with the distribution for $k^2/a_1 = 0$, indicating zero effect of advection of turbulent energy and equivalent to making the mixing-length assumption that

$$\partial U/\partial z = \tau^{1/2}/(kz).$$

It is interesting to notice how advection of energy reduces the thickness of the modified layer and that not even the implausibly large value of the diffusion constant does much more than smooth over the sharp edge predicted for zero diffusion.[†]

Velocity distributions after a change of roughness have been measured in the atmospheric boundary layer (e.g. Bradley 1968) and in the wind tunnel (Antonia & Luxton 1971), but not the more distinctive stress distributions. It appears that the profile shapes fall somewhere between the simple logarithmic form of Elliott and the calculated form for zero diffusion.

Self-preserving development means that the profile shapes are approached asymptotically as forms for a moving equilibrium, and the shapes will be little changed by small deviations from the ideal external conditions or failure to satisfy strongly the condition of small change of friction velocity. Then a better approximation than (7.15.8) for the dependence of layer thickness on fetch may be found by assuming a distribution function for the velocity change (e.g. the Elliott profile or a solution of (7.15.14)) and using the condition for overall conservation of momentum in the modified layer. For not too large changes of roughness and large values of $\log(l_0/z_0)$, the relation is

$$\log(l_0/z_0) - \tfrac{1}{2}M - C_0 - 2 = 2k^2(x - x_0), \quad (7.15.15)$$

where

$$C_0 = -\int_0^\infty f(\eta) \log(\eta)\, d\eta \Big/ \int_0^\infty f(\eta)\, d\eta$$

(Townsend 1966a).

If the change of friction velocity is not small, the development may still be nearly self-preserving but the profiles will be different. An extreme example is the transition from a smooth surface to an extremely rough one. Compared with the wall stress after the

[†] Application of the diffusion assumption (7.15.13) to the energy balance in the zero-stress equilibrium layer leads to the result,

$$k/k_0 = 1 - \tfrac{3}{2}\alpha k^2/(2a_1).$$

The available measurements suggest that k/k_0 cannot be much less than 0.8, i.e. α is not more than 0.3.

transition, the stress before is so small that the velocities in the un-modified flow are nearly constant on all streamlines that lie outside the equilibrium layer of the internal boundary layer. The flow is then nearly the same as in a boundary layer on the rough surface with uniform stream velocity, and it is expected to have a typical boundary-layer profile.

7.16 Boundary layers with three-dimensional mean flow

It is not uncommon for boundary layers to have distributions of velocity that are not two-dimensional, one example being the layer on a swept-back wing where the pressure gradients are not in the same direction as the free stream velocity. Using the boundary-layer approximation, only the two shear components of the Reynolds stress tensor are significant, i.e.

$$\tau_{13} = -\overline{uw}, \qquad \tau_{23} = -\overline{vw},$$

and the equations of mean flow are

$$U\frac{\partial U}{\partial x} + V\frac{\partial U}{\partial y} + W\frac{\partial U}{\partial z} = -\frac{\partial P}{\partial x} + \frac{\partial \tau_{13}}{\partial z}, \left. \right\}$$
$$U\frac{\partial V}{\partial x} + V\frac{\partial V}{\partial y} + W\frac{\partial V}{\partial z} = -\frac{\partial P}{\partial y} + \frac{\partial \tau_{23}}{\partial z}, \left. \right\} \qquad (7.16.1)$$

where $P + \frac{1}{2}(U_1{}^2 + V_1{}^2)$, the total head in the free stream, is every-where the same if the free stream flow is irrotational.

Integrating the equations through the layer and using the condition of incompressibility leads to

$$\frac{\partial}{\partial x}\int_0^\infty U(U-U_1)\,\mathrm{d}z + \frac{\partial}{\partial y}\int_0^\infty V(U-U_1)\,\mathrm{d}z$$
$$+ \frac{\partial U_1}{\partial x}\int_0^\infty (U-U_1)\,\mathrm{d}z + \frac{\partial U_1}{\partial y}\int_0^\infty (V-V_1)\,\mathrm{d}z = -\tau_x, \left. \right\}$$
$$\frac{\partial}{\partial x}\int_0^\infty U(V-V_1)\,\mathrm{d}z + \frac{\partial}{\partial y}\int_0^\infty V(V-V_1)\,\mathrm{d}z \qquad (7.16.2)$$
$$+ \frac{\partial V_1}{\partial x}\int_0^\infty (U-U_1)\,\mathrm{d}z + \frac{\partial V_1}{\partial y}\int_0^\infty (V-V_1)\,\mathrm{d}z = -\tau_y, \left. \right\}$$

where τ_x, τ_y are the components of the wall stress. If the free stream flow is everywhere parallel to Ox, $V_1 = 0$ and the equations simplify to

$$\left.\begin{array}{c}\dfrac{\partial}{\partial x}\displaystyle\int_0^\infty U(U-U_1)\,\mathrm{d}z+\dfrac{\partial}{\partial y}\displaystyle\int_0^\infty V(U-U_1)\,\mathrm{d}z\\[2mm]+\dfrac{\mathrm{d}U_1}{\mathrm{d}x}\displaystyle\int_0^\infty (U-U_1)\,\mathrm{d}z = -\tau_x,\\[3mm]\dfrac{\partial}{\partial x}\displaystyle\int_0^\infty UV\,\mathrm{d}z+\dfrac{\partial}{\partial y}\displaystyle\int_0^\infty V^2\,\mathrm{d}z = -\tau_y.\end{array}\right\}\qquad (7.16.3)$$

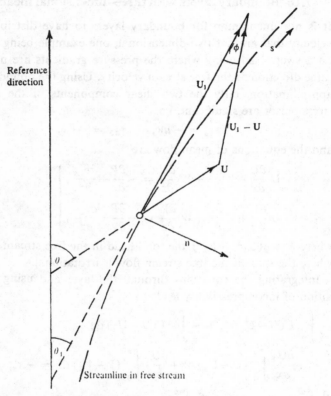

Fig. 7.16. Diagram showing notation and co-ordinates for three-dimensional mean flow in a boundary layer.

Not much is known about the turbulent motion in three-dimensional layers but, remembering that the observed structure of two-dimensional layers is very similar to that produced by a finite rapid shear of simple isotropic turbulence, a relation between the directions of velocity shear and shear stress can be formulated. Since we are concerned only with plane shearing and two shear components of Reynolds stress, both may be represented by two-dimensional vectors. The nature of the approximations in the rapid distortion theory is such that the structure of the turbulence depends only on the total distortion and is not affected by the way in which the distortion is achieved. The distortion is always a plane shear of magnitude $\alpha(t)$ and the (vector) Reynolds shear stress is in the same direction, say making an angle θ with the Ox axis. Then if ϕ is the direction of the current (vector) rate of shear $\dot{\alpha}$, consideration of the vector diagram shows that

$$\frac{d\theta}{dt} = \sin(\phi - \theta)\frac{1}{|\alpha|}\left|\frac{d\alpha}{dt}\right|$$
$$= \sin(\phi - \theta)\frac{1}{|\alpha|}\left[\left(\frac{\partial U}{\partial z}\right)^2 + \left(\frac{\partial V}{\partial z}\right)^2\right]^{1/2} \tag{7.16.4}$$

since $\dot{\alpha}$ has components $(\partial U/\partial z, \partial V/\partial z)$.

Earlier consideration of the relation between observations and the predictions of the rapid-distortion theory suggested that the actual stress–intensity ratios are near those for effective total strains which vary from one flow to another and from one part of a flow to another. For wall flows, effective strains are probably in the range 6 to 12, and putting a value in equation (7.16.4) provides a relation between the directions of Reynolds stress and rate of shear. If the angle difference is small, the equation is

$$\frac{D}{Dt}(\theta - \phi) + \frac{[(\partial U/\partial z)^2 + (\partial V/\partial z)^2]^{1/2}}{|\alpha|}(\theta - \phi) = -\frac{D\phi}{Dt}, \tag{7.16.5}$$

where D/Dt indicates variation following the mean flow. The simple meaning is that the stress direction lags behind the direction of rate of shear with a characteristic time for adjustment of

$$t_\theta = |\alpha|\left/\left[\left(\frac{\partial U}{\partial z}\right)^2 + \left(\frac{\partial V}{\partial z}\right)^2\right]^{1/2}\right. . \tag{7.16.6}$$

If the change in direction of the rate of shear is considerable along a streamline during a time of flight comparable with t_θ, the direction of Reynolds stress will be appreciably different from the current direction of shear. Some idea of the magnitude of the difference can be obtained by calculating the characteristic time for a constant-stress layer. It is

$$t_\theta = k|\alpha|z/\tau_0^{1/2}, \tag{7.16.7}$$

and it is of similar magnitude to the adjustment time for the turbulent
energy and stresses,

$$t_d = \tfrac{1}{2}\overline{q^2}/\varepsilon = \frac{1}{2a_1}\frac{kZ}{\tau_0^{1/2}}.$$

The condition for an equilibrium layer is that surface conditions
should be homogeneous over distances with times of flight long
compared with t_d, and so the angle difference cannot be appreciable
in an equilibrium layer. In general, the difference may be appreciable
only in parts of the layer where the effects of energy advection are
large, normally only in the outer region of entrainment. It should
be emphasised that flow conditions leading to large differences
between the directions of shear and stress are commonly those in
which the Reynolds stresses have little influence on the flow
development.

J. P. Johnston (1970) has measured mean velocities, turbulent
intensities and stresses in a boundary layer approaching a swept-
back step, and finds angular differences up to fifteen degrees. He
compares the angles of shear and of stress with Bradshaw's (1971)
method for calculating the development of three-dimensional
boundary layers, which equates the angular adjustment time to t_d,
the adjustment time for turbulent energy. Except in the equilibrium
layer, the calculated angles of stress are within three degrees of the
original direction and use of the adjustment time given by (7.16.6)
would make the change even less. The observed values over the
outer part of the flow are in the opposite sense to the calculated
ones, but it is possible that the differences are within experimental
uncertainty. It appears likely that the vector shear stress remains
nearly constant along streamlines of the mean flow that lie in the
outer layer, and that the observed distributions of mean velocity are
nearly those for constant total head along streamlines.

7.17 Three-dimensional flow with negligible Reynolds stresses

In the three-dimensional flow of a boundary layer approaching a
swept-back step, parcels of turbulent fluid in the outer flow are
carried from the upstream, two-dimensional flow into the three-
dimensional flow in times that are short compared with the adjustment times for Reynolds stresses, and the development of the outer

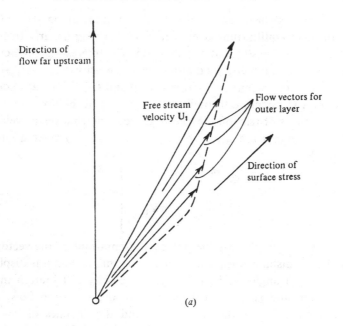

Direction of
flow far upstream

Free stream
velocity U_1

Flow vectors for
outer layer

Direction of
surface stress

(a)

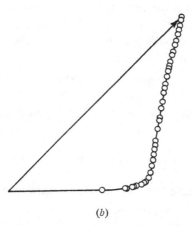

(b)

Fig. 7.17. (a) Plots of mean flow vectors for yawing flow, showing the 'triangle'
effect in the outer flow: (b) typical plot (Perry & Joubert 1965).

flow is nearly that of an inviscid fluid with the same initial velocity
distribution. The situation resembles that discussed in § 7.13, and an
an approximate description could be obtained by assuming an outer

layer with total head and vector stress constant along streamlines and an inner equilibrium layer with stress varying linearly from its wall value to the stress at the junction. An interesting prediction, made by Johnston (1960), is that the vector velocity defect $(\mathbf{U}_1 - \mathbf{U})$, where $\mathbf{U}_1$ is the velocity in the free stream and $\mathbf{U}$ is the local velocity, is constant in direction through the outer part of the flow.

If the effects of friction may be neglected, the changes of velocity along streamlines are given to the usual boundary-layer approximation, by

$$\left. \begin{aligned} \mathscr{U}\, \frac{\partial \mathscr{U}}{\partial s} &= \mathscr{U}_1\, \frac{\partial \mathscr{U}_1}{\partial s}\,, \\ \mathscr{U}^2\, \frac{\partial \theta}{\partial s} &= \mathscr{U}_1\, \frac{\partial \mathscr{U}_1}{\partial n}\,, \end{aligned} \right\} \tag{7.17.1}$$

where $(\mathscr{U}, \theta)$, $(\mathscr{U}_1, \theta_1)$ are the polar representations of the vectors $\mathbf{U}$ and $\mathbf{U}_1$, s is distance measured along a streamline, and n is displacement at right angles to it. For a given velocity distribution in the free stream and specified values of $\mathbf{U}$ in the upstream flow, it is possible to calculate the streamlines and the changes of velocity along them, and so to verify that the direction of the velocity defect at one position is the same for defects up to about 50%.

For small defects, the streamlines in the layer are nearly co-incident in plan with the streamlines of the free stream, and the second equation (7.17.1) becomes

$$\mathscr{U}^2\, \frac{\partial \theta}{\partial s} = \mathscr{U}_1{}^2\, \frac{\partial \theta_1}{\partial s}\,. \tag{7.17.2}$$

since by the first equation $(\mathscr{U}^2 - \mathscr{U}_1{}^2)$ is constant along a streamline, the angular difference is

$$\theta - \theta_1 = (\mathscr{U}_1{}^2 - \mathscr{U}^2) \int_0 \frac{1}{\mathscr{U}^2}\, d\theta_1. \tag{7.17.3}$$

Then ϕ, the angle between the free stream velocity and the velocity defect (fig. 7.16), is given by

$$\tan \phi = \frac{\mathscr{U}_1}{\mathscr{U}_1 - \mathscr{U}} (\theta - \theta_1)$$

$$= 2\mathscr{U}_1{}^2 \int_0 \frac{1}{\mathscr{U}_1{}^2}\, d\theta_1 \tag{7.17.4}$$

to the approximation that $\mathscr{U}_1 - \mathscr{U} \ll \mathscr{U}_1$. Since the magnitude of the free stream velocity usually decreases as the flow becomes three-dimensional, $\tan \phi$ should be a little less than $2\theta_1$.

The variation of flow direction with height may be displayed by plotting velocity vectors on a diagram, and, if the flow has undergone rapid changes of direction, the ends of the vectors define a nearly triangular area (fig. 7.17). The vectors of the outer flow have their ends along a nearly straight line with a slope that approximates to the angle given by (7.17.4), and, except for a small blending region, the vectors of the inner flow are all nearly in the same direction and parallel to the surface stress. Examples are given by Johnston (1960) and by Perry & Joubert (1965).

7.18 Homogeneous three-dimensional flow — the Ekman layer

If the flow boundaries and the undisturbed fluid are rotating uniformly but otherwise the surface and flow are homogeneous over planes parallel to the surface, a turbulent Ekman layer of uniform thickness forms that is of importance as an ideal form of the boundary layers in the ocean and in the atmosphere.

If flow is described in co-ordinates rotating with angular velocity Ω with respect to an inertial frame, the equations of motion include additional terms for the Coriolis force $2U \times \Omega$ and the centrifugal force $(\Omega \times x) \times \Omega$. With an incompressible fluid, the centrifugal force is of the form $\mathrm{grad}(\frac{1}{2}(\Omega \times x)^2)$ and merely changes the reference hydrostatic pressure, but Coriolis forces alter the nature of the flow. Assuming steady flow parallel to the surface with mean values dependent only on height above the surface, the equations for the mean velocity are

$$
\left.
\begin{aligned}
0 &= -\frac{\partial P}{\partial x} + \frac{\partial \tau_{13}}{\partial z} + fV, \\[2mm]
0 &= -\frac{\partial P}{\partial y} + \frac{\partial \tau_{23}}{\partial z} - fU, \\[2mm]
0 &= -\frac{\partial P}{\partial z} + \frac{\partial \tau_{33}}{\partial z} + 2\Omega_2 U - 2\Omega_1 V,
\end{aligned}
\right\}
\tag{7.18.1}
$$

where τ_{ij} is the Reynolds stress tensor, and $f = 2\Omega_3$ is the Coriolis parameter for horizontal flow. Since U, V and τ_{33} depend only on height, the third equation shows that the horizontal components of $\mathrm{grad}\, P$ are also independent of height, and the first two equations relate the mean velocity to the constant pressure gradient and to the stresses. Writing two-dimensional vectors as complex quantities, $\mathbf{U} = U + iV$ and $\tau = \tau_{13} + i\tau_{23}$, the equations become

$$\frac{d\tau}{dz} = if(\mathbf{U} - \mathbf{G}), \qquad (7.18.2)$$

where $\mathbf{G} = -(\partial P/\partial y - i\,\partial P/\partial x)/f$ is the gradient wind outside the boundary layer where stresses are negligible. The condition for overall conservation of momentum is expressed by integrating (7.18.2) over the whole layer, and it gives the surface stress as

$$\tau_0 = -if \int_0^\infty (\mathbf{U} - \mathbf{G})\,dz. \qquad (7.18.3)$$

Near the surface, rates of shear are expected to be large compared with the rate of rotation and the flow there is essentially that of a constant-stress equilibrium layer. In the outer flow, rotation makes the flow change in direction with height and the Reynolds stresses vary in magnitude and direction, but it appears possible that the distributions of velocity defect and turbulent stress have universal forms analogous to those for channel flow. That is,

$$\begin{aligned}
\mathbf{U} - \mathbf{G} &= \mathbf{u}_0 F(z/h) \\
\tau &= \tau_s G(z/h),
\end{aligned} \qquad (7.18.4)$$

where h is an effective depth, and $\mathbf{u}_0$, τ_s are vector scales of velocity and stress. Since the universal forms are valid near the surface in the constant-stress layer, they must be consistent with the appropriate forms,

$$\mathbf{U} = \frac{\tau_0}{k\tau_0^{1/2}} \log(z/z_0) \quad \text{and} \quad \tau = \tau_0$$

The conditions are that

$$\left.\begin{aligned}
&F(z/h) \to \log(z/h) + C \text{ for small } z/h, \\
&\mathbf{u}_0 = \frac{\tau_0}{k\tau_0^{1/2}} \quad \text{and} \quad \tau_s = \tau_0, \\
&\mathbf{G} = \frac{\tau_0}{k\tau_0^{1/2}}[\log(h/z_0) - C].
\end{aligned}\right\} \qquad (7.18.5)$$

and

A further condition is provided by (7.18.3), that

$$\tau_0 = -if \int_0^\infty (\mathbf{U} - \mathbf{G})\,dz = -ifhI_1\tau_0/(k\tau_0^{1/2}), \qquad (7.18.6)$$

where

$$I_1 = \int_0^\infty F(z/h)\, dz/h.$$

Dividing by the surface stress,

$$fh/\tau_0^{1/2} = ik/I_1. \tag{7.18.7}$$

The last equation shows that universal forms for velocity and stress distributions are not only consistent with the equation of mean flow but that they imply that the Rossby number for turbulent flow $fh/\tau_0^{1/2}$ would be the same in all layers. Unless the horizontal components of the rotation influence the motion, the hypothesis of universal forms is dynamically self-consistent.

From the last of the conditions (7.18.5), the dependence of the surface stress on the external conditions of the flow, G, f and z_0, can be found in terms of the flow constants, C and I_1. The magnitude is given by

$$k^2 G^2/\tau_0 = [\log(h/z_0) - C_r]^2 + C_i^2 \tag{7.18.8}$$

and θ, the difference of angle between the gradient wind and the surface stress, by

$$\sin\theta = -C_i\tau_0^{1/2}/(kG), \tag{7.18.9}$$

where $C = C_r + iC_i$.

To establish values for the flow constants requires a knowledge of the velocity profile, obtained either from observations or from some theoretical model of the flow. Following the procedure used for other wall flows, we suppose the profile to be approximately the composite form,

$$\begin{aligned}
U - G &= A\exp[-(1+i)z/h] \quad \text{for } z > \alpha h, \\
&= \frac{\tau_0}{k\tau_0^{1/2}}\log(z/z_0) - G \quad \text{for } z < \alpha h,
\end{aligned} \tag{7.18.10}$$

where α and A are flow constants. If velocity across the junction is continuous,

$$A\exp[-(1+i)\alpha] = \frac{\tau_0}{k\tau_0^{1/2}}\log(h/z_0) - G \tag{7.18.11}$$

but it is not possible to require that the gradients be continuous. Assume that the magnitude of the gradient does not change across the junction. Then

$$\frac{\tau_0}{k\tau_0^{1/2}} = -(1+i)\alpha A\exp[-(1+i)\alpha + i\phi], \tag{7.18.12}$$

where ϕ is the change of direction across the junction, and the flow integral may be calculated to be

$$I_1 = -[1+\alpha(1+i)+2i\alpha^2 e^{i\phi}]/(2ie^{i\phi}).$$

If equation (7.18.7) is valid, $\mathbf{I}_1$ must be a positive imaginary number, possible if

$$\tan \phi = \alpha/(1 + \alpha). \qquad (7.18.13)$$

Then

$$I_1 = \frac{i}{2\alpha}\frac{1+\alpha+2\alpha^2 \cos \phi}{\cos \phi} \qquad (7.18.14)$$

and

$$\frac{fh}{k\tau_0^{1/2}} = \frac{2\alpha}{1+\alpha}+O(\alpha^3). \qquad (7.18.15)$$

Using (7.18.11, 7.18.12) and the value of ϕ, the relation between surface stress and gradient wind is

$$G = \frac{\tau_0}{k\tau_0^{1/2}}\left[\log(h/z_0)+\log \alpha+\frac{e^{-i\phi}}{\alpha(1+i)} \right] \qquad (7.18.16)$$

of the same form as the third of the conditions (7.18.5).

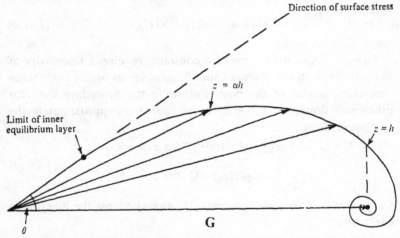

Fig. 7.18. Plot of mean flow vectors for a turbulent Ekman layer (diagrammatic only).

The predictions based on the assumed profile may be compared with atmospheric observations given by Plate (1971). Measurements of the angle between gradient wind and surface stress are well described by

$$\sin \theta = -10.7\,\tau_0^{1/2}/G$$

corresponding to $\alpha = 0.128$. The magnitudes of surface stresses are less dependent on the value of α, and equation (7.18.6) provides a good description of the somewhat scattered observations, particularly those analysed by Blackadar (1962).

The Ekman layer is of special interest as a turbulent flow with zero net rate of entrainment, a property that depends on the rotation. The basic limit to the upward growth of the layer depends on the rotation of direction of the absolute momentum being transferred to the fluid, which causes an evanescent wave to propagate upwards and which, if the effective eddy viscosity were constant, would have the exponential form of (7.18.10). In reality, interaction between the turbulent motion and the mean flow and effects of the rotation on the turbulence lead to a much more sharply defined top to the layer. In the first place, using the equation for turbulent energy with similarity assumptions of the kind used by Bradshaw *et al.* (see § 7.12) leads to predicted profiles with a sharp limit, but it is likely that the vector stress is at right angles to the vector velocity gradient at the upper edge of the layer. There the adjustment time for direction of stress becomes large compared with the period of rotation and so, viewed from an inertial frame, the stress lags nearly a quarter-period behind the rate of shear and very little energy can be transferred to the turbulence. Energy may also be lost by radiation of inertial waves with periods longer than the reciprocal of the Coriolis parameter.

7.19 Secondary flow in a boundary layer with a free edge

Turbulent channel flow with mean velocity everywhere in the same direction is possible only between parallel planes or in channels with axisymmetric boundaries, i.e. pipes of circular section or between concentric circular cylinders. With any other section, a pattern of secondary flow in the transverse yOz-plane is superimposed on the axial flow. The secondary flow is a consequence of the non-Newtonian form of the Reynolds stress tensor and it does not occur for laminar flow of a Newtonian fluid along a uniform channel of any section. Secondary flow of similar origin will occur in a boundary layer on a flat plate if the surface stress is not homogeneous normal to the stream direction, and a simple example is the flow over a flat plate with free edges parallel to the direction of flow.

Let the flat plate be of width D and occupy the plane $z = 0$ between the edges, $x = 0$ and $y = 0, D$. The equations for the mean velocity components in the Oy- and Oz- directions are

$$\left.\begin{aligned}
\left(U\frac{\partial}{\partial x}+V\frac{\partial}{\partial y}+W\frac{\partial}{\partial z}\right)V &= -\frac{\partial P}{\partial y}+\frac{\partial\tau_{12}}{\partial x}+\frac{\partial\tau_{22}}{\partial y}+\frac{\partial\tau_{23}}{\partial z}, \\
\left(U\frac{\partial}{\partial x}+V\frac{\partial}{\partial y}+W\frac{\partial}{\partial z}\right)W &= -\frac{\partial P}{\partial z}+\frac{\partial\tau_{13}}{\partial x}+\frac{\partial\tau_{23}}{\partial y}+\frac{\partial\tau_{33}}{\partial z}.
\end{aligned}\right\} \quad (7.19.1)$$

Consider flow on the upper surface and take line integrals of the equations from a point in the layer to the free stream along paths of

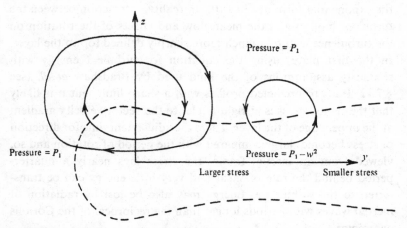

Fig. 7.19. Section of a boundary layer near a free edge, showing co-ordinate system and conjectural streamlines of the secondary flow.

constant x and y and constant x and z (fig. 7.19). If the point is distant from the free edge, to the boundary layer approximation

$$\left.\begin{aligned}
\int_{-\infty}\left(U\frac{\partial V}{\partial x}+V\frac{\partial V}{\partial y}+W\frac{\partial V}{\partial z}\right)&\mathrm{d}y \\
&= P_0-P+\tau_{22}+\int_{-\infty}\frac{\partial\tau_{23}}{z}\,\mathrm{d}y, \\
0 &= P_0-P+\tau_{33}+\int_{-\infty}\frac{\partial\tau_{23}}{\partial y}\,\mathrm{d}z,
\end{aligned}\right\} \quad (7.19.2)$$

where P_0 is the pressure in the free stream, and so

$$\tau_{33}-\tau_{22} = \int_{-\infty}\frac{\partial\tau_{23}}{\partial z}\,\mathrm{d}y-\int_{\infty}\frac{\partial\tau_{23}}{\partial y}\,\mathrm{d}z$$

$$-\int_{-\infty} \left(U \frac{\partial V}{\partial x} + V \frac{\partial V}{\partial y} + W \frac{\partial V}{\partial z} \right) dy. \qquad (7.19.3)$$

The characteristically 'turbulent' term is

$$\tau_{33} - \tau_{22} = \overline{v^2} - \overline{w^2}$$

the difference of the normal stresses, which is not zero in wall flows and which describes the driving force of the secondary flow. The equation balances the driving force by the sum of forces developed by transverse shear stresses, τ_{23}, and the transverse flow accelerations. Since transverse shear stresses are most likely to be generated by gradients of the transverse flow, the inequality of the normal stresses requires a cross-flow at least in the neighbourhood of the free edge.

For an estimate of the magnitude of the secondary flow, neglect the term involving flow accelerations and suppose that L, the lateral extent of edge-influenced flow is considerably more than the layer thickness. Then, nearly,

$$\int_{-\infty} \frac{\partial \tau_{23}}{\partial z} dy = \overline{v^2} - \overline{w^2} \qquad (7.19.4)$$

and a typical value of $\partial \tau_{23}/\partial z$ is $(v^2 - w^2)/L$. In an equilibrium layer, $\overline{v^2} - \overline{w^2} \simeq 1.5\,\tau$ and, assuming a stress distribution far from the edge, $\tau = \tau_0(1 - z/\delta)$, a typical distribution of transverse stress near the edge is

$$\tau_{23} = -0.75\tau_0 \frac{\delta}{L}(1-z/\delta)^2. \qquad (7.19.5)$$

At the wall, the vector stress is directed toward the free edge at angles of about $0.75\,\delta/L$ with the mean flow in the free stream, and flow near the wall will be nearly in the same direction. The outflow towards the edge by inflow normal to the surface and the consequent momentum flux leads to an increase in the wall stress in addition to any arising from the flow geometry.

The secondary flow has been studied in some detail by Elder (1960), using flat plates at Reynolds numbers up to 10^6. Measuring the longitudinal vorticity of the secondary flow, he finds that it is concentrated near the free edges, and so the associated circulation extends one or two boundary-layer thicknesses into the free stream and on to the plate (fig. 7.20). More than a few thicknesses from the edge, the flow is unaffected by the secondary flow and the motion is very nearly that of a truly two-dimensional layer. The conclusion that the central flow is unaffected by the edges is supported by careful measurement by Davies & Young (1964) at larger Reynolds numbers. They find that the lateral distributions of surface stress have the similarity form,

$$\tau_0(y) - \tau_0(\infty) = \tau_0(\infty)G(y/\delta), \qquad (7.19.6)$$

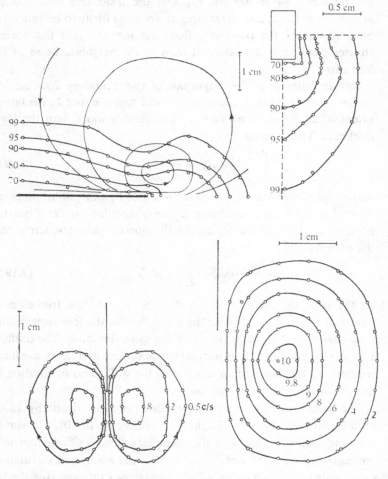

Fig. 7.20. Observations of secondary flow near the free edge of a thin plate by Elder (1960).

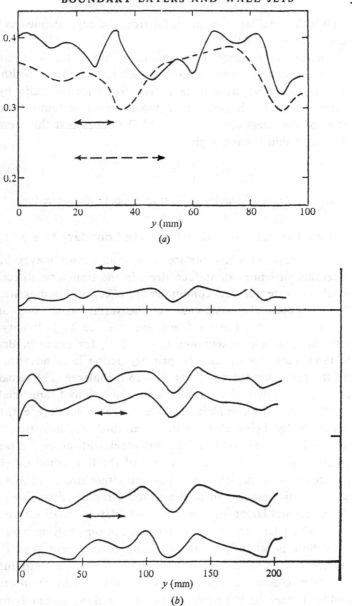

Fig. 7.21. Lateral variations of surface stress and mean velocity in boundary layers. (a) Wall stress variations at two distances from layer origin. (b) Mean velocity variations across layer. (Arrow-ended lines indicate approximate total thickness of the boundary layers)

where δ is the total layer thickness far from the edge (defined as the distance where $U = 0.99\, U_1$).

A practical consequence of the secondary flow is that the average skin friction on a flat plate of finite aspect ratio (ratio of width to length) exceeds that on a plate of effectively infinite ratio by a quantity that is much larger than would occur in laminar flow. Validity of the stress distribution (7.19.5) implies that the average coefficient of skin friction is given by

$$C_f = (C_f)_\infty\left(1 + A\frac{\delta}{D}\right). \tag{7.19.7}$$

The measurements of Davies & Young indicate that $A \simeq 0.4$.

7.20 Lateral variations of stress in boundary layers

Boundary layers on a flat surface in a wind tunnel may exhibit considerable variations of surface stress in the transverse direction although the external flow conditions are effectively homogeneous in that direction. The basic cause of the variations is commonly small spatial variations in the free stream, caused by instability of flow through turbulence-damping screens (see, for example, Bradshaw 1965, Patel 1964), and the primary action is to advance or retard the transition from laminar flow to turbulence. The remarkable feature is that the variations persist without appreciable attenuation over the complete observed range of development, the principal change being that, as the layer thickness increases, the profiles of lateral variation become smoother with much the same amplitude (fig. 7.21). Even if each slice of the flow could develop independently of its neighbours, slices with larger stress and smaller thickness would increase in thickness more rapidly than those of smaller stress and larger thickness, and the relative variations would diminish with fetch. The fact that they do not diminish must mean that the flow is laterally unstable during development, and that small variations of suitable scale increase up to a critical amplitude.

A likely mechanism of instability is the induction by the normal Reynolds stresses of a pattern of secondary flow, directed from regions of large stress to ones of small stress (fig. 7.22). Consider the equation for the longitudinal component of the mean vorticity, $\Omega_1 = \partial W/\partial y - \partial V/\partial z$

$$\frac{D\Omega_1}{Dt} = \Omega_1 \frac{\partial U}{\partial x} + \Omega_2 \frac{\partial V}{\partial y} + \Omega_3 \frac{\partial U}{\partial z}$$

$$+ \frac{\partial^2 \tau_{33}}{\partial y\, \partial z} - \frac{\partial^2 \tau_{23}}{\partial z^2} - \frac{\partial^2 \tau_{22}}{\partial y\, \partial z} + \frac{\partial^2 \tau_{23}}{\partial y^2}$$

$$= \Omega_1 \frac{\partial U}{\partial x} - \frac{\partial W}{\partial x} \frac{\partial U}{\partial y} + \frac{\partial V}{\partial x} \frac{\partial U}{\partial z}$$

$$+ \frac{\partial^2}{\partial y\, \partial z}(\tau_{33} - \tau_{22}) - \frac{\partial^2 \tau_{23}}{\partial z^2} + \frac{\partial^2 \tau_{23}}{\partial y^2} \,. \tag{7.20.1}$$

If the vorticity changes slowly in the stream direction, the terms describing advection and production of vorticity by straining are small and, nearly,

$$\frac{\partial^2}{\partial y\, \partial z}(\tau_{33} - \tau_{22}) = \left(\frac{\partial^2}{\partial z^2} - \frac{\partial^2}{\partial y^2}\right)\tau_{23} \tag{7.20.2}$$

expressing a balance between the force generated by the normal stresses and the effect of the transverse stresses generated by the secondary flow. Near the wall,

$$\tau_{33} - \tau_{22} = \overline{v^2} - \overline{w^2} = a\tau_{13} \tag{7.20.3}$$

where a is about two.

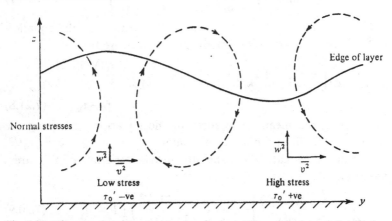

Fig. 7.22. Diagram showing a mechanism for instability of the lateral distribution of surface stress.

The possibility of persistent secondary flow may be investigated by assuming a small periodic deviation from the ideal two-dimensional flow in which the lateral variations of mean values are periodic with wave number m. Approximating the distributions of shear stress by

$$\tau \equiv \tau_{13} = \tau_0(1 - z/\delta) \tag{7.20.4}$$

and imposing the conditions that both τ_{23} and $\partial\tau_{23}/\partial z$ are zero at $z = \delta$, the edge of the layer, the solution of equation (7.20.2) is

$$\tau_{23} = -\frac{ia\tau_0'}{m\delta}\{1 - \cos[m(z-\delta)]\}. \tag{7.20.5}\dagger$$

If $m\delta$ is not large, a sufficiently accurate form is

$$\tau_{23} = -\tfrac{1}{2}ia\tau_0'm\delta(1 - z/\delta)^2. \tag{7.20.6}$$

Assuming that the direction of the velocity gradient vector coincides with the direction of shear stress, i.e.

$$\frac{\partial V}{\partial z} = \frac{\tau_{23}}{\tau_{13}}\frac{\partial U}{\partial z},$$

and approximating the velocity distributions by

$$U = \frac{\tau_0^{1/2}}{k}\log(z/z_0) \quad \text{(for } z < \delta),$$

the distribution of transverse velocity is, to first order,

$$V = \tfrac{1}{2}iam\delta\frac{\tau_0^{1/2}}{k}\left[\frac{\log(z/z_0)}{\log(\delta/z_0)} - \frac{z}{\delta}\right]\left(\frac{\delta'}{\delta} - \frac{\tau_0'}{\tau_0}\right) \tag{7.20.7}$$

if $V = 0$ at $z = \delta$. Using the velocity distributions in the equation (7.16.3) for the total flux of momentum defect, we find

$$U_1\frac{d}{dx}\left[-\frac{\bar{\tau}_0^{1/2}\delta}{k} + 2\frac{\bar{\tau}_0\delta}{k^2U_1}\right] + \tfrac{1}{2}am^2\delta^2\frac{\bar{\tau}_0}{k^2}\left(\frac{\delta'}{\delta} - \frac{\tau_0'}{\tau_0}\right)\left(\frac{3}{4} - \frac{2\bar{\tau}_0^{1/2}}{kU_1}\right)$$
$$= -\bar{\tau}_0 - \tau_0'. \tag{7.21.8}$$

Subtracting the mean value form of the equation and using the roughness length appropriate to a smooth surface, $z_0 = e^{-A}v/\tau_0^{1/2}$, we obtain an equation for the streamwise variation of the stress variation,

$$\frac{1}{2}\left[1 + \frac{x\bar{\tau}_0}{\tau_0'}\frac{d(\tau_0'/\bar{\tau}_0)}{dx}\right] - \tfrac{1}{4}am^2\delta^2\left(\frac{3}{4} - \frac{2\bar{\tau}_0^{1/2}}{kU_1}\right) + 1 = 0. \tag{7.20.9}$$

The critical condition for zero growth of the variations is that

$$\tfrac{1}{4}am^2\delta^2 = \left(\frac{1}{2} + \frac{\bar{\tau}_0^{1/2}}{kU_1}\right)\bigg/\left(\frac{3}{4} - \frac{2\bar{\tau}_0^{1/2}}{kU_1}\right) \tag{7.20.10}$$

† Dashes denote fluctuations from mean values over a complete spatial period.

and using the typical value, $\tau_0^{1/2}/(kU) = 0.1$, we find
$$(m\delta)_{\text{crit}} = 1.5$$
or the ratio of wavelength to layer thickness is 4.2.

The calculation indicates that lateral variations of surface stress with characteristic wavelengths less than about four layer thicknesses sustain themselves through the induced secondary flows, but the approximations are not likely to be valid for variations of much smaller wavelengths. Only a limited range of wavelengths can be self-sustaining and, in a developing layer, the dominant wavelength is that which has been able to grow most rapidly and will be proportional to the layer thickness. The tendency of wall layers to develop periodic lateral variations of surface stress through secondary flows is evident in some measurements of secondary flow near the corners of a channel of rectangular section.

7.21 Periodic structure of flow near the viscous layer

The lateral instability of two-dimensional flow in a boundary layer arises from the contrast of the inequality of the normal Reynolds stresses in the flow with their equality in the free stream. Since Reynolds stresses become small in the viscous layer as well, it appears possible that a similar lateral instability could exist in the transition layer between the fully turbulent flow and the viscous layer with negligible Reynolds stresses. Gupta, Laufer & Kaplan (1971), using a technique for measuring values of the velocity product,
$$u(x, y, z)u(x, y + r, z)$$
simultaneously for a range of separations r and calculating the average values for various time intervals, find strong but transitory, periodic lateral variations. The period varies from one occurrence to the next and the long-term averages give a correlation function that is the superposition of many periodic components and displays no obvious periodicity.

The relevant length scale for the transverse flow is the thickness of the viscous layer which, in zero longitudinal pressure gradient, changes very slowly in the stream direction, and the dominant wavelength will remain nearly unchanged over fetches long compared with the total thickness of the layer and very long compared with the length scale of the motion. With such a long time for development, it is not surprising that the periodicity is so strong and that the duration of the patterns is long. It is likely that the apparent duration depends mostly on the effect of the outer flow in changing slightly the alignment of the axes of circulation from the direction of the free stream.

Since the transverse flow patterns are changing, their velocity fields appear as fluctuations and affect the components of the correlation function. For a nearly stationary pattern, the equation for the streamwise component of 'mean velocity' is

$$V \frac{\partial U}{\partial y} + W \frac{\partial U}{\partial z} = \frac{\partial \tau}{\partial z} + \nu \frac{\partial^2 U}{\partial z^2} \qquad (7.21.1)$$

and, supposing the scales of lateral and normal variation of transverse velocities to be comparable, variations of wall stress are related to variations of normal 'mean velocity' by

$$\tau_0' \approx -W'U. \qquad (7.21.2)$$

To an order of magnitude, the change of longitudinal 'mean velocity' produced by the change of wall stress is

$$U' = \frac{1}{2} \frac{\tau_0'}{\tau_0^{1/2}} \left(\log \frac{\tau_0^{1/2} z}{\nu} + A + 1 \right) \bigg/ k \qquad (7.21.3)$$

compared with changes in transverse 'mean velocities' of order

$$W' = -\frac{\tau_0'}{\tau_0^{1/2}} k \bigg/ \left(\log \frac{\tau_0^{1/2} z}{\nu} + A \right). \qquad (7.21.4)$$

Using an electrochemical method, Mitchell & Hanratty (1966) have measured fluctuations of wall stress and find that

$$(\overline{\tau_0'^2})^{1/2}/\overline{\tau_0} = 0.32.$$

If all the variation were due to transverse circulations, the contribution to the root-mean-square of the longitudinal fluctuation would be

$$(\overline{U'^2})^{1/2}/\tau_0^{1/2} = 2.2$$

while the contributions to the transverse fluctuations are

$$(\overline{W'^2})^{1/2}/\tau_0^{1/2} \quad \text{and} \quad (\overline{V'^2})^{1/2}/\tau_0^{1/2} \approx 0.02$$

(taking a value for $\tau_0^{1/2} z/\nu$ of 50). Hence the effect of transverse flow is considerable for the longitudinal fluctuations but nearly negligible for transverse fluctuations. These arguments apply to the velocity fluctuations induced by variations of wall stress, whatever the cause, and it is likely that the anomalous behaviour of the $R_{11}(r, 0, 0)$ component of the correlation, i.e. its comparative invariance through the whole of a boundary layer or channel flow, is a consequence of the longitudinal coherence of the stress variations on the wall. The coherent stress variations are probably caused by transitory patterns of transverse secondary flow, induced by the non-Newtonian nature of turbulent flow either at the inner or outer boundary to the region of fully turbulent flow.

CHAPTER 8

TURBULENT CONVECTION OF HEAT AND PASSIVE CONTAMINANTS

8.1 Governing equations and dimensional considerations

In the Boussinesq approximation to the equations of motion, the fields of velocity, temperature and concentration of a passive contaminant, $\mathbf{u}(\mathbf{x}, t)$, $T(\mathbf{x}, t)$, $c(\mathbf{x}, t)$, are described by the equations,

$$\left.\begin{aligned}
\partial u_k/\partial x_k &= 0, \\[4pt]
\partial u_i/\partial t + u_k\,\partial u_i/\partial x_k &= -\partial p/\partial x_i - \frac{\rho - \rho_a}{\rho_a}g_i + \nu\nabla^2 u_i, \\[4pt]
\partial T/\partial t + u_k\,\partial T/\partial x_k &= k\nabla^2 T, \\[4pt]
\partial c/\partial t + u_k\,\partial c/\partial x_k &= D\nabla^2 c, \\[4pt]
\rho - \rho_a &= \left(\frac{\partial \rho}{\partial T}\right)_{p,c}(T - T_a) + \left(\frac{\partial \rho}{\partial c}\right)_{p,T}(c - c_a),
\end{aligned}\right\} \qquad (8.1.1)$$

where T is the potential temperature, i.e. the thermodynamic temperature increased by ϕ/c_p, ϕ being the gravitational potential,

T_a, c_a are average values over a horizontal surface through the position,

$p\rho_a$ is the difference of (mechanical) pressure from the 'hydrostatic' pressure defined by grad $p_h = -\rho_a\mathbf{g}$,

$\mathbf{g} = \operatorname{grad}\phi$, the negative of the gravitational acceleration,

κ is the thermometric conductivity,

and D is the diffusion coefficient for the contaminant. The equations are a valid approximation if the density variation in the flow is small and if the Mach number of the velocity variation is small.

An important feature is that the equations for temperature and concentration are linear in these quantities, and that they appear in the equation of motion only in the buoyancy term, as the relative density defect multiplied by the gravitational acceleration. Consider flows with geometrically similar boundaries with length scale l_0, with similar distributions on them of velocity, temperature and concentration proportional to the scales u_0, θ_0 and c_0. If the equations are put into non-dimensional form with these scales, constant coefficients appear in the equations which are combinations of the five independent parameters,

Reynolds number $R_0 = u_0 l_0 / \nu$,

Prandtl number ν/κ,

Schmidt number ν/D,

Richardson number $Ri = g\theta_0 l_0 / (u_0{}^2 T_r)$,

Density change ratio $S = c_0 \left(\dfrac{\partial \rho}{\partial c}\right)_{p,T} \bigg/ \left(\theta_0 \left(\dfrac{\partial \rho}{\partial c}\right)_{p,c}\right)$,

and the solutions are of the forms,

$$\left.\begin{aligned}
\mathbf{u} &= u_0 \mathbf{F}(\mathbf{x}/l_0, u_0 t/l_0, R_0, \nu/\kappa, Ri, S, \nu/D), \\
T - T_r &= \theta_0 G(\mathbf{x}/l_0, u_0 t/l_0, R_0, \nu/\kappa, Ri, S, \nu/\Delta), \\
c - c_r &= c_0 H(\mathbf{x}/l_0, u_0 t/l_0, R_0, \nu/\kappa, Ri, S, \nu/D),
\end{aligned}\right\} \qquad (8.1.2)$$

where T_r, c_r are average or representative values over the whole flow. Notice that gravitation only appears in the Richardson number, essentially as a scale of buoyancy force $g\theta_0/T_r$ where T_r is the reciprocal of the thermal expansion coefficient $-(\partial\rho/\partial T)_{p,c}/\rho$ and is equal to the Kelvin temperature for a perfect gas.

If the buoyancy forces are small compared with the inertial and viscous terms in the equation of motion, the motion is not affected by the variations of temperature and concentration and the fields have the forms,

$$\left.\begin{aligned}
\mathbf{u} &= u_0 \mathbf{F}(\mathbf{x}/l_0, u_0 t/l_0, R_0) \\
T - T_r &= \theta_0 G(\mathbf{x}/l_0, u_0 t/l_0, R_0, \nu/k) \\
c - c_r &= c_0 H(\mathbf{x}/l_0, u_0 t/l_0, R_0, \nu/D)
\end{aligned}\right\} \qquad (8.1.3)$$

Unless the Reynolds number is small, the condition of *forced convection* is expected if both the Richardson number for temperature, Ri, and the Richardson number for concentration, Ri S, are sufficiently small. Then not only are the distributions of temperature and concentration proportional to the scales of their boundary variations but the linearity of the governing equations means that solutions for different kinds of boundary distributions can be added to give a solution for the composite distribution.

In *natural convection*, energy is supplied to the flow almost entirely through the buoyancy forces and to a negligible extent by mechanical working on the flow boundaries. The condition is that one Richardson number should be large and the scale of velocity variation over the boundaries becomes irrelevant. Introducing a new buoyancy parameter that does not involve the velocity scale, the *Rayleigh number*

$$\lambda = g\theta_0 l_0^3/(\nu\kappa T_r), \tag{8.1.4}$$

the solutions have the forms,

$$\left.\begin{array}{l} \mathbf{u} = u_t\mathbf{F}(\mathbf{x}/l_0, \, u_t t/l_0, \, \lambda, \, \nu/\kappa, \, \nu/D, \, S), \\[4pt] T - T_r = \theta_0 G(\mathbf{x}/l_0, \, u_t t/l_0, \, \lambda, \, \nu/\kappa, \, \nu/D, \, S), \\[4pt] c - c_r = c_0 H(\mathbf{x}/l_0, \, u_t t/l_0, \, \lambda, \, \nu/\kappa, \, \nu/D, \, S), \end{array}\right\} \tag{8.1.5}$$

where $u_t = (g\theta_0 l_0/T_r)^{1/2}$ is a measure of the velocity that would be acquired by fluid of typical buoyancy rising through the depth of the flow. If the convection is driven entirely by concentration variations, i.e. $\theta_0 = 0$, the velocity scale should be replaced by

$$u_c = [gc_0 l_0(\partial\rho/\partial c)_{p,T}]^{1/2}. \tag{8.1.6}$$

For most aspects of turbulent convective flows, the influence of molecular diffusivity is confined to small-scale variations that do not affect distributions of mean values of quantities such as temperature and heat flux. By defining an effective temperature,

$$T_e = T + \left(\frac{\partial\rho}{\partial c}\Big/\frac{\partial\rho}{\partial T}\right)(c - c_r) \tag{8.1.7}$$

the equations for a flow with fluctuations of both temperature and concentration are identical with those for temperature variations in a perfect gas, if molecular diffusion can be neglected. In most of the following account, the discussion refers to temperature variation but most of the results may be used for concentration variations by using the effective temperature.

8.2 Diffusion by continuous movements: effect of molecular diffusive transport

If the molecular diffusivity is very small, i.e. if the Péclét number, $u_0 l_0/\kappa$, is large, it would appear that the equation for the temperature reduces to

$$\frac{DT}{Dt} \equiv \frac{\partial T}{\partial t} + (\mathbf{u} \cdot \nabla)T = 0 \tag{8.2.1}$$

and the temperature at the current position of a particular fluid particle does not change. Then, a knowledge of the particle movements is sufficient information to determine the changes in the temperature field, given its initial form. The displacements of single particles from their original positions are specified statistically by the Lagrangian auto-correlation function,

$$R_{ij}(\tau; \mathbf{x}_s, t_s) = \langle v_i(t_s; \mathbf{x}_s, t_s)v_j(\tau + t_s; \mathbf{x}_s, t_s)\rangle, \qquad (8.2.2)$$

where $v_i(t; \mathbf{x}_s, t_s)$ is the velocity of a particle that was at position $\mathbf{x}_s$ at time t_s. The function was introduced by G. I. Taylor (1921) in his theory of turbulent diffusion by continuous movements, and he showed that the mingling of isothermal particles, which retain their original temperatures, leads to distributions of mean temperature that are characteristically diffusive in appearance. For diffusion from an instantaneous point source in a homogeneous, stationary turbulent flow, he showed that the dispersion at time t after release is given by

$$\langle X_1^2 \rangle = 2 \int_0^t R_{11}(\tau)(t-\tau)\, d\tau. \qquad (8.2.3)$$

If the auto-correlation function becomes negligible for large time intervals, for long dispersion times

$$\langle X_1^2 \rangle = 2\kappa_T t + \text{constant}, \qquad (8.2.4)$$

where $\kappa_T = \int_0^\infty R_{11}(\tau)\, d\tau$ is an eddy coefficient of diffusion, defined without any reference to a mixing process.

Just as with the influence of viscosity on the velocity field, the influence of molecular diffusivity cannot be neglected for all aspects of the flow although it may be small for the large-scale features. If equation (8.2.1) is valid, isothermal surfaces are material surfaces always composed of the same fluid particles, and they are convected and distorted by the fluid motion. Batchelor (1952) shows from the theory of diffusion by continuous movements that each small element of a material surface, δA, increases in (scalar) area at a mean rate given by

$$\frac{1}{\delta A}\frac{d(\delta A)}{dt} = \beta(\varepsilon/\nu)^{1/2}, \qquad (8.2.5)$$

where β is an absolute constant of order one, and $(\varepsilon/\nu)^{1/2}$ measures

the root-mean-square rate of distortion of the fluid. From observation and from the theory of local similarity, it is known that
rates of deformation are nearly constant over volumes with sides
less than about $v^{3/4}\varepsilon^{-1/4}$, and so elements of material surfaces separated by smaller distances are distorted similarly and increase in
area at the same rates. Since the fluid is incompressible, the volume
between adjacent elements in the two surfaces does not change and
the separation must decrease at the same logarithmic rate as the
area. Hence, temperature gradients increase at the same rate as
areas of material surfaces and, in the absence of molecular diffusion,
would become indefinitely large.

With small but non-zero diffusivity, the growth of temperature
gradients is eventually limited by diffusion when the spatial scale of
the gradient variations has been reduced by crowding together so
much that they are being destroyed by diffusion as fast as they are
being amplified by turbulent distortion (for a fuller account, refer
to a review in Batchelor & Davies (1956)). The important point
is that the irreversible flow of heat down temperature gradients
occurs almost entirely on the smallest scales of variation, and all
but the fine detail of the temperature field is unaffected by the
magnitude of the molecular diffusivity. The principle that turbulent
diffusion is independent of the magnitude of the diffusivity is
analogous to the principle of Reynolds number similarity for
turbulent motion and, like it, it applies to all aspects but those
concerned directly with the irreversible processes of the flow.

8.3 Eulerian description of convective flows: mean value equations and correlation functions

In general, the Lagrangian theory of diffusion by continuous movements is more useful for discussing the nature of convective diffusion
than for the practical purpose of predicting its magnitude and
properties, and the Eulerian specification used in most experimental
studies allows the application to convective flow of the general
concepts of turbulent shear flow. The basic equations for a flow
with temperature fluctuations are readily derived from the Boussinesq equations (8.1.1) and, making the usual distinction between
the mean values, T and U_i, and their fluctuations, θ and u_i, mean

values may be derived for quantities of physical significance. The more important ones are:

Mean flow

$$\frac{\partial U_i}{\partial t} + U_k \frac{\partial U_i}{\partial x_k} = -\frac{\partial P}{\partial x_i} - \frac{\overline{\partial u_i u_k}}{\partial x_k} + g_i \frac{T - T_a}{T_a} + \nu \frac{\partial^2 U_i}{\partial x_k^2}. \tag{8.3.1}$$

Mean temperature

$$\frac{\partial T}{\partial t} + U_k \frac{\partial T}{\partial x_k} = -\frac{\overline{\partial u_k \theta}}{\partial x_k} + \kappa \frac{\partial^2 T}{\partial x_k^2}. \tag{8.3.2}$$

Kinetic energy of velocity fluctuations

$$\frac{\partial}{\partial t}(\tfrac{1}{2}\overline{q^2}) + U_k \frac{\partial}{\partial x_k}(\tfrac{1}{2}\overline{q^2}) + \frac{\partial}{\partial x_k}(\overline{pu_k + \tfrac{1}{2}q^2 u_k})$$
$$= -\overline{u_i u_k} \frac{\partial U_i}{\partial x_k} + \frac{g_i}{T_a}\overline{\theta u_i} + \overline{\nu u_i \frac{\partial^2 u_i}{\partial x_k^2}}. \tag{8.3.3}$$

Intensity of temperature fluctuations

$$\frac{\partial}{\partial t}(\tfrac{1}{2}\overline{\theta^2}) + U_k \frac{\partial}{\partial x_k}(\tfrac{1}{2}\overline{\theta^2}) + \frac{\partial}{\partial x_k}(\tfrac{1}{2}\overline{\theta^2 u_k})$$
$$= -\overline{\theta u_k} \frac{\partial T}{\partial x_k} + \kappa \frac{\partial^2}{\partial x_k^2}(\tfrac{1}{2}\overline{\theta^2}) - \kappa \overline{\left(\frac{\partial \theta}{\partial x_k}\right)^2}. \tag{8.3.4}$$

Mean heat flux

$$\frac{\partial}{\partial t}(\overline{\theta u_i}) + U_k \frac{\partial}{\partial x_k}(\overline{\theta u_i}) + \frac{\partial}{\partial x_k}(\overline{\theta u_i u_k})$$
$$= -\overline{u_i u_k} \frac{\partial T}{\partial x_k} - \overline{\theta u_k} \frac{\partial U_i}{\partial x_k} - \overline{\theta \frac{\partial p}{\partial x_i}} + \frac{g_i}{T_a}\overline{\theta^2} + \tag{8.3.5}$$
$$+ \frac{\partial}{\partial x_k}\left(\overline{\kappa u_i \frac{\partial \theta}{\partial x_k}} + \overline{\nu \theta \frac{\partial u_i}{\partial x_k}}\right) - (\nu + \kappa) \overline{\frac{\partial u_i}{\partial x_k} \frac{\partial \theta}{\partial x_k}}.$$

It should be noticed that $-c_p \tfrac{1}{2}\overline{\theta^2}/T_a^2$ is the entropy density associated with the temperature fluctuations, and so the quantity,

$$\varepsilon_\theta = \overline{\kappa(\partial \theta / \partial x_k)^2} \tag{8.3.6}$$

which appears in (8.3.4) as the rate of destruction of fluctuation intensity by molecular diffusion, is a multiple of the local rate of entropy increase by irreversible conduction of heat down temperature gradients. The equation for the fluctuation intensity

12

expresses the balance of fluctuation entropy between net gain by advective, convective and diffusive transfers, transfer from the mean temperature field at a rate $\overline{\theta u_k}\, \partial T/\partial x_k$, and the irreversible rate of increase ε_θ (omitting for clarity the conversion factor c_p/T_a^2). Only the last term involves the magnitude of the thermal diffusivity and if the large-scale aspects of the velocity and temperature fields are independent of the magnitudes of viscosity and diffusivity, it follows that

$$\varepsilon_\theta = \overline{\theta^2}(\overline{q^2})^{1/2}/L_\theta, \tag{8.3.7}$$

where L_θ is a length determined by the large-scale properties and similar in magnitude to the integral length-scale and the dissipation length-scale.

To describe the spatial properties of the fluctuations, correlation functions may be defined, in particular,

$$R_\theta(\mathbf{r}; \mathbf{x}) = \overline{\theta(\mathbf{x})\theta(\mathbf{x} + \mathbf{r})},$$

and

$$R_{\theta, i}(\mathbf{r}; \mathbf{x}) = \overline{\theta(\mathbf{x})u_i(\mathbf{x} + \mathbf{r}}. \tag{8.3.8}$$

Both correlation functions are expected to become small for large values of the separation $|\mathbf{r}|$, that is to say, large compared with the integral scale of the turbulent motion. Since the velocity fluctuations satisfy the continuity condition, $\partial u_k/\partial x_k = 0$, the cross-correlation $R_{\theta, i}$ satisfies

$$\frac{\partial}{\partial r_i} R_{\theta, i}(\mathbf{r}; \mathbf{x}) = 0 \tag{8.3.9}$$

and integration over a plane in $\mathbf{r}$-space leads to

$$\iint_{-\infty}^{\infty} R_{\theta, 3}(r_1, r_2, r_3)\, \mathrm{d}r_1\, \mathrm{d}r_2 = 0.$$

For large values of r_3, the correlation is small for all values of r_1 and r_2, and so

$$\iint_{-\infty}^{\infty} R_{\theta, 3}(r_1, r_2, r_3)\, \mathrm{d}r_1\, \mathrm{d}r_2 = 0 \tag{8.3.10}$$

for all r_3. It follows that $R_{\theta, i}$ (the direction of Ox_3 is arbitrary) cannot be everywhere of the same sign over any plane normal to Ox.

8.4 Local forms of the Richardson number

The relative magnitude of the buoyancy forces may be measured by non-dimensional parameters that are essentially ratios of typical buoyancy forces to typical 'inertial' forces, i.e. accelerations of the flow. For flows with similar boundary conditions, the scale Richardson number $g\theta_0 l_0/(T_r u_0^2)$ is such a parameter of the whole flow, but its magnitude depends on how the scales are chosen and the values in different kinds of flows cannot be compared easily. One way of comparing buoyancy forces in different flows is to define local Richardson numbers, such as that introduced by L. F. Richardson (1920)

$$Ri = \frac{g_i}{T_r}\frac{\partial T}{\partial x_i}\bigg/\left(\frac{\partial U_i}{\partial x_k}\right)^2. \tag{8.4.1}$$

It is the ratio of the quantity $(g_i/T_r)(\partial T/\partial x_i)$ to the square of the rate of shear, and it compares the rate of shear with the rate of variation of small local disturbances in still fluid with the same density stratification. With stable stratification, $[(g_i/T_r)(\partial T/\partial x_i)]^{1/2}$ is the maximum radian frequency of internal gravity waves while, for unstable stratification, $[(-g_i/T_r)(\partial T/\partial x_i)]^{1/2}$ is the maximum (exponential) rate of growth of unstable perturbations of the still fluid.

A number of more direct relevance to the turbulent flow is the *flux Richardson number*,

$$R_f = \frac{(g_i/T_r)\overline{\theta u_i}}{\overline{u_i u_k}\,\partial U_i/\partial x_k}. \tag{8.4.2}$$

The kinetic energy equation (8.3.3) shows it to be the ratio of the rate of energy loss by working of the velocity fluctuations against buoyancy forces to the rate of energy production by working of the mean velocity against Reynolds stresses. If the first three terms of the energy equation (which describe energy transport across the flow) can be ignored, it follows that

$$R_f = 1 - \varepsilon/(-\overline{u_i u_k}\,\partial U_i/\partial x_k). \tag{8.4.3}$$

The denominator of the last term is the rate at which energy is passing from the mean flow to the turbulence, and it is negative only in exceptional circumstances. It appears that steady conditions are possible only if

$$R_f \leqslant 1. \tag{8.4.4}$$

Using the coefficients of eddy viscosity and eddy diffusion, v_T and κ_T, to describe the relation between transport rates and gradients, the ordinary Richardson number is related to the flux number by

$$Ri = \frac{v_T}{\kappa_T} R_f \qquad (8.4.5)$$

but, particularly in flows with stable stratification, the ratio v_T/κ_T is strongly dependent on the stability and no upper limit can be set on Ri.

8.5 Spectrum functions and local similarity

In the equation for the intensity of the temperature fluctuations, three kinds of term appear – terms describing transport of fluctuations from one part of the flow to another, a term $-\overline{\theta u_i}\, \partial T/\partial x_i$ that is the local rate of production, and a term $\kappa(\partial\theta/\partial x_i)^2$ that is the rate of destruction. The last two terms must be similar in magnitude and so the root-mean-square gradient must be large compared with the mean gradient if the Péclet number is large. The reason for the difference is to be found in the folding and stretching of isothermal surfaces by the turbulent motion, but the process of generating large local temperature gradients can be discussed in terms of the Fourier components of the temperature field.

If a spectrum function is to have a local significance, it must be restricted to wave numbers of magnitude large compared with the reciprocal of the scale of inhomogeneity of the flow, so that a spectrum may be defined using the fluctuations within a volume V small enough for the fields to be considered statistically homogeneous in space. With that restriction, the local fields of velocity and temperature fluctuations may be expressed as Fourier sums,

$$\left. \begin{aligned} u_i(\mathbf{x}, t) &= \sum_{\mathbf{k}} a_i(\mathbf{k}, t) \exp(i\mathbf{k} . \mathbf{x}), \\ \theta(\mathbf{x}, t) &= \sum_{\mathbf{k}} b(\mathbf{k}, t) \exp(i\mathbf{k} . \mathbf{x}), \end{aligned} \right\} \qquad (8.5.1)$$

where $k_i a_i(\mathbf{k}, t) = 0$, and the gradients of mean velocity and mean temperature are nearly constant over the volume. Substitution in the equations of motion and temperature leads to the equations for the coefficients,

$$\left(\frac{\partial}{\partial t} + v k^2\right) a_i = -\left(\frac{\partial U_i}{\partial x_j} - 2\frac{k_i k_m}{k^2}\frac{\partial U_m}{\partial x_j}\right) a_j(\mathbf{k}) + \left(\frac{g_i}{T_a} - \frac{k_i k_m}{k^2}\frac{g_m}{T_a}\right) b(\mathbf{k})$$

$$-ik_l \sum_{\mathbf{k'}+\mathbf{k''}=\mathbf{k}} \left(\delta_{im} - \frac{k_i k_m}{k^2} \right) a_l(\mathbf{k'}) a_m(\mathbf{k''}) \qquad (8.5.2)$$

and

$$\left(\frac{\partial}{\partial t} + \kappa k^2 \right) b(\mathbf{k}) = -\frac{\partial T}{\partial x_i} a_i(\mathbf{k}) - ik_m \sum_{\mathbf{k'}+\mathbf{k''}=\mathbf{k}} a_m(\mathbf{k'}) b(\mathbf{k''}), \qquad (8.5.3)$$

where $dk_i/dt = -k_j\, \partial U_j/\partial x_i$. It is then possible to write down equations for the three-dimensional spectrum functions of $\overline{q^2}$, the total turbulent intensity, and of $\overline{\theta^2}$, the temperature fluctuation intensity,

and
$$\left.\begin{aligned} \Phi_{ii}(\mathbf{k};\, \mathbf{x}) &= (2\pi)^{-3/2} \int R_{ii}(\mathbf{r};\, \mathbf{x})\, e^{i\mathbf{k}\cdot\mathbf{r}}\, dV(\mathbf{r}), \\ \Phi_\theta(\mathbf{k};\, \mathbf{x}) &= (2\pi)^{-3/2} \int R_\theta(\mathbf{r};\, \mathbf{x})\, e^{i\mathbf{k}\cdot\mathbf{r}}\, dV(\mathbf{r}). \end{aligned}\right\} \qquad (8.5.4)$$

They are

$$\left(\frac{\partial}{\partial t} + 2vk^2 \right) \Phi_{ii}(\mathbf{k}) = -\frac{\partial U_i}{\partial x_j} \Phi_{ij}(\mathbf{k}) + \frac{g_i}{T_a} \Phi_{\theta,i}(\mathbf{k}) - \partial S_i(\mathbf{k})/\partial k_i, \qquad (8.5.5)$$

where

$$\partial S_i(\mathbf{k})/\partial k_i = ik_l \frac{V}{8\pi^3} \sum_{\mathbf{k'}+\mathbf{k''}=\mathbf{k}} \langle a_l(\mathbf{k'}) a_i(\mathbf{k''}) a_i^*(\mathbf{k}) - a_i^*(\mathbf{k'}) a_i^*(\mathbf{k''}) a_i(\mathbf{k}) \rangle$$

and

$$\left(\frac{\partial}{\partial t} + 2\kappa k^2 \right) \Phi_\theta(\mathbf{k}) = -\frac{\partial T}{\partial x_i} \Phi_{\theta,i} - \partial S_{\theta,i}(\mathbf{k})/\partial k_i, \qquad (8.5.6)$$

where

$$\partial S_{\theta,i}(\mathbf{k})/\partial k_i = ik_l \frac{V}{8\pi^3} \sum_{\mathbf{k'}+\mathbf{k''}=\mathbf{k}} \langle a_l(\mathbf{k'}) b(\mathbf{k''}) b^*(\mathbf{k}) - a_i^*(\mathbf{k'}) b^*(\mathbf{k''}) b(\mathbf{k}) \rangle.$$

The quantities, S_i and $S_{\theta,i}$, are fluxes of intensity in wave-number space, an interpretation that is possible since their divergences measure transfer of intensity between Fourier components without net gain or loss of the physical quantities, kinetic energy and entropy.

The equation for the velocity spectrum function, Φ_{ii}, contains two terms, $-\Phi_{ij}\, \partial U_i/\partial x_j$ and $\Phi_{\theta,i} g_i/T_a$, that describe transfer of energy from the mean flow and from buoyancy forces. If the spectrum functions for velocity and temperature are such that most of the total intensities comes from components of small wave number, these sources of intensity are small compared with the intensity flux $S_i(\mathbf{k})$ for wave numbers greater than k_e, selected so that

$$\int_{k_e}^{\infty} \Phi_{ii}(\mathbf{k})\, dV(\mathbf{k}) \text{ is small compared with } \overline{q^2} \text{ and } \int_{k_e}^{\infty} \Phi_{\theta}(\mathbf{k})\, dV(\mathbf{k})$$

is small compared with $\overline{\theta^2}$. Then the arguments of §3.13, based on the cascade concept of transfer between eddies of different sizes, can be used to conclude:

(1) that the turbulent motion described by components with wave numbers greater than k_e is not affected by buoyancy forces and is effectively in a state of forced convection,

(2) that the velocity spectrum function depends only on the rate of energy dissipation and the fluid viscosity. For the same range of large wave numbers, the term $-\Phi_{\theta, i}\, \partial T/\partial x_i$ represents a transfer of temperature fluctuation intensity at rates that are negligible compared with the flux $S_{\theta, i}$, and similar arguments suggest that the temperature spectrum function depends only on the velocity spectrum function, on ε_{θ} the rate at which fluctuation intensity is being transmitted to the similarity range, and on the fluid thermal diffusivity.

Since the motion is determined by the dissipation and is not directly dependent on the temperature field, the temperature spectrum function has the form,

$$\Phi_{\theta}(\mathbf{k}) = \varepsilon_{\theta} F[k, \Phi_{ii}]. \tag{8.5.7}$$

If both the Reynolds and Péclét numbers of the flow are large, it is likely that an inertial–convective range can be distinguished, in which the effects of viscosity on the motion are small and the effects of diffusivity on the temperatures are small. In that range, the function depends only on wave number, on dissipation and, linearly, on entropy production, and so

$$\Phi_{\theta}(\mathbf{k}) = \frac{1}{4\pi} C_{\theta} \varepsilon_{\theta} \varepsilon^{-1/3} k^{-11/3}, \tag{8.5.8}$$

where C_{θ} is an absolute constant.

The upper limit to the convective spectrum (8.5.8) cannot exceed the viscous limit to the velocity spectrum, near $k_s = \varepsilon^{1/4} \nu^{-3/4}$, and occurs at a smaller wave number if the Prandtl number ν/κ is small.

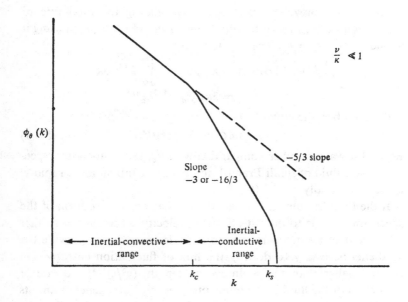

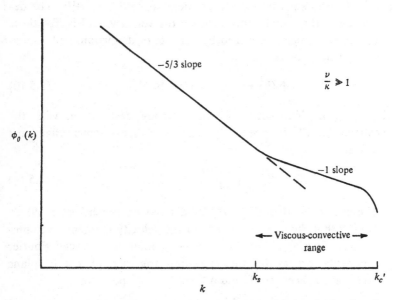

Fig. 8.1. Equilibrium spectra of temperature fluctuations in fluids of various Prandtl numbers. (The scales are logarithmic to display power-law variations.)

Its magnitude may be estimated by calculating the total rate of production of entropy in the inertial–convective range, supposing it to end where $|\mathbf{k}| = k_c$. The rate is nearly

$$\kappa \int k^2 \Phi_\theta(\mathbf{k}) \, dV(\mathbf{k}) = \kappa C_\theta \varepsilon_\theta \varepsilon^{-1/3} \int_0^{kc} k^{1/3} \, dk$$
$$= 3\kappa C_\theta \varepsilon_\theta \varepsilon^{-1/3} k_c^{4/3}$$

and would become equal to ε_θ if

$$k_c = (3/4C_\theta)^{-3/4}(\varepsilon/\kappa^3)^{1/4}. \tag{8.5.9}$$

Beyond a wave number comparable with k_c, the temperature spectrum for a fluid of small Prandtl number must fall off more rapidly, at least as rapidly as k^{-5} (fig. 8.1).

If the Prandtl number is large, the inertial–convective form of the spectrum extends to the cut-off of the velocity spectrum but further refinement of the temperature structure is necessary to provide the gradients necessary to destroy the flux of fluctuation entropy. On scales smaller than the Kolmogorov length, $(\nu^3/\varepsilon)^{1/4}$, the rate of distortion of the fluid is instantaneously nearly homogeneous and its magnitude and rate of variation with time are both described by the root-mean-square velocity gradient, of order $(\varepsilon/\nu)^{1/2}$. The dependence of the temperature field on the velocity field is, for these large wave numbers, described by the rate of distortion, and dimensional considerations lead to

$$\Phi_\theta(\mathbf{k}) = \frac{1}{4\pi} C_\theta' \varepsilon_\theta (\nu/\varepsilon)^{1/2} k^{-3}. \tag{8.5.10}$$

Repeating the calculation of the entropy destruction, with the spectrum (8.5.10), it appears that the convective–viscous range must end near

$$k_c' = \left(\frac{\nu}{\kappa}\right)^{1/2} \varepsilon^{1/4} \nu^{-3/4}. \tag{8.5.11}$$

The spectrum form (8.5.10) was derived by Batchelor (1959) by assuming that the rates of change of the velocity gradients are comparatively slow and that the transfer of intensity between Fourier components occurs by the crowding together of nearly plane isothermal surfaces. He found a form for the spectrum,

$$\Phi_\theta(\mathbf{k}) = \frac{1}{4\pi} C_\theta' \varepsilon_\theta \left(\frac{\nu}{\varepsilon}\right)^{1/2} k^{-3} \exp[-2k^2/k_c'^2] \tag{8.5.12}$$

which specifies the form of the cut-off at the end of the convective range. Observational evidence for the validity of the spectrum forms (8.5.8) and (8.5.10), i.e. in the convective–inertial range and in the convective–viscous range, is nearly conclusive, although the difficulty of calculating the two dissipation rates makes the values of C_θ and C_θ' uncertain. For the convective–inertial range, measurements of C_θ have been made for temperature and salt distributions in a water tunnel by Gibson & Schwarz (1963), for temperature fluctuations in air by Mills, Kistler, O'Brien & Corrsin (1958) and for salinity and temperature fluctuations in the ocean by Grant, Hughes, Voget & Moilliet (1968), with values in the range 0.52 to 0.9. For the convective–viscous range, the measurements of Gibson & Schwarz indicate a value of C_θ' near 2, while salinity measurements in the ocean (Grant et al.) may imply a rather larger value.

The form of the spectrum in the diffusive–inertial range, i.e. beyond the end of the convective–inertial range for small values of the Prandtl number, is not yet established beyond doubt. Batchelor, Howells & Townsend (1959) argue that components of the temperature field with wave numbers large compared with $k_c \approx \varepsilon^{1/4}\kappa^{-3/4}$ are formed by distortion of temperature gradients of scales near k_c^{-1} by velocity components of similar magnitudes, and arrive at

$$\Phi_\theta(\mathbf{k}) = \tfrac{1}{3}C\varepsilon_\theta\varepsilon^{2/3}\kappa^{-3}k^{-23/3}. \qquad (8.5.13)$$

Gibson (1968) considers the distribution and motion of temperature maxima and minima, and arrives at

$$\Phi_\theta(\mathbf{k}) = C_\theta''\varepsilon_\theta\kappa^{-1}k^{-5} \qquad (8.5.14)$$

for values of wave number both large compared with k_c and small compared with $[\varepsilon/(\nu\kappa^2)]^{1/4}$. Unlike the Batchelor, Howells & Townsend spectrum, the Gibson result indicates that entropy production does not fall off sharply with wave number in the range. Temperature spectra for mercury with a Prandtl number of 0.02 have been measured by Rust & Sesonske (1966), and the measurements in the diffusive–inertial range show a variation that is not inconsistent with the Gibson prediction but seems to fall off somewhat more rapidly with wave number.

8.6 Scattering of light by density fluctuations in a turbulent flow

Light and other electromagnetic waves passing through turbulent flows may be scattered by spatial variations of optical density, a phenomenon that sets a limit to the definition of stellar objects obtainable with terrestial telescopes, that permits long-range propagation of high frequency radio waves by atmospheric scattering in the stratosphere, and that is the basis of the shadowgraph technique for making visible regions of turbulent flow. The theory of the scattering has been developed in detail for turbulent flows with small variations of refractive index (see Tatarski 1961), and it is of interest to discuss the results for a simple but practically useful situation.

Both in the shadowgraph arrangement and in the illumination of the earth's surface by a single star, a nearly parallel beam of light is incident on a layer of density fluctuations which are so weak that the emergent beam is nearly unchanged in amplitude but has lateral variations of phase dependent on the variations of optical path along wave normals of the incident wave. After emergence, lateral fluctuations of amplitude develop as the beam travels from the scattering region, and it may be shown that the spatial spectrum of the lateral variations of light intensity, $F(m, n)$, is related to the spectrum of the optical density fluctuations in the turbulent layer, $\Phi(l, m, n)$, by

$$F(m, n) = 2\pi h H^2 (m^2 + n^2)^2 \times$$

$$\times \Phi(0, m, n) \left\{ \sin\left[\frac{(m^2 + n^2)H}{2N} \right] \Big/ \frac{(m^2 + n^2)H}{2N} \right\}^2, \quad (8.6.1)$$

where h is the thickness of the scattering layer, H is the distance of the shadow-plane from the layer, N is the wave number of the light, supposed large compared with wave numbers of the fluctuations. The spectrum is the product of two factors, the first one that could be obtained using the approximation of geometrical optics, and the second a diffraction factor that limits the intensity spectrum to Fourier components with wavelengths exceeding the diameter of the first Fresnel zone.

Both for laboratory shadowgraphs and for star illumination at ground level, the dimensions of the turbulent flow are large compared with diffraction limit at a wave number of roughly

$$k_d = (2\pi N/H)^{1/2} \qquad (8.6.2)$$

and nearly all the fluctuations in the shadow-pattern come from components in the similarity range of the spectrum of density variations. The effect of the weighting factor, $(m^2 + n^2)^2$, is to concentrate contributions within a comparatively small range of wave numbers just below the end of the convective range, i.e. near $(\varepsilon/\kappa^3)^{1/4}$ if $\nu/\kappa \ll 1$ or $(\varepsilon/(\nu\kappa^2))^{1/4}$ if $\nu/\kappa \gg 1$, or just below the diffraction limit, whichever is the less. In either case, the pattern of intensity variations arises from statistically isotropic fluctuations and is itself isotropic, and the limited range of wave numbers implies a quasi-periodic variation with a definite scale.

Most of the variations of illumination of the ground produced by a single star arise from regions of temperature fluctuation at heights of order 8 km and they are observed to have a well-defined wavelength of about 0.15 m, very near the calculated diffraction limit for visible light and the height. It is probable that the fluctuations are in regions of clear-air turbulence and from a study of the temporal variation of the shadow-bands, it is possible to estimate the characteristics of the turbulence (Protheroe 1964, Townsend 1965d). They are consistent with measurements in clear-air turbulence by aircraft.

While the dominant scale of stellar shadow-bands is determined by diffraction, the diffraction limit of wave number is usually beyond the end of the density fluctuation spectrum in laboratory shadowgraph systems. If the elements of the intensity pattern are of simple structure, the distribution of mean square fluctuation among elements of different sizes is nearly the contribution from Fourier components with scalar wave numbers in unit range of $\log k = \log(m^2 + n^2)^{1/2}$, where k^{-1} is a measure of their size. Then within an inertial–convective range where $\Phi_\theta(0, m, n) \propto k^{-11/3}$, the size distribution of the intensity pattern is

$$2\pi k^3 F(k) \propto k^{10/3} \qquad (8.6.3)$$

and, within a viscous–convective range $(\nu/\kappa \gg 1)$ where $\Phi_\theta(0, m, n) \propto k^{-3}$, it is

$$2\pi k^3 F(k) \propto k^4. \qquad (8.6.4)$$

At the end of the convective ranges, the spectrum functions for density variations fall off rapidly and so the dominant elements of the intensity pattern have scales determined by the dissipation rate of the turbulent motion and by the Prandtl number of the fluid. The scales are well defined and are proportional to the larger of k_c^{-1} and $k_c'^{-1}$, where k_c and k_c' are the spectrum limits for $\nu/\kappa \ll 1$ and for $\nu/\kappa \gg 1$.

8.7 Self-preserving development of temperature fields in forced convection flows

If small amounts of heat are present in a turbulent flow that is developing in self-preserving fashion, the temperature distributions may also be of self-preserving form, that is

$$T - T_a = \theta_0 f_\theta(z/l_0), \quad \overline{\theta^2} = \theta_0{}^2 g_\theta(z/l_0),$$
$$\overline{\theta w} = \theta_0 u_0 g_{\theta,3}(z/l_0), \text{ etc.,} \tag{8.7.1}$$

where u_0, l_0, θ_0 are scales dependent only on downstream position, and f_θ, g_θ, $g_{\theta,3}$ are distribution functions characteristic of the flow and, possibly, of the thermal boundary conditions. Substituting the similarity forms in the equation for the mean temperature leads to

$$(U_1 + u_0 f)\frac{\mathrm{d}T_a}{\mathrm{d}x} + U_1 \frac{\mathrm{d}\theta_0}{\mathrm{d}x} f_\theta - \frac{\theta_0}{l_0}\frac{\mathrm{d}(U_1 l_0)}{\mathrm{d}x}\eta f_\theta{}' + u_0\frac{\mathrm{d}\theta_0}{\mathrm{d}x} f f_\theta$$
$$- \frac{\theta_0}{l_0}\frac{\mathrm{d}(u_0 l_0)}{\mathrm{d}x} f_\theta{}' \int_0^\cdot f\,\mathrm{d}\eta + \frac{u_0\theta_0}{l_0} g_{\theta,3}{}' = 0 \tag{8.7.2}$$

and the necessary condition for the development is that the co-efficients of the distribution functions should either be negligible or remain in constant ratios to each other for all values of downstream displacement. If one of the flow boundaries is unrestricted, non-turbulent fluid, the equation reduces to

$$U_1\,\mathrm{d}T_a/\mathrm{d}x = 0 \tag{8.7.3}$$

and either the ambient temperature is constant or the ambient fluid is at rest.

For a free turbulent flow with no variation of ambient temperature, it may appear that self-preserving development of the temperature and velocity fields is possible if the scale θ_0 varies as any power of $(x - x_0)$, where x_0 is the effective origin of the flow. However, the condition for overall conservation of heat in a free turbulent flow without rigid boundaries is

$$\frac{\mathrm{d}}{\mathrm{d}x} \int_{-\infty}^{\infty} U(T - T_a)\, \mathrm{d}z + \frac{\mathrm{d}T_a}{\mathrm{d}x} \int_{-\infty}^{\infty} U\, \mathrm{d}z = 0 \qquad (8.7.4)$$

or, after inserting the similarity forms and integrating,

$$U_1 \theta_0 l_0 \int_{-\infty}^{\infty} f_\theta\, \mathrm{d}\eta + u_0 \theta_0 l_0 \int_{-\infty}^{\infty} f f_\theta\, \mathrm{d}\eta = \text{constant.} \qquad (8.7.5)$$

In self-preserving flows of constant u_0/U_1, the development is possible only if θ_0 is proportional to the velocity scale u_0. In approximately self-preserving, 'small-defect' flows with $|u_0/U_0| \ll 1$, the condition is that $\theta_0 \propto (U_1 l_0)^{-1}$, but the momentum condition leads to $U_1 u_0 l_0 = \text{constant}$, and again the temperature scale must be proportional to the velocity scale.

If a boundary layer has self-preserving temperature distributions, the equation for conservation of heat,

$$\frac{\mathrm{d}}{\mathrm{d}x} \int_0^{\infty} U(T - T_a)\, \mathrm{d}z = Q_0(x), \qquad (8.7.6)$$

where $Q_0(x)$ is the rate of heat transfer from the surface, shows that

$$\frac{\mathrm{d}}{\mathrm{d}x}(U_1 \theta_0 l_0) \propto Q_0(x). \qquad (8.7.7)$$

For self-preserving development of the flow, the flow width must increase nearly linearly with distance, and to that approximation, the surface flux Q_0 may vary as any power of l_0 [or $(x - x_0)$] and then

$$\theta_0 \propto Q_0/U_1. \qquad (8.7.8)$$

The ambient temperature need not be constant if the ambient fluid is at rest, i.e. the flow is a simple jet. Inspection of (8.7.2) shows that the conditions are that

$$\frac{\mathrm{d}T_a}{\mathrm{d}x} \propto \frac{\mathrm{d}\theta_0}{\mathrm{d}x} \propto \frac{\theta_0}{l_0} \qquad (8.7.9)$$

or that the ambient temperature gradient, $\mathrm{d}T_a/\mathrm{d}x$, should vary as a power of distance from the flow origin.

8.8   Forced convection in wall flows

The considerations of wall similarity that lead to predictions of the relation between mean velocity gradient and Reynolds stress (§ 5.4) are readily extended to the forced convection of heat through similarity layers. In fully turbulent and fully convective flow, rates of entropy production and dissipation near the wall are large compared with the rates of gain by mean flow advection, and the basic requirement for an equilibrium layer – negligible influence of advective processes on the balance of fluctuation intensity – is satisfied for the temperature field. Then, the thermal structure as well as the motion is independent of conditions in the viscous layer or around roughness elements, and the local gradient of mean temperature depends only on distance from the wall and on the distributions of heat flux and Reynolds stress in the equilibrium layer. For the simple but important case of nearly constant stress and constant heat transfer, dimensional considerations show that

$$\frac{dT}{dz} = -Q_0/(k_\theta \tau_0^{1/2} z) \tag{8.8.1}$$

where k_θ is the Kármán constant for diffusion. Notice that the eddy diffusivity is

$$\kappa_T = k_\theta \tau_0^{1/2} z \tag{8.8.2}$$

and that the ratio, κ_T/ν_T, is the ratio of the Kármán constants, k_θ/k.

Integrating (8.8.1) leads to the logarithmic distribution of mean temperature,

$$T - T_0 = -\frac{Q_0}{k_\theta \tau_0^{1/2}} \log(z/z_t), \tag{8.8.3}$$

where T_0 is the wall temperature, and z_t is a thermal 'roughness length', dependent on the nature of the surface and on the fluid viscosity and thermal diffusivity. On a smooth surface,

$$T - T_0 = -\frac{Q_0}{k_\theta \tau_0^{1/2}}[\log(\tau_0^{1/2} z/\nu) + A_\theta], \tag{8.8.4}$$

where the additive constant A_θ describes the temperature difference across the conduction layer and is a function of the Prandtl number,

ν/κ. As with the velocity profile, the form is determined by the wall values of stress and heat flux, and changes of fluid viscosity and conductivity only change the reference level of temperature (fig. 8.2).

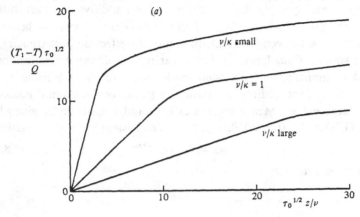

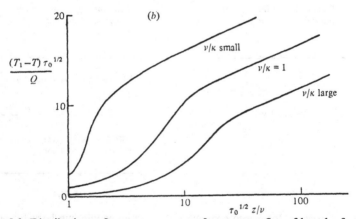

Fig. 8.2. Distributions of mean temperature for constant flux of heat by forced convection through an equilibrium layer in fluids of various Prandtl numbers: (a) linear scale of non-dimensional separation, $\tau_0^{1/2} z/\nu$; (b) logarithmic scale of separation.

If the fluid Prandtl number is large, three distinct regions can be distinguished within the equilibrium layer, an inner conductive layer across which most of the heat flux is conveyed by molecular conduction, the remainder of the viscous layer where most of the flux is transferred through convection by the small but not negligible velocity fluctuations, and the fully turbulent, convective flow outside the viscous layer (fig. 8.3). To determine the additive constant in the profile equation, knowledge of the motion in the viscous layer is necessary and, even then, calculation of the effect on heat transfer is not simple. Calculations using a variety of plausible assumptions and simplifications have been made, most of them leading to a variation of the additive constant as a power of the Prandtl number for large values. Measurements of heat and mass transfer given by R. G. Deissler (Lin 1959, p. 297) are consistent with the variation,

$$A_\theta = 4.5 \, (\nu/\kappa)^{3/4} \qquad (8.8.5)$$

for large values of ν/κ.

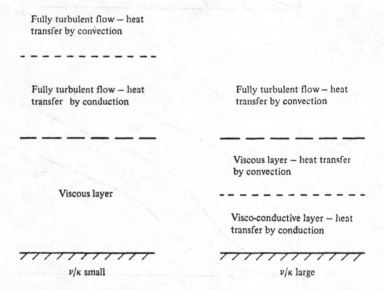

Fig. 8.3. Regions for forced convection through an equilibrium layer for large and small Prandtl number.

With small Prandtl numbers, conductive transfer of heat remains the dominant process for some distance beyond the viscous layer and into the fully turbulent flow. Once convective transfer is established, the eddy diffusivity is expected to be independent of molecular diffusivities and the boundary between the conduction layer and the fully convective region must be near the position where the eddy diffusivity, $\kappa_T = k_\theta \tau_0^{1/2} z$, would equal the molecular diffusivity, κ. Assuming constant temperature gradient of Q_0/κ in the conduction layer and a logarithmic profile outside it, the additive constant is found to be

$$A_\theta = (1 + \log k_\theta) + \log(\nu/\kappa) \qquad (8.8.6)$$

for small values of the Prandtl number. More careful treatment of the blending region between the conduction and convective layers would change the additive constant from the value $(1 + \log k_\theta)$ but preserve the dependence on the logarithm of the Prandtl number.

If mean temperature distributions in a wall flow are of the similarity form,

$$T - T_1 = \theta_0 f_\theta(z/\delta), \qquad (8.8.7)$$

independently of Reynolds number and possibly of downstream position, the distribution function must be consistent with the logarithmic distribution (8.8.4) (T_1 is a reference temperature, the temperature of the free stream for boundary layers or the axial temperature for channel and pipe flows). The conditions that the similarity distribution should be identical with the logarithmic distribution are that

$$\theta_0 = -Q_0/(k_\theta \tau_0^{1/2}) \quad \text{and} \quad f_\theta \to \log(z/\delta) + C_\theta \qquad (8.8.8)$$

for small values of z/δ. Here C_θ is a constant characteristic of the flow and not dependent on Prandtl number. It follows that heat flux and temperature difference across the flow are related by

$$T_0 - T_1 = \frac{Q_0}{k_\theta \tau_0^{1/2}}[\log(\tau_0^{1/2}\delta/\nu) + A_\theta - C_\theta], \qquad (8.8.9)$$

an equation similar to that connecting flow velocity and surface stress,

$$U_1 = \frac{\tau_0^{1/2}}{k}[\log(\tau_0^{1/2}\delta/\nu) + A - C_0] \qquad (8.8.10)$$

(§ 5.6). The relative rates of heat and momentum transfer may be compared using the ratio,

$$\frac{Q_0 U_1}{\tau_0(T_0 - T_1)} = \frac{k_\theta}{k}\frac{[\log(\tau_0^{1/2}\delta/\nu) + A - C_0]}{[\log(\tau_0^{1/2}\delta/\nu) + A_\theta - C_\theta]} \qquad (8.8.11)$$

which is of order k_θ/k unless the Prandtl number is large.

In general, most of the temperature change occurs in the constant-stress, constant-flux layer (particularly if the Prandtl number is large), and the heat transfer coefficients for large flow Reynolds numbers depend most critically on the surface stress and on the Prandtl number if it is large. The value of the flow thickness is not at all critical and neither is the variation of the flow constant C_θ between different flows.

8.9 Rates of heat transfer in forced convection

If an Eulerian treatment of turbulent diffusion is to lead to predictions of transfer rates and mean profiles, the equation for the mean temperature,

$$U \frac{\partial T}{\partial x} + W \frac{\partial T}{\partial z} + \frac{\partial \overline{\theta w}}{\partial z} = \kappa \frac{\partial^2 T}{\partial z^2} \qquad (8.9.1)$$

to the usual boundary-layer approximation for flow homogeneous in the Oy direction, must be supplemented with a second relation between the convective flux $\overline{\theta w}$, the field of mean temperature and the turbulent motion. The simplest of several proposals is the Reynolds analogy between the convective transports of heat and momentum. If heat and momentum were convected in the same way by the turbulent motion, velocity and temperature profiles should be proportional to each other, i.e.

$$(T - T_1) \propto (U - U_1)$$

in flows with similar boundary conditions for velocity and temperature and without longitudinal pressure gradients (which amount to sources of momentum with no thermal counterpart). Then, the formal coefficients of eddy transfer for heat and for momentum would be equal,

$$\kappa_T = -\overline{\theta w} \bigg/ \frac{\partial T}{\partial z} = \nu_T = -\overline{uw} \bigg/ \frac{\partial U}{\partial z} . \qquad (8.9.2)$$

Although the analogy gives an order of magnitude, the ratio κ_T/ν_T (K_h/K_m to meteorologists) is observed to vary between 1.2 and 2.0 in forced convection flows, and it varies significantly within individual flows such as jets and wakes.

For the corresponding problem of the second relation between

Reynolds stress and the mean velocity field, a starting point is an assumption of similarity of the local turbulent motion, in particular the approximate constancy of the stress–intensity ratio, $a_1 = \overline{|uw|/q^2}$, except near positions of zero velocity gradient. It is natural to ask if a corresponding quantity, the heat transport coefficient,

$$a_\theta = \overline{|\theta w|}/(\overline{\theta^2 q^2})^{1/2} \tag{8.9.3}$$

also has nearly the same value in convective flows. Far less is known about the temperature fluctuations than about the velocity fluctuations but, for forced convection in constant-stress layers, the correlation coefficient, $\overline{|\theta w|}/(\overline{\theta^2 w^2})^{1/2}$, appears to lie in the range 0.50–0.55. The value is rather larger than the shear correlation coefficient, $\overline{|uw|}/(\overline{u^2 w^2})^{1/2}$, and indicates that

$$a_\theta \approx 0.18.$$

If the heat-transport coefficient is constant in flows with unidirectional heat flux, assumptions of the kind made by Bradshaw *et al.* (1967) allow the equation for $\frac{1}{2}\overline{\theta^2}$,

$$\left(U \frac{\partial}{\partial x} + W \frac{\partial}{\partial z}\right)\frac{1}{2}\overline{\theta^2} + \overline{w\theta}\frac{\partial T}{\partial z} + \frac{\partial}{\partial z}(\frac{1}{2}\overline{\theta^2 w}) + \varepsilon_\theta = 0 \tag{8.9.4}$$

to be used as the second relation. Suitable assumptions are (1) that the thermal dissipation length is in constant ratio to the dissipation length parameter, i.e.

$$L_\theta = bL_\varepsilon \tag{8.9.5}$$

and (2) that the diffusive lateral flux of $\frac{1}{2}\overline{\theta^2}$ is described by the same convection velocity $\mathscr{V}$,

$$\frac{1}{2}\overline{\theta^2 w} = \frac{1}{2}\overline{\theta^2}\mathscr{V} = \frac{1}{2}\overline{\theta^2}(\overline{pw} + \frac{1}{2}\overline{q^2 w})/\frac{1}{2}\overline{q^2}. \tag{8.9.6}$$

It is supposed that the flow is known in sufficient detail that the distributions of the quantities, L_ε, $\overline{q^2}$ and $\overline{pw} + \frac{1}{2}\overline{q^2 w}$, are known.

Some theoretical justification for the assumption of constant heat-transfer coefficient might be sought by supposing that the apparent ability of the rapid-distortion theory to predict the velocity structure of shear turbulence extends to the temperature field, and using the theory to calculate the development of temperature fluctuations and heat flux in uniform shear and uniform temperature gradient. For distributions of mean velocity and temperature given by

$$\frac{\partial U_1}{\partial x_3} = \alpha, \qquad \frac{\partial T}{\partial x_3} = \alpha_\theta$$

equation (8.5.3) for the variation of amplitude of a Fourier component of the temperature field becomes

$$\left(\frac{\partial}{\partial t} + \kappa_T k^2\right) b(\mathbf{k}) = -\alpha_\theta a_3(\mathbf{k}), \qquad (8.9.7)$$

where κ_T is introduced to model transfer from the large-scale components to smaller-scale components by the non-linear interactions neglected in the rapid-distortion theory. Using the previous solution for the variation of the velocity coefficient $a_3(\mathbf{k})$, the solution is (for $\kappa_T = \nu_T$)

$$b(\mathbf{k}, t) = D\left(b_0(\mathbf{k}) - \alpha_\theta a_{30}(\mathbf{k})\frac{k_0{}^2 \psi}{\alpha k_1(k_1{}^2 + k_2{}^2)^{1/2}}\right), \qquad (8.9.8)$$

where

$$D = \exp[-\kappa_T t(k_0{}^2 - \alpha t k_1 k_{30} + \tfrac{1}{3}\alpha^2 t^2 k_1{}^2)]$$

and

$$\psi = \arctan\left[\frac{\alpha t k_1(k_1{}^2 + k_2{}^2)}{k_0{}^2 - \alpha t k_1 k_{30}}\right].$$

Combining the expression for the temperature coefficients with those for the velocity coefficients given in § 3.12, intensities of temperature fluctuations and heat flux may be calculated as functions of the total strain and the initial configuration of the velocity and temperature fluctuations. If the initial fluctuations are isotropic and uncorrelated, only the initial intensities and, if the transfer terms are appreciable, the scales need be known. In particular, the initial intensity of the temperature fluctuations has no effect on the heat transport which depends only on the temperature gradient and the velocity field, and the ratio of the eddy diffusivities,

$$\kappa_T/\nu_T = (\overline{\theta w}\, \partial U/\partial z)/(\overline{uw}\, \partial T/\partial z)$$

is nearly a function only of the total strain (fig. 8.4). On the other hand, the values of the heat-transport and correlation coefficients do depend on the initial intensity of the fluctuations (fig. 8.5).

Ratios of the eddy diffusivities are plotted in fig. 8.4 as functions of the total strain for several rates of (simulated) transfer, and it is seen that the ratio is 2.5 for small strains and appears to approach one for large total strains. The trend is in good agreement with measurements of the ratios in wakes, jets and boundary layers, using the effective strains found in the way described in § 6.8 or, for equilibrium layers, selected to force agreement with the predicted value of the shear coefficient.

To bring observed values of the heat transport coefficient into agreement with the calculations, it is necessary to assume that the contribution to the current intensity of temperature fluctuations from the original isotropic fluctuations is nearly equal to the intensity generated by interaction with the temperature gradient. In free turbulent flows, both the initial velocity and temperature

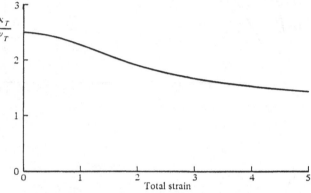

Fig. 8.4. Convective structure of turbulence calculated from the rapid-distortion assumption: dependence of turbulent Prandtl number on total strain.

fluctuations arise from the entrainment process whose nature changes from one kind of flow to another, and it is more likely that the coefficient depends critically on the entrainment process than that it has the same value in all flows. However, the possibility of variation between different kinds of flow does not mean that the hypothesis of a constant coefficient is not useful for wall flows, where most of the interest is in heat transport within the equilibrium layer at the wall.

A conclusion might be that calculations of forced convection in a specified flow should use values of eddy diffusivity based on a constant value of the ratio, κ_T/ν_T, appropriate to the kind of flow, and on the known distribution of eddy viscosity. Good results might be expected if the boundary conditions for velocity and temperature are not too dissimilar, but the rapid distortion calculations are relevant only if the parcel of turbulent fluid has been subjected to rates of distortion that were always in nearly constant ratio to the mean temperature gradient.

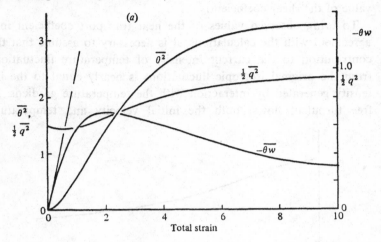

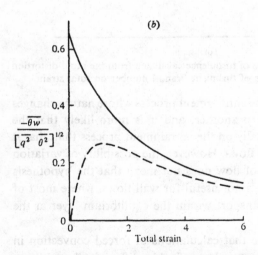

Fig. 8.5. Convective structure of turbulence calculated from the rapid-distortion assumption. (a) Fluctuation intensities and heat flux as a function of total strain, $N = 0.2$. (b) Transport correlation coefficient as a function of total strain for two values of initial intensity of temperature fluctuation: ——, generated $\overline{\theta^2}$; – –, with unit initial $\overline{\theta^2}$.

8.10 Convection in a constant-stress layer after an abrupt change in wall flux or temperature

A situation in which the boundary conditions for velocity and temperature are quite dissimilar arises if the wall temperature or heat flux changes abruptly, and it occurs frequently in the atmospheric boundary layer, both for diffusion of heat and for diffusion of matter. Consider the basic situation in which the surface heat flux changes discontinuously across a line normal to the flow direction, say along the Ox axis, from zero upstream to a constant value Q_0 downstream. With the atmospheric applications in mind, the temperature variations are assumed to be confined within a growing thermal boundary layer inside a thicker constant-stress layer with the velocity distribution,

$$U = \frac{\tau_0^{1/2}}{k} \log(z/z_0). \qquad (8.10.1)$$

Upstream of the discontinuity (for negative x), the temperature is everywhere T_a.

It is natural to ask whether the distributions of mean temperature and heat flux can have the self-preserving forms,

$$\left. \begin{array}{l} T - T_a = \theta_0 f(z/l_0), \\ \overline{\theta w} = Q_0 g(z/l_0), \end{array} \right\} \qquad (8.10.2)$$

where θ_0 is a scale of temperature variation, l_0 is a measure of the thickness of the thermal layer, and both may depend on x. For small values of $\eta = z/l_0$, the heat flux is nearly equal to Q_0, and the convection is that of a constant-stress, constant-flux layer with the temperature distribution,

$$T - T_0 = -\frac{Q_0}{k_\theta \tau_0^{1/2}} \log(z/z_t). \qquad (8.10.3)$$

To be consistent with the distribution of (8.10.2), it is necessary that

$$\theta_0 = Q_0/(k_\theta \tau_0^{1/2}) \quad \text{and} \quad f(\eta) \to -(\log \eta + C_1) \qquad (8.10.4)$$

for small values of η. Then the surface temperature is given by

$$T_0 - T_a = \frac{Q_0}{k_\theta \tau_0^{1/2}}[\log(l_0/z_t) - C_1]. \qquad (8.10.5)$$

The conditions (8.10.4) are sufficient to reconcile the self-preserving development with the properties of the new thermal, constant-flux layer, but the effects of advection are certainly important for larger values of z/l_0. The possibility of the development may be tested here in the usual way by substituting the distributions in the equation for the mean temperature. Then,

$$\frac{\tau_0^{1/2}}{k}\left(\log\frac{l_0}{z_0}+\log\eta\right)\left(\frac{d\theta_0}{dx}f-\frac{\theta_0}{l_0}\frac{dl_0}{dx}\eta f'\right)+\frac{Q_0}{l_0}g'=0. \quad (8.10.6)$$

First, if $\log(l_0/z_0)$ is large, $|\log\eta| \ll \log(l_0/z_0)$ everywhere except within the constant-flux layer, and the equation takes the self-preserving form,

$$\eta f' - g' = 0 \quad (8.10.7)$$

if the length-scale satisfies the equation,

$$\log(l_0/z_0)\,dl_0/dx = kk_\theta. \quad (8.10.8)$$

Notice that the choice of the length scale to satisfy equation (8.10.8) implies that

$$\int_0^\infty f(\eta)\,d\eta = g(0) = 1. \quad (8.10.9)$$

A better approximation to the layer thickness may be found from the condition of conservation of heat, i.e.

$$Q_0 x = \int_0^\infty U(T-T_a)\,dz$$

$$= \frac{\tau_0^{1/2}}{k}\frac{Q_0 l_0}{k_\theta \tau_0^{1/2}}\left[\log l_0/z_0 + \int_0^\infty f(\eta)\log\eta\,d\eta\right] \quad (8.10.10)$$

i.e.

$$l_0\left(\log l_0/z_0 + \int_0^\infty f(\eta)\log\eta\,d\eta\right) = kk_\theta x,$$

a result consistent with (8.10.8) if $\log(l_0/z_0)$ is large.

Diffusion from a line source of heat can be treated by super-imposing the two distributions of surface flux,

(1) $Q_0(x) = 0$ for $x < 0$, $\qquad Q_0(x) = Q_1$ for $x > 0$,
(2) $Q_0(x) = 0$ for $x < \Delta x$, $\qquad Q_0(x) = -Q_1$ for $x > \Delta x$.

Then the composite distribution of mean temperature is

$$T-T_a = \frac{Q_1}{k_\theta \tau_0^{1/2}}[f(z/l_0)-f(z/l_0')], \quad (8.10.11)$$

where l_0' is the thickness for the second distribution, starting a distance Δx downstream of the first. If Δx is small,

$$T - T_a = -\frac{Q_1 \Delta x}{k_\theta \tau_0^{1/2}} \frac{1}{l_0} \frac{\mathrm{d}l_0}{\mathrm{d}x} \eta f', \qquad (8.10.12)$$

where $Q_1 \Delta x$ is the strength of the line source.

Like the function for a step-change, the distribution function for a line source has unit area, i.e.

$$\int_0^\infty (-\eta f') \, \mathrm{d}\eta = 1,$$

and is equal to one for $\eta = 0$. Certainly, its gradient is everywhere negative or zero, and the thinnest possible profile is one with $-\eta f' = 1$ for η less than one and zero beyond, i.e. the extreme profile for the diffusion is the Elliott profile,

$$\left. \begin{aligned} f(\eta) &= -\log \eta \quad (\eta < 1), \\ &= 0 \quad\quad\;\; (\eta > 1). \end{aligned} \right\} \qquad (8.10.13)$$

Measurements of the concentration of chemical tracers emitted by a line source are well described by the analysis, and they indicate that the distribution function is probably intermediate between an exponential form,

$$-\eta f' = \exp(-\eta),$$

and a linear form,

$$\begin{aligned} -\eta f' &= 1 - \tfrac{1}{2}\eta \quad (\eta < 2), \\ &= 0 \quad\quad\quad\;\; (\eta > 2) \end{aligned}$$

(Townsend 1965b).

It is interesting to interpret these results in terms of the eddy diffusivity. For the exponential profile, the eddy diffusivity is simply

$$\kappa_T = k_\theta \tau_0^{1/2} z$$

the same as in an equilibrium layer, and, for the linear profile,

$$\kappa_T = k_\theta \tau_0^{1/2} z (1 + \tfrac{1}{2}z/l_0) \quad (z < 2l_0).$$

If anything, the ratio of the eddy diffusivities, κ_T/ν_T, exceeds the normal value over most of the thermal internal layer, very much in contrast to the coefficient of eddy viscosity that describes the flow modification when a boundary layer flows across a change of surface roughness (§ 7.15). For the Elliott velocity profile,

$$\nu_T = k\tau_0^{1/2} z (1 - z/l_0)$$

(l_0 in this equation is greater than the scale-length of this section in the ratio $2k/k_\theta$), showing it to become less in the internal boundary layer.

It appears that the interaction between a growing thermal boundary layer and an already existing shear flow is not essentially different from the interaction with random isotropic turbulence, and that the initial turbulent Prandtl numbers of the outer flow are similar in magnitude to those calculated using the total effective strain since inception of the temperature gradient.

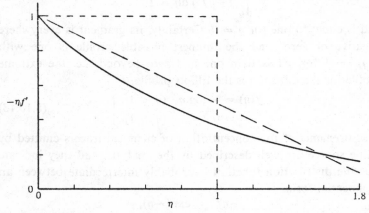

Fig. 8.6. Profiles of change of heat flux following an abrupt change of surface flux, showing the extreme Elliott profile, the log-linear profile and the mixing-length profile.

Nicholl (1970) and Johnson (1959) have measured the temperature fluctuations in boundary layers flowing from a cold surface to a hot one. Perhaps the most remarkable feature is the pronounced intermittency of the temperature fluctuations near the edge of the heated region, although the velocity fluctuations are normal and continuous.

8.11 Longitudinal diffusion in pipe flow

If the composition of the fluid flowing along a pipe is suddenly changed, the interface between the two fluids moves along the pipe and becomes diffuse. In forced convection, solutions may be superimposed and the basic diffusion problem is that of a thin diametral section of marked fluid, distinct from the fluid up and down stream. We need to find the motion of the centroid of the marked fluid and the dispersion around the centroid as a function of time.

The problem has been treated by the theory of diffusion by continuous movements (Batchelor, Binnie & Phillips, 1955) which assumes that the bulk diffusion is independent of the molecular diffusivity. With time, a marked particle migrates across the pipe section, spending time in all parts, and its velocity is a stationary random function of time with a non-zero mean value. After a sufficient time, the position and velocity is independent of its place of release, and the statistical mean velocity is the same for all particles. Consequently, it is the same as the mean velocity taken over all fluid particles in the pipe and is equal to the mean flow velocity,

$$U_m = \frac{1}{\pi R^2} \int_0^R U 2\pi r \, dr. \tag{8.11.1}$$

The Lagrangian autocorrelation function for a marked particle initially depends on the radial position of release,

$$\langle (v(t) - U_m)(v(t + \tau) - U_m) \rangle = R(\tau, t; r_0), \tag{8.11.2}$$

where r_0 is the initial radial position at time $t = 0$, but, after a long diffusion time, the position is independent of r_0 and the correlation depends only on the flow and the time interval τ. Then, for long diffusion times,

$$\Delta^2 = \langle (X - U_m t)^2 \rangle = 2(t - t_0) \int_0^\infty R_\infty(\tau) \, d\tau, \tag{8.11.3}$$

where R_∞ is the autocorrelation function for large values of time, and t_0 is a time displacement. It is known that the scales of the turbulent motion are the friction velocity, $\tau_0^{1/2}$, and the pipe radius, R, and dimensional reasoning shows that

$$\tfrac{1}{2}\Delta^2 = A_d \tau_0^{1/2} R t - O(R^2) \tag{8.11.4}$$

since t_0 must be of order $R/\tau_0^{1/2}$. The quantity A_d is to be regarded as a coefficient of longitudinal diffusion.

The problem was first considered by Taylor (1954) using the Eulerian equation for concentration,

$$\frac{\partial C}{\partial t} + U \frac{\partial C}{\partial x} + \frac{1}{r} \frac{\partial (\overline{cw}r)}{\partial r} + \frac{\partial \overline{cu}}{\partial x} = \frac{1}{r} \frac{\partial}{\partial r}\left(Dr \frac{\partial c}{\partial r} \right), \tag{8.11.5}$$

where c is the concentration of the marking material. By taking the first moment with respect to downstream position, it can be shown that X_m, the position of the centroid, varies as

$$\frac{\mathrm{d}X_m}{\mathrm{d}t} = \iint (CU + \overline{cu})r \, \mathrm{d}r \, \mathrm{d}x \Big/ \iint Cr \, \mathrm{d}r \, \mathrm{d}x \qquad (8.11.6)$$

which is equal to the mean velocity over the section, if (1) concentration varies little over any section, and (2) longitudinal transport by the *turbulent* motion has a negligible effect. Assuming that the eddy coefficients of viscosity and diffusion are everywhere the same, Taylor showed that the coefficient of longitudinal diffusion would be 10.1, which may be compared with a value of 10.4 measured for diffusion of salt in water flow.

8.12 Natural convection and energy transfer

Natural convection flows derive their energy from the action of buoyancy forces that convert gravitational potential energy into kinetic energy of movement, but the transfer can take place in two ways. In buoyant heat plumes and in other flows with nearly vertical mean flow, nearly all the energy is transferred by working of buoyancy forces on the mean flow and very little by direct action on the velocity fluctuations. The relation between the turbulent motion and the mean flow in these flows is very much the same as in ordinary shear flows, the principal difference being that they are driven by body forces rather than inertia or pressure gradients.

If the mean flow is either zero or substantially horizontal, little energy can be transferred to the mean flow and the turbulence is strongly influenced by the buoyancy. Natural convection of heat between parallel horizontal planes is the best known example, and the buoyancy-driven turbulent motion is considerably different in nature from shear-generated turbulence.

8.13 Buoyant plumes and thermals

Continuous release of heat from a line or point source in an unrestricted fluid leads to the development of a rising plume, and sudden release to the formation of a rising volume of heated fluid, a thermal. If the surrounding fluid is either unstratified or stratified in particular ways, plumes and thermals may develop in self-preserving fashion and they have much in common with turbulent jets and puffs of fluid (fig. 8.7).

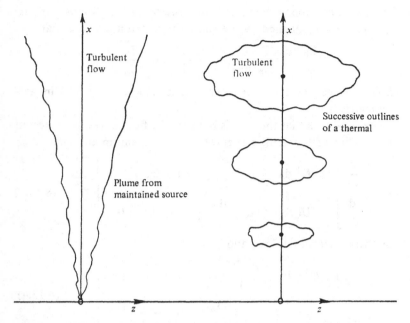

Fig. 8.7. Sketch of developments of a plume from a maintained source and a thermal from a transient source.

To conform with the notation used by meteorologists, the Oz axis is vertical and in the direction of motion and, for two-dimensional flows, Ox is in the horizontal direction of shear. The development of a plume is self-preserving if mean values have the forms,

$$
\left.
\begin{aligned}
W &= u_0 f(x/l_0) && \text{or} && u_0 f(r/l_0), \\
T - T_a &= \theta_0 f_\theta(x/l_0) && \text{or} && \theta_0 f_\theta(r/l_0), \\
\overline{uw} &= u_0{}^2 g(x/l_0) && \text{or} && u_0{}^2 g(r/l_0), \\
\overline{\theta w} &= \theta_0 u_0 g_\theta(x/l_0) && \text{or} && \theta_0 u_0 g_\theta(r/l_0)
\end{aligned}
\right\}
\qquad (8.13.1)
$$

as the plume is two-dimensional or axisymmetric. The equations for mean velocity and mean temperature are then

$$
\left.
\begin{aligned}
u_0 \frac{du_0}{dz} f^2 - \frac{u_0}{l_0} \frac{d(u_0 l_0)}{dz} f' \int_0^{} f\, d\eta + \frac{u_0{}^2}{l_0} g' &= \frac{g}{T_a} \theta_0 f_\theta, \\
u_0 \frac{d\theta_0}{dz} f f_\theta - \frac{\theta_0}{l_0} \frac{d(u_0 l_0)}{dz} f_\theta' \int_0^{} f\, d\eta + \frac{u_0 \theta_0}{l_0} g_\theta' + u_0 f \frac{dT_a}{dz} &= 0
\end{aligned}
\right\}
\qquad (8.13.2)
$$

for two-dimensional flow, and the necessary conditions for the equations to be satisfied by the similarity distributions are that

$$\frac{\mathrm{d}l_0}{\mathrm{d}z}, \qquad \frac{l_0}{\theta_0}\frac{\mathrm{d}T_a}{\mathrm{d}z}, \qquad \frac{gl_0\theta_0}{u_0^2 T_a}, \qquad \frac{l_0}{u_0}\frac{\mathrm{d}u_0}{\mathrm{d}z}$$

should be constant or zero. The same conditions apply in axisymmetric flow.

The condition that the flow is bounded by fluid at rest is expressed by equations for overall conservation of momentum and heat,

$$\left.\begin{array}{l} \dfrac{\mathrm{d}}{\mathrm{d}z}\displaystyle\int_{-\infty}^{\infty} U^2\,\mathrm{d}x = \dfrac{g}{T_a}\displaystyle\int_{-\infty}^{\infty}(T-T_a)\,\mathrm{d}x, \\[3mm] \dfrac{\mathrm{d}}{\mathrm{d}z}\displaystyle\int_{-\infty}^{\infty} U(T-T_a)\,\mathrm{d}x + \dfrac{\mathrm{d}T_a}{\mathrm{d}z}\displaystyle\int_{-\infty}^{\infty} U\,\mathrm{d}x = 0, \end{array}\right\} \tag{8.13.3}$$

for two-dimensional flow, and

$$\left.\begin{array}{l} \dfrac{\mathrm{d}}{\mathrm{d}z}\displaystyle\int_{0}^{\infty} U^2 r\,\mathrm{d}r = \dfrac{g}{T_a}\displaystyle\int_{0}^{\infty}(T-T_a)r\,\mathrm{d}r, \\[3mm] \dfrac{\mathrm{d}}{\mathrm{d}z}\displaystyle\int_{0}^{\infty} U(T-T_a)r\,\mathrm{d}r + \dfrac{\mathrm{d}T_a}{\mathrm{d}z}\displaystyle\int_{0}^{\infty} Ur\,\mathrm{d}r = 0. \end{array}\right\} \tag{8.13.4}$$

In terms of the profile integrals,

$$I_n = \int_0^\infty [f(\eta)]^n \eta^\alpha\,\mathrm{d}\eta, \qquad I_{n,\theta} = \int_0^\infty [f(\eta)]^n f_\theta(\eta)\eta^\alpha\,\mathrm{d}\eta$$

they are

$$\left.\begin{array}{l} I_2 \dfrac{\mathrm{d}}{\mathrm{d}z}(u_0^2 l_0^{1+\alpha}) = I_{0,\theta}\dfrac{g}{T_a}\theta_0 l_0^{1+\alpha}, \\[3mm] \end{array}\right.$$

and

$$\left.\begin{array}{l} I_{1,\theta}\dfrac{\mathrm{d}}{\mathrm{d}z}(u_0\theta_0 l_0^{1+\alpha}) + I_1\dfrac{\mathrm{d}T_a}{\mathrm{d}z}u_0 l_0^{1+\alpha} = 0, \end{array}\right\} \tag{8.13.5}$$

where $\alpha = 0$ or 1 as the flow is two-dimensional or axisymmetric.

For both kinds of symmetry, the conditions for self-preserving development are met if the scales and ambient temperature gradient vary as

$$\left.\begin{array}{ll} \theta_0 \propto (z-z_0)^a, & u_0 \propto (z-z_0)^{\frac{1}{2}(a+1)}, \\[2mm] l_0 \propto (z-z_0), & \mathrm{d}T_a/\mathrm{d}z \propto (z-z_0)^{(a-1)}, \end{array}\right\} \tag{8.13.6}$$

where the exponent a lies in a range that allows acceptable values of the flow parameters. With the variations (8.13.6), the conservation equations can be put into the forms,

$$Ri = \frac{g\theta_0(z - z_0)}{T_a u_0{}^2}$$

$$= \frac{I_2}{I_\theta}(a + \alpha + 2) \qquad (8.13.7)$$

$$= -\frac{I_{1,\theta}}{I_1}(\tfrac{3}{2}(a + 1) + \alpha).$$

For buoyant plumes, the Richardson number Ri must be positive, and so the exponent a must be more positive than $-(2 + \alpha)$. If a lies in the range, $-(2 + \alpha) < a < -(1 + \tfrac{2}{3}\alpha)$, the ambient temperature gradient is positive and plume momentum decreases with development. If a exceeds $-(1 + \tfrac{2}{3}\alpha)$, plume buoyancy increases from its initial value and plumes are self-generating in the unstably-stratified ambient fluid.

A finite volume of heated fluid may rise as a turbulent thermal whose development is self-preserving in time, i.e. the distributions of velocity and temperature have the forms,

$$\left. \begin{array}{l} \mathbf{U} = u_0\mathbf{f}[(\mathbf{x} - \mathbf{X}_m)/l_0], \\ T - T_a = \theta_0 f_\theta[(\mathbf{x} - \mathbf{X}_m)/l_0], \end{array} \right\} \qquad (8.13.8)$$

where the scales, u_0 and l_0, are functions of time, and $\mathbf{X}_m$ is the centroid of the temperature distribution. The ambient fluid is at rest, and the velocity of rise, dZ_m/dt, is a constant multiple of the velocity scale if the distribution remains similar. Substitution in the equations for mean velocity and mean temperature gives as conditions for self-preserving development that the ratios,

$$\frac{1}{u_0}\frac{dl_0}{dt}, \qquad \frac{l_0}{u_0\theta_0}\frac{d\theta_0}{dt}, \qquad \frac{l_0}{u_0{}^2}\frac{du_0}{dt},$$

$$\frac{g\theta_0 l_0}{T_a u_0{}^2}, \qquad \frac{l_0}{\theta_0}\frac{dT_a}{dz},$$

must be constant or negligible. It should be noted that

$$\frac{dl_0}{dZ_m} = \frac{dl_0}{dt} \Big/ \frac{dZ_m}{dt} \propto \frac{1}{u_0}\frac{dl_0}{dt} \qquad (8.13.9)$$

and the angle of spread is constant.

The rate of increase of the impulse of the motion equals the total buoyancy force, that is,

$$\frac{d}{dt} \int \tfrac{1}{2}(x \times \Omega)\, dV = \int g(T - T_a)/T_a\, dV$$

or, in terms of the distribution functions,

$$M^* \frac{d}{dt}(u_0 l_0{}^3) = I_\theta g \theta_0 l_0{}^3/T_a, \qquad (8.13.10)$$

where M^* is a non-dimensional coefficient of virtual mass, determined by the distribution function for velocity. The equation for overall conservation of heat is

$$\frac{d}{dt} \int (T - T_a)\, dV + \frac{dT_a}{dz} \int W\, dV = 0,$$

or

$$I_\theta \frac{d}{dt}(\theta_0 l_0{}^3) + \frac{dT_a}{dz} I_1 u_0 l_0{}^3 = 0. \qquad (8.13.11)$$

Self-preserving development is then possible if the scales vary as

$$u_0 \propto t^a, \qquad \theta_0 \propto t^{(a-1)}, \qquad l_0 \propto t^{(a+1)} \qquad (8.13.12)$$

and if the gradient of ambient temperature varies as

$$dT_a/dz \propto z^{-2/(a+1)} \qquad (8.13.13)$$

In order that the buoyancy and impulse be both directed upwards, equation (8.13.10) requires that the exponent a should be more positive than $-\tfrac{3}{4}$.

The most interesting case is the thermal in a neutrally stratified environment, i.e. $dT_a/dz = 0$, for which the exponent is $-\tfrac{1}{2}$ (see equation (8.13.11)). Measurements of the rate of rise and rate of spread of individual thermals are well described by the variations (8.13.12), but different thermals with the same total buoyancy rise and spread at different rates (Scorer 1957, Richards 1961). If the velocity distributions are the same in all the thermals, the rate of rise is

$$dZ_m/dt = W_m^* u_0, \qquad (8.13.14)$$

where W_m^* is a non-dimensional constant, and the angle of spread, l_0/Z_m, is related to the rate of rise by

$$\frac{dZ_m{}^2}{dt} \bigg/ (gB)^{1/2} = (2W_m{}^*/M^*)^{1/2}(Z_m/l_0)^{3/2}, \qquad (8.13.15)$$

where $B = \int (T - T_a)/T_a \, dV = I_\theta \theta_0 l_0{}^3/T_a$ is the total buoyancy. The relation is well confirmed by the measurements. Variability of the rate of rise means that, although the ratio of the axial circulation to the square root of the total buoyancy force,

$$K^* = (gB)^{-1/2} \int_{-\infty}^{\infty} w(0, 0, z) \, dz$$

$$= \int_{-\infty}^{\infty} f_3(0, 0, \eta) \, d\eta \, u_0 l_0/(gB)^{1/2} \qquad (8.13.16)$$

remains the same during rise of a particular thermal, its value depends on initial conditions at release. If thermals have the form of turbulent vortex rings with non-turbulent fluid along the axis, the axial circulation cannot change and the observed variability is easily explained. (Turner 1957), but thermals generated by simple release of buoyant fluid are observed to be turbulent along the axis. It is most likely that the circulation ratio, K^*, does approach a universal value after a sufficient time of rise but the rate of approach is too slow to be easily detectable in the experiments.

Two-dimensional thermals have scales that vary with time in the ways given by equations (8.13.12), but now the exponent a must be more positive than $-\frac{2}{3}$. In neutral surroundings, its value is $-\frac{1}{3}$ and

$$u_0 \propto t^{-1/3}, \qquad \theta_0 \propto t^{-4/3}, \qquad l_0 \propto t^{2/3}. \qquad (8.13.17)$$

The circulation does not remain constant and the turbulent transfer of heat and vorticity across the plane of symmetry must be appreciable to maintain the similarity of the distributions.

Attempts have been made to calculate the development of plumes and thermals in conditions of ambient fluid stratification that do not conform to the strict requirements for self-preserving development, usually by assuming similarity of the distributions of velocity and temperature and a universal value of the entrainment constant,

$$\beta = \frac{1}{2}\frac{dl_0}{dx} \quad or \quad \frac{1}{u_0}\frac{dl_0}{dt}. \qquad (8.13.18)$$

Taken with the equations for overall conservation of momentum (or impulse) and heat, the motion can be found for arbitrary external stratification (for example, Morton, Taylor & Turner 1956).

13

8.14 The effect of buoyancy forces on turbulent motion

If the direction of mean flow is nearly horizontal, the buoyancy forces may act directly on the turbulence and transfer more energy to the fluctuations than to the mean flow. The effect might be simply to transfer energy to the turbulence without changing its structure, but it is likely that the strongly directional forces will change the geometrical form of the eddies. In spite of possible changes of form, a good description of some aspects of flow behaviour can be obtained by ignoring them and considering the energy balance of the motion.

For nearly horizontal mean flow, the equation for the turbulent kinetic energy is

$$U\frac{\partial}{\partial x}(\tfrac{1}{2}\overline{q^2}) + W\frac{\partial}{\partial z}(\tfrac{1}{2}\overline{q^2}) + \overline{uw}\frac{\partial U}{\partial z} + \frac{\partial}{\partial z}(\overline{pw} + \tfrac{1}{2}\overline{q^2 w}) = g\frac{\overline{\theta w}}{T_a} - \varepsilon \quad (8.14.1)$$

to the boundary-layer approximation. If the turbulent level is not too large, the advection terms and the lateral diffusion terms may be omitted, and the equation expresses a balance between

(1) energy generation by Reynolds stresses,

(2) energy generation by buoyancy forces, and

(3) viscous energy dissipation.

If the structure of the turbulence is unaffected by the buoyancy forces,

$$\left.\begin{array}{ll} |\overline{uw}| = a_1\overline{q^2}, & |\overline{\theta w}| = a_\theta(\overline{\theta^2}\,\overline{q^2})^{1/2} \\ \varepsilon = (\overline{q^2})^{3/2}/L_\varepsilon, & \varepsilon_\theta = (\overline{q^2})^{1/2}\overline{\theta^2}/L_\theta \end{array}\right\} \quad (8.14.2)$$

where a_1, a_θ and L_ε/L_θ are constants. Substitution in the energy equation leads to

$$a_1\overline{q^2}\frac{\partial U}{\partial z} - a_\theta\frac{g}{T_a}(\overline{q^2\theta^2})^{1/2}\,\mathrm{sgn}\left(\frac{\partial T}{\partial z}\right) - (\overline{q^2})^{3/2}/L_\varepsilon = 0 \quad (8.14.3)$$

and, in the equation for the fluctuation entropy to

$$a_\theta(\overline{q^2\theta^2})^{1/2}\frac{\partial T}{\partial z} = -(\overline{q^2})^{1/2}\overline{\theta^2}/L_\theta. \quad (8.14.4)$$

Eliminating $\overline{\theta^2}$ from the first equation, we find

$$a_1\overline{q^2}\frac{\partial U}{\partial z} - a_\theta{}^2\frac{g}{T_a}(\overline{q^2})^{1/2}L_\theta\frac{\partial T}{\partial z} - (\overline{q^2})^{3/2}/L_\varepsilon = 0 \quad (8.14.5)$$

an equation for the turbulent intensity whose only physical solution is

$$\frac{(\overline{q^2})^{1/2}}{a_1 L_\varepsilon \, \partial U/\partial z} = \frac{1}{2} + \frac{1}{2}\left[1 - 4\frac{a_\theta^2 L_\theta}{a_1^2 L_\varepsilon}Ri\right]^{1/2} \tag{8.14.6}$$

where $Ri = (g/T_a)(\partial T/\partial z)/(\partial U/\partial z)^2$ is the local Richardson number. The important feature of (8.14.6) is that, if the Richardson number exceeds the critical value of

$$R_c = a_1^2 L_\varepsilon/(4a_\theta^2 L_\theta) \tag{8.14.7}$$

there is no solution with a physical meaning. At the critical Richardson number, the non-dimensional intensity $(\overline{q^2})^{1/2}/(a_1 L_\varepsilon \, \partial U/\partial z)$ has half the value it has for zero Richardson number, and the implication is that the equilibrium intensity drops suddenly to zero as

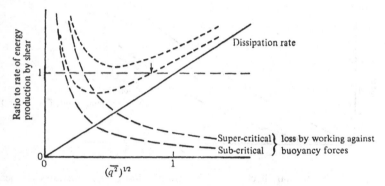

Fig. 8.8. Relative magnitudes of terms in the turbulent energy equation for shear flow with stable density stratification.

the critical condition is exceeded. The reason can be understood by considering the ratios of the terms in equation (8.14.5) to the rate of energy generation by Reynolds stresses as functions of the non-dimensional intensity (fig. 8.8). In stable conditions, the sum of the ratios for buoyancy loss and viscous dissipation has a minimum value which exceeds one if the Richardson number exceeds the critical value. The conclusion is that, in a flow for which the local Richardson numbers increase in the flow direction (or with time), normal but somewhat weakened turbulence persists until the critical number is exceeded. Then the energy supply is no longer sufficient and the motion collapses to almost laminar flow. The behaviour is

easily observed in jets and mixing layers established along an interface between fluid layers of different densities.

Although these considerations lead to a good qualitative description of the effect of buoyancy forces and even predict with
some success the critical Richardson number, the structure of the
turbulence is changed considerably if the buoyancy forces are large.
For example, the ratio of the eddy diffusivities should be

$$\frac{\kappa_T}{\nu_T} = \frac{\overline{\theta w}}{\overline{uw}} \frac{\partial U/\partial z}{\partial T/\partial z} = \frac{a_\theta^2}{a_1^2} \frac{a_1 L_\varepsilon}{(\overline{q^2})^{1/2}} \frac{\partial U/\partial z}{} \qquad (8.14.8)$$

which increases with flow stability, in contrast to observation and
to common intuition.

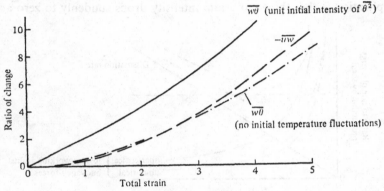

Fig. 8.9. Incremental effects of density stratification on transfer rates for heat
and momentum, from the rapid-distortion assumption.

Use of the rapid-distortion approximation allows calculation of
changes in turbulence structure for a dynamically consistent flow,
even if it is not completely realistic. In fig. 8.9, the calculated effects
of Richardson number on transport rates are given in the form of
the coefficients β in relations of the kind,

$$\overline{u^2} = (\overline{u^2})_{Ri=0}(1 - \beta Ri) \qquad (8.14.9)$$

The coefficients depend strongly on the total strain of the turbulence,
the sensitivity to Richardson number increasing rapidly with increasing strain, but the coefficients for a strain ratio of about four
and with a plausible choice of the initial intensity of temperature
fluctuations are in order of magnitude agreement with measurements

in the atmospheric boundary layer (see Lumley & Panofsky 1964, Plate 1971). It will be noticed that the quantities most strongly affected by buoyancy are the transport rates, $\overline{uw}$ and $\overline{\theta w}$, and the vertical fluctuation intensity $\overline{w^2}$. The longitudinal intensity changes much less, and the lateral intensity is relatively unaffected.

In the discussion above, it is assumed implicitly that one Richardson number describes the whole flow. For this reason, the results are applicable only to flows with not too large variations of Richardson number, in particular to the outer flow in boundary layers and to free turbulent flows in stratified fluid.

8.15 Horizontal wall layers with heat transport

The results of the previous section cannot be used deep within a boundary layer where the local Richardson number depends strongly on position in the flow. Given a sufficient degree of horizontal homogeneity, it seems possible that a stratified equilibrium layer may be distinguished in which the Reynolds stress and vertical heat flux are nearly independent of height. Assuming that it resembles an ordinary equilibrium layer in that its motion is unaffected either by conditions at the upper boundary of the flow or directly by the surface roughness, dimensional considerations lead to the results,

$$
\left.
\begin{aligned}
dU/dz &= \frac{\tau_0^{1/2}}{kz}F(z/L), \\
dT/dz &= -\frac{Q_0}{k_\theta \tau_0^{1/2}z}F_\theta(z/L),
\end{aligned}
\right\} \tag{8.15.1}
$$

where $L = \tau_0^{3/2}T_a/(kgQ_0)$ is the Monin–Obukhov length, and τ_0, Q_0 are the constant values of the Reynolds stress and upward heat flux. If the effect of buoyancy is negligible, the relations (8.15.1) must reduce to the forced convection forms and so both functions are nearly one. Then the local Richardson number is given by

$$
Ri = -\frac{k}{k_\theta}z/L \tag{8.15.2}
$$

if its magnitude is small (in practice less than about 0.02), and the Monin–Obukhov length is a measure of the height at which buoyancy forces become dominant.

The equation for the turbulent kinetic energy is

$$\tau_0 \frac{\partial U}{\partial z} + \frac{g}{T_a} Q_0 = \varepsilon + \frac{\partial}{\partial z}(\overline{pw} + \tfrac{1}{2}\overline{q^2 w})$$

and can be written in terms of the distribution function of equation (8.15.1) as

$$F(z/L) + z/L = \left[\varepsilon + \frac{\partial}{\partial z}(\overline{pw} + \tfrac{1}{2}\overline{q^2 w}) \right] \Big/ (\tau_0\, \partial U/\partial z) \qquad (8.15.3)$$

If the heat flux is downward, the left-hand terms become very large compared with those on the right (supposing the vertical diffusive transport of energy to be moderate), and it appears that

$$F(z/L) \rightarrow -z/L \quad \text{for large } -z/L, \qquad (8.15.4)$$

indicating that nearly all the energy generated by shear is used in working against the buoyancy forces. Observations in conditions of extreme stability, either in the atmosphere or in the laboratory, are very difficult, but the scattered measurements could be reconciled with the predicted values of the velocity gradient, $-gQ_0/(\tau_0 T_a)$.

If the heat flux is directed upwards, buoyancy forces transfer energy to the turbulence at a rate that is independent of height, while transfer from the mean flow is proportional to the velocity gradient and decreases with height. As the contribution from buoyancy becomes relatively larger, the ratio of turbulent intensity to shear stress increases and the eddy diffusivity is more than in a constant-density flow.

Measurement in the atmospheric boundary layer (Lumley & Panofsky 1964, Plate 1971) give an indication of the variation of the distribution functions of equations (8.15.1) (fig. 8.10). For small values of z/L, the functions have the forms,

$$\left. \begin{array}{l} F(z/L) = 1 - \beta z/L, \\ F_\theta(z/L) = 1 - \beta_\theta z/L, \end{array} \right\} \qquad (8.15.5)$$

where $\beta \simeq 3.8$ and $\beta_\theta \simeq 8.2$, showing that the initial effect of buoyancy on heat convection is considerably greater than that on momentum transport. For larger values of z/L in strongly unstable conditions, both profile functions approach zero and the mean velocity and temperature approach limiting values, given by

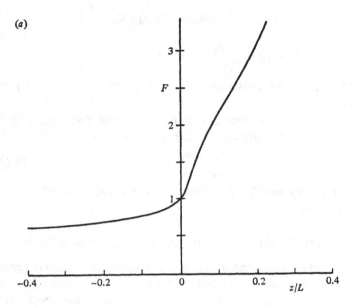

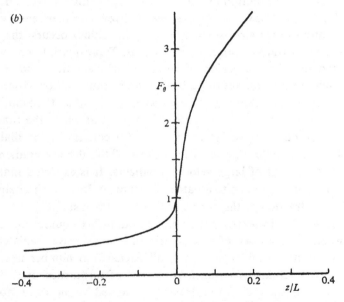

Fig. 8.10. Dependence of transfer rates for heat and momentum on stability in a constant-stress, constant-flux layer.

$$U_\infty = \frac{\tau_0^{1/2}}{k}[\log(L/(\beta z_0))] + C_u),$$

$$T_\infty - T_0 = \frac{Q_0}{k_\theta \tau_0^{1/2}}[\log(L/(\beta_\theta z_0))] + C_\theta, \quad \Big\}$$ (8.15.6)

where C_u, C_θ are constants depending on the forms of the profile functions.

In conditions of strong stability (z/L large and negative), the ratio of the eddy diffusivity to the eddy viscosity,

$$\frac{\kappa_T}{\nu_T} = \frac{k_\theta}{k}\frac{F}{F_\theta}$$ (8.15.7)

becomes very small, while the value for strongly unstable flow is around four.

8.16 Nature of turbulence in strongly stable flows

If the Richardson number of a flow is greater than 0.25, energy may be propagated in the form of internal gravity waves whose displacements may be random and aperiodic in form but which do not form a turbulent motion in the sense of a highly dissipative, diffusive motion. Like most wave motions, they convey with them energy and momentum but, unless some secondary instability occurs, they are incapable of transferring matter and heat. Typically, internal waves are radiated from a region of strongly turbulent flow and spread throughout the volume of stably stratified fluid. If the volume is finite and viscous damping of the waves is negligible, the amplitude of the wave motion will increase to the point at which the random superposition of waves gives rise to local concentrations that are unstable either through local inversions of the density gradient or the development of large velocity gradients. It is expected that the instability will give rise to a patch of turbulent flow which dissipates at least a fraction of the energy released by the instability.

The motion in a strongly stable, constant-flux equilibrium layer is probably composed of a basic field of internal waves radiated by the vigorous turbulent flow of small Richardson number near the surface, punctuated by transient patches of turbulence. It is likely that most of the momentum transfer is carried by the wave motion but heat transfer depends on movement of matter and only occurs within the turbulent patches. In a flow of this kind, the conditions for horizontal homogeneity are difficult to satisfy since internal waves propagate horizontally as well as vertically.

Over most of the earth's atmosphere, the density stratification is stable and Richardson numbers based on mean gradients are large except in the boundary layer, in clouds and in jet streams. That mixing does occur, even much higher than the limit of active convection, is shown by the substantial uniformity of the atomic composition of the atmosphere at all heights to at least 100 km without any sign of the separation that would appear by molecular diffusion in an unstirred atmosphere. In fact, very little mixing is necessary to maintain the well-mixed state, and the agent above the tropopause is probably randomly occurring patches of sporadic turbulence (clear-air turbulence) generated by local instabilities in a system of random internal waves. The patches may be remote from their source, though consideration of their propagation in an atmosphere with vertical wind-shear and observation show that they are more likely to occur in regions of smaller Richardson number (Townsend 1968). One source of wave energy is the jet stream near the tropopause, and clear-air turbulence is most common there although it is found at much greater heights. In particular, observations of stellar shadow-bands and of the scatter-propagation of short waves show that at least one patch usually lies along a line of sight through a clear atmosphere.

8.17 Transient behaviour of boundary layers with heat transfer

If an adiabatic boundary layer with no heat flux flows onto a hotter or cooler surface, the development may be strongly influenced by the buoyancy forces both while the temperature change is spreading from the surface and afterwards when some form of equilibrium development is re-established. Similar effects are found in the earth's boundary layer if the surface heat flux is changed considerably, for example, by the removal of incoming radiation at sunset in clear weather or, most dramatically, at a solar eclipse.

Initially, the temperature changes spread from the surface in an internal, thermal boundary layer of forced convection whose development is described by the analysis of § 8.10, but when the thickness of the internal layer reaches an appreciable fraction of the

Monin–Obukhov length, buoyancy forces begin to influence the flow. If the heat flux is upwards, additional turbulence is generated and the internal layer spreads more rapidly, but the horizontal gradients of mean temperature cause acceleration of the mean flow that may lead to the transient formation of a sea-breeze, resembling a wall jet in a moving stream. (Townsend 1972).

If the surface is colder than the flow and the heat flux is downward, there is a tendency to deceleration of the mean flow but the most dramatic effect is a collapse of both the turbulent motion and the mean flow below the level at which the Richardson number first attains a value of about 0.1. When the internal layer has spread so far that the Richardson number passes the critical value, the turbulent motion at this level can no longer be maintained at a level sufficient to transmit mean flow energy towards the wall by the Reynolds stress working against the mean velocity. With no energy supply, the lower part of the flow dissipates rapidly all its energy and comes nearly to rest. The collapse of the flow persists for some time with an upper layer of warm, fast-moving fluid overlying colder fluid of low velocity. Measurements by Nicholl (1970) show the effect very clearly.

The effect is also to be observed at sunset with clear skies, and the characteristic reduction in surface wind is commonly confined to a layer whose thickness is of order 10 m. Above the collapsed layer, the wind is nearly unaffected as can sometimes be confirmed by observing the top branches of trees. After collapse, the flow is accelerated by the synoptic pressure gradient at a rate that is nearly fU (f is the Coriolis parameter) since the daytime flow is roughly in balance with the pressure gradient and the Coriolis force. In a time of order f^{-1} (about three hours), the ground flow may be moving rapidly enough to become turbulent once more, to renew the heat transfer, and, probably, to suffer a second collapse.

8.18 Convective turbulence

The turbulent flows described in previous chapters are essentially constant-density flows without appreciable curvature of the streamlines of the mean flow, and their properties were related to two ideal kinds of turbulence, free and wall turbulence. Both kinds receive energy from the mean flow through the medium of eddies of specialised structure, adapted to the boundary restriction in wall flows but not unrecognisably so. In a third kind of turbulence, convective turbulence, the source of energy is an actual or virtual store of potential energy and, in consequence, the mechanism of energy release is much simpler and can be carried through by eddies of simple structure. The two simplest examples of convective turbulence are

(1) heat convection between parallel, horizontal planes in a gravitational field, and

(2) Couette flow between concentric, rotating cylinders with circulation decreasing outwards.

The forms of the mean value profiles, of temperature and circulation respectively, reflect the efficiency of transfer. At very moderate values of the Reynolds number or its analogue, the mixing is so complete that the gradients are too small to measure over the greater part of the flows.

8.19 Heat convection between horizontal, parallel planes

The natural convection that develops between horizontal plane surfaces is a pure form of convective turbulence if the Rayleigh number of the flow is large. Consider the motion between surfaces at $z = 0$ and $z = 2D$, the lower one being at a constant temperature T_1 and the upper one at temperature T_2. The boundary conditions are that the velocity is zero on the bounding surfaces and that the temperature fluctuation is zero there. Stability analysis shows that the flow is unstable to small disturbances from the time-independent state,

$$\theta = u_i = 0, \qquad T = T_1 - (T_1 - T_2)z/(2D) \qquad (8.19.1)$$

if the Rayleigh number,

$$\lambda = \frac{g}{T_a} \frac{(T_1 - T_2)8D^3}{\nu\kappa},$$

exceeds 1708. The initial instability takes the form of regular, hexagonal convection cells, and the motion preserves a considerable degree of regularity to Rayleigh numbers of at least 10^6.

For horizontally homogeneous turbulent flow, the mean value equations are, (1) for temperature,

$$\frac{\partial \overline{\theta w}}{\partial z} = \kappa \frac{\partial^2 T}{\partial z^2} \qquad (8.19.2)$$

or, after integration,

$$Q_0 = \overline{\theta w} - \kappa \, \partial T / \partial z \qquad (8.19.3)$$

where Q_0 is the constant upwards flux of (thermometric) heat, (2) for turbulent kinetic energy,

$$\frac{\partial}{\partial z}(\overline{pw} + \tfrac{1}{2}\overline{q^2 w}) = \frac{g}{T_a}\overline{\theta w} - \varepsilon \qquad (8.19.4)$$

and (3) for the fluctuation entropy, $\tfrac{1}{2}\overline{\theta^2}$,

$$\overline{\theta w}\frac{\partial T}{\partial z} + \frac{\partial}{\partial z}(\tfrac{1}{2}\overline{\theta^2 w}) = -\varepsilon_\theta. \qquad (8.19.5)$$

The equations have points of similarity with the equations for pressure flow between parallel planes, and, by arguing that eddy transfer must be much greater than conduction in the central turbulent flow, it follows from (8.19.3) that the temperature gradient over most of the flow is small compared with its value near the boundaries, $-Q_0/\kappa$.

Differences from the channel flow appear when the equations for kinetic energy and fluctuation entropy are examined. Except within the thin layers at the walls where heat is transferred by molecular conduction, the buoyancy forces generate energy at a uniform local rate of gQ_0, and, unlike the channel flow, most of the energy transfer to the turbulent motion takes place in the central region and not in wall layers of relatively small thickness (fig. 8.11). On the other hand, equation (8.19.5) for the temperature fluctuation intensity shows that it is generated for the most part in the wall layers where the mean temperature gradients are large, and so, to provide the fluctuations necessary for the heat flux, in the central region, it must be transported there by diffusive movements.

Because of the necessity for the diffusion of temperature fluctuations and the uniformity of the rate of energy production, it is not possible to assert with confidence that an equilibrium layer exists near the wall with properties independent both of the central flow and of the conductive–viscous layers on the wall. Visualisation studies of the convection indicate that much of the heat transfer is carried by rising columns of hot fluid (and falling columns of cold fluid), extending from the edge of the conduction layer nearly to the upper surface. If they retain a memory of their thermal past, it is likely that they retain also a memory of their dynamical origin and that conditions in the conduction layer have some influence on the convection in the fully turbulent flow.

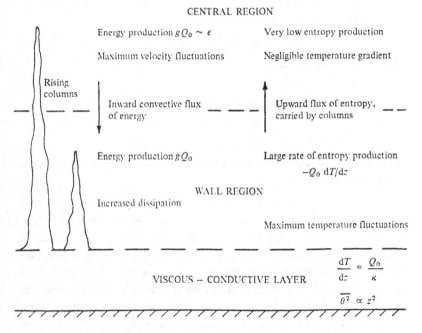

CENTRAL REGION

Energy production $gQ_0 \sim \epsilon$ · · · · · Very low entropy production

Maximum velocity fluctuations · · · · · Negligible temperature gradient

Rising columns

Inward convective flux of energy · · · · · Upward flux of entropy, carried by columns

Energy production gQ_0 · · · · · Large rate of entropy production $-Q_0\, \mathrm{d}T/\mathrm{d}z$

WALL REGION

Increased dissipation

Maximum temperature fluctuations

VISCOUS – CONDUCTIVE LAYER $\dfrac{\mathrm{d}T}{\mathrm{d}z} = \dfrac{Q_0}{\kappa}$

$\overline{\theta^2} \propto z^2$

Fig. 8.11. Production and transfer of kinetic energy and entropy of fluctuations in natural convection over a horizontal plane and a Prandtl number near one.

One reason for the predominance of long, rising columns is that vertical motions alone are effective in converting potential to kinetic energy and that the magnitude of the lateral flow is least in vertically elongated velocity patterns. For example, in uniform temperature gradient, a periodic disturbance of wave number, $\mathbf{k} = (l, m, n)$, grows at an exponential rate of $(-g/T_a\, \mathrm{d}T/\mathrm{d}z)^{1/2}(l^2 + m^2)/k^2$, which has its maximum value if $n = 0$ and the motion is entirely vertical.

8.20 Heat transfer in Bénard convection

For Bénard convection between parallel planes, the 'mean flow' problem is to find a second relation between heat flux and the distribution of mean temperature, but the distributions of turbulent energy and entropy production are so different that the notions of similarity and wall layers cannot be used with the comparative certainty of the application to channel flow. Perhaps the only firm prediction is the dimensional one that, if the conditions for validity of the Boussinesq approximation are met, the distribution of mean temperature has the form,

$$T - T_c = \theta_0 fn[z/z_0, D/z_0, \lambda, \nu/\kappa], \qquad (8.20.1)$$

where $\theta_0 = Q_0^{3/4}(\kappa g/T_a)^{-1/4}$

$z_0 = (\kappa^3 T_a/gQ_0)^{1/4}$

T_c is the temperature on the central plane, $z = D$,

$$\lambda = g/T_a(T_1 - T_2)8D^3/(\nu\kappa)$$

is the Rayleigh number of the convection.

The relation between heat flux and overall temperature difference is of the form,

$$T_1 - T_2 = \theta_0 fn[D/z_0, \lambda, \nu/\kappa]. \qquad (8.20.2)$$

For further progress, some physical assumptions must be made. If all the intense temperature gradients and fluctuations occur in thin layers near the surfaces, it appears plausible that the convection within each surface layer depends only on heat flux and fluid properties within the layer and not on the central convection which merely transfers heat with a negligible gradient of mean temperature. The assumption of surface similarity ignores the essential transport of temperature fluctuations to the central region, but its predictions receive some experimental support. If convection in the surface layer is independent of the width, D, the distributions of mean temperature and fluctuation intensity have the forms,

$$\left. \begin{array}{l} T_1 - T = \theta_0 F(z/z_0, \nu/\kappa) \\ \\ \overline{\theta^2} = \theta_0^2 G(z/z_0, \nu/\kappa) \end{array} \right\} \qquad (8.20.3)$$

and

and, since temperature variation in the central region is very small, the relation between heat flux and temperature difference is

$$T_1 - T_2 = 2C\theta_0 = 2CQ_0{}^{3/4}(\kappa g/T_a)^{-1/4}, \qquad (8.20.4)$$

where the factor C depends on Prandtl number.

After assuming the convection in the surface layer to be uninfluenced by the central flow and the distant boundary, it is tempting to suppose further that the convection in the fully turbulent region is both independent of the fluid viscosity and conductivity and of the nature and motion in the conductive-viscous layer at the surface. That the convection is independent of the fluid properties can hardly be doubted, but the penetrative nature of heat convection almost certainly makes the scale of the convection strongly dependent on conditions at the origin of the rising columns. The 'similarity theory' of convection does assume independence, and makes the predictions that

$$\left. \begin{aligned} \mathrm{d}T/\mathrm{d}z &= -\tfrac{1}{3}C'Q_0(gQ_0/T_a)^{-1/3}z^{-4/3}, \\ \overline{\theta^2} &= C''Q_0{}^{4/3}(g/T_a)^{-2/3}z^{-2/3}. \end{aligned} \right\} \qquad (8.20.5)$$

Similar results are obtained from the mixing-length theory, supposing mixing length to be proportional to distance from the surface.

Many measurements of mean temperatures and heat flux have been made in the range of Rayleigh numbers 10^6–10^8, and, while some of them support the assumption that the surface convection does not depend on separation, the predictions of the 'similarity theory' (8.20.5) are clearly wrong. Nearly all measurements show the temperature gradient to diminish with height at least as rapidly as z^{-2}. Evidence that the assumption of independence of separation is a good approximation is to be obtained from confirmation of the heat transfer relation (8.20.4). In a common non-dimensional form, it is

$$k_{\mathrm{eff}}/k = \frac{2Q_0 D}{\kappa(T_1 - T_2)} = (2C)^{-4/3}(\nu/\kappa)^{1/3}\lambda^{1/3} \qquad (8.20.6)$$

but some of the measurements indicate that the ratio k_{eff}/k varies as a power of λ smaller than $-\tfrac{1}{3}$. The measured values depend to a considerable extent on measurements at Rayleigh numbers less than 10^7, where the effects of transition to turbulence from periodic eddies are still apparent, and it is not certain that the results are representative of very large Rayleigh numbers.

In the central region, temperature gradients are very small indeed, and examples of slight reversal of gradient are not uncommon. Whether reversed gradients persist as the Rayleigh number is increased is not certain, but their probable cause is the retardation of rising (or falling) columns as they approach the opposite surface. If dissipation of the columns is more rapid at larger Rayleigh numbers, it is possible that the reversed gradients may not persist.

8.21 Similarity and structure of Bénard convection

The most detailed study of the convective motion and temperature fluctuations in Bénard convection is that by Deardorff & Willis (1967a) for Rayleigh numbers up to 10^7, a value not really large enough to extrapolate the results with confidence to very large Rayleigh numbers. Some information about the convection at Rayleigh numbers up to perhaps 10^9 is to be found in measurements

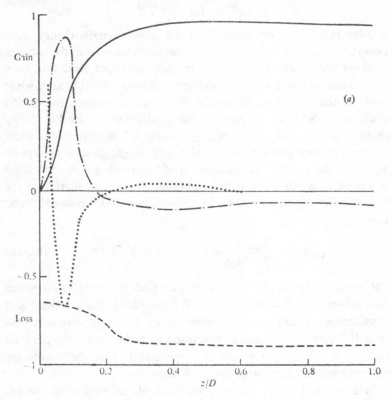

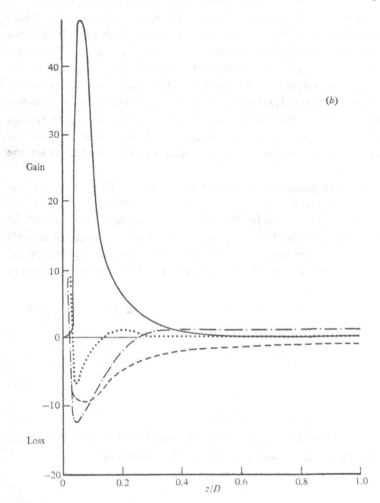

Fig. 8.12. Distributions of terms in the balance equations for kinetic energy and entropy of fluctuations in natural convection over a horizontal plane. (From Deardorff & Willis 1967a, but using the scales of (8.21.1).) (a) Energy balance, (b) entropy balance. —— generation; − − − destruction; − · − · turbulent diffusion; . . . . conductive diffusion.

over a single heated surface with inflow restricted by non-conducting vertical walls (Townsend 1959), but the upper boundary conditions are those at the base of a rising heat plume and the results can be applied to Bénard convection only with caution.

The energy and entropy balances in the flow can be appreciated from diagrams showing the terms in equations (8.19.4, 8.19.5) for the kinetic energy and the fluctuation entropy (fig. 8.12). Outside the wall layer, energy is generated at a uniform rate and it is dissipated everywhere at nearly the same rate, so that turbulent diffusion of energy is small. On the other hand, temperature gradients are appreciable only in the wall layers and, in the central region, there is no generation of fluctuation entropy and the entropy production by conduction is balanced by net diffusion from the wall layers.

From the balance diagrams, it appears possible that the central region of the flow may have nearly similar structures at all Rayleigh numbers large enough for the wall layer to form a small part of the flow volume. Since the convection is fully turbulent and not directly affected by molecular diffusivity, the controlling quantities are the surface separation and the buoyancy flux which determine the scales as

$$D, u_c = (gQ_0 D/T_a)^{1/3} \quad \text{and} \quad \theta_c = Q_0^{2/3}(gD/T_a)^{-1/3} \qquad (8.21.1)$$

and the various distributions are of the forms,

$$\left.\begin{aligned}
\overline{q^2} &= u_c^2 f(z/D), \\
\overline{\theta^2} &= \theta^2 f_\theta(z/D), \\
\varepsilon &= (u_c^3/D)h(z/D), \\
\varepsilon_\theta &= (u_c\theta_c^2/D)h_\theta(z/D).
\end{aligned}\right\} \qquad (8.21.2)$$

Notice that $u_c^3/D = gQ_0/T_a$ so that $h(z/D) = 1$ except in the wall layers. Also the entropy equation (8.19.5) becomes

$$\frac{\partial}{\partial z}(\tfrac{1}{2}\overline{\theta^2 w}) = -\varepsilon_\theta$$

and the entropy diffusion term is

$$\tfrac{1}{2}\overline{\theta^2 w} = u_c\theta_c^2 \int_{z/D}^{1} h_\theta(z/D)\,\mathrm{d}(z/D). \qquad (8.21.3)$$

Figs. 8.13, 8.14 show some of the measurements by Deardorff & Willis with appropriate scaling, and the central similarity expressed by the relations (8.21.2, 8.21.3) does seem to exist. Within the wall layers, considerable deviations are expected and do occur, even outside the regions of appreciable conduction of heat. The deviations are small for the velocity fluctuations, but they are large and vary systematically with Rayleigh number for the temperature fluctuations and for the mean temperature gradient.

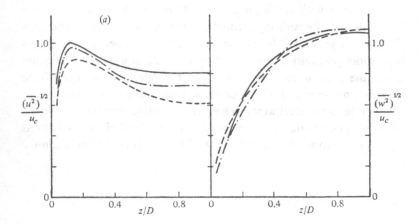

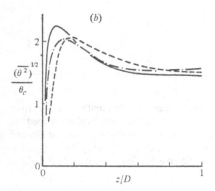

Fig. 8.13. Distributions of intensities of velocity fluctuations and temperature fluctuations above a heated horizontal plane (from Deardorff & Willis 1967a). —— $\lambda = 10 \times 10^6$; $-\cdot-\cdot$ $\lambda = 2.5 \times 10^6$; $---$ $\lambda = 0.63 \times 10^6$.

If the convection in the wall layer is, as may appear plausible, independent of the flow width, it should be in a state of similarity with scales of length, velocity and temperature, z_0, u_0 and θ_0, given in (8.20.1). A fundamental objection to wall similarity described by these scales is that the ratio of the velocity scale of the central flow to the wall scale u_0 is

$$u_c/u_0 = (gQ_0 D^4/k^3)^{1/12} \qquad (8.21.4)$$

and increases with Rayleigh number. It is not conceivable that the convection in the wall layer can be independent of the central flow if velocities generated by buoyancy in the wall layer are much less than those just outside generated by the central flow. On the other hand, the ratio u_c/u_0 varies nearly as the one-ninth power of the Rayleigh number and effects of its variation are small over ranges obtainable in the laboratory. Then, it is possible that measurements of the wall convection may be consistent with similarity distributions (as appears from the measurements of fig. 8.13) only over a limited range.

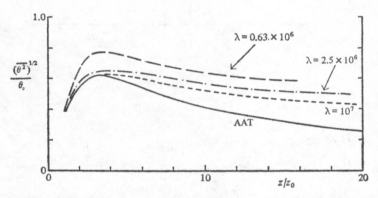

Fig. 8.14. Distributions of intensities of temperature fluctuations close to a heated horizontal plane, made non-dimensional with the scales defined in (8.20.1). (The upper curves are taken from Deardorff & Willis 1967a, for Benard convection between two planes, the lower from Townsend 1959, for convection with no upper surface and probably an effective Rayleigh number near 10^8.)

8.22 Natural convection in wall layers

The penetrative nature of convective motions has as a consequence that the whole region of variation of mean temperature is strongly influenced by the conductive-viscous layer, and the existence of wall similarity cannot be discussed profitably without knowledge of the convection. The fluctuations in the wall layer are remarkable in that the temperature fluctuations are strongly intermittent while velocity fluctuations are nearly continuous. The reason is simply that the principal velocity fluctuations are those of the central flow while the large temperature fluctuations are caused by plumes of hot fluid coming from the edge of the conduction layer.

By measuring the probability distribution functions for fluctuations of temperature, temperature gradient and time rate of temperature change, it can be shown that

(1) the rising columns of hot and thermally active fluid occupy a fraction of the flow that diminishes with height above the surface,

(2) their maximum temperatures also diminish with height,

(3) the remainder of the flow has considerably weaker fluctuations of temperature of constant intensity,

(4) the mean temperature of the inactive fluid is nearly independent of height, and

(5) the intensity of velocity fluctuations is roughly the same in the active and inactive regions.

It appears that the columns are emitted from the conductive layer into a strongly turbulent environment and so, instead of spreading as they rise, they are eroded by the surrounding turbulence and merge into it at a height of around fifteen layer thicknesses.

Evidently the effective thickness of the conductive layer is of first importance for convection in the wall layer. Once its thickness is set, the probable penetration of the columns is determined and also the initial temperature difference between the column and ambient inactive fluid. Particularly for the mean temperature distribution, the scale of temperature variation will be the product of the temperature gradient in the conduction layer and its thickness, say z_1, and the temperature distribution is expected to be nearly of the form,

$$T - T_c = \theta_1 F(z/z_1), \tag{8.22.1}$$

where $\theta_1 = Q_0 z_1/(\kappa T_a)$. The distribution of intensity of temperature fluctuations within the thermally active regions may be expected to have a similar form, but the total intensity is made up of a wall contribution of form,

$$\overline{\theta^2} = \theta_1{}^2 G(z/z_1) \tag{8.22.2}$$

added to the intensity in the ambient fluid which is proportional to the square of the temperature scale for the central region. It is likely that the considerable differences between the distributions for Bénard convection (fig. 8.14) and for convection over a horizontal plane arise from the difference in relative level of the two kinds of fluctuation.

On a smooth surface, the thickness of the conduction layer will be determined by two effects, (1) the degree of buoyant instability of the stratification in the layer, measured non-dimensionally by a Rayleigh number, and (2) the destabilising influence of velocity gradients induced by the turbulent motion of the central region. If only the first is active, the conditions for wall similarity are met and the thickness z_1 may be identified with the wall scale,

$$z_0 = (\kappa^3 T_a/g Q_0)^{1/4}.$$

If the central flow affects the thickness, the scales of length and temperature variation will depend to some extent on the flow width, but the influence of the central flow is measured by the changes in the ratio u_c/u_0 (see equation (8.21.4)) and is difficult to detect unless measurements can be made over a very large range of Rayleigh numbers.

Over a rough surface or a shear layer of forced convection, the initial scale of the columns is determined not by conditions in a viscous layer but by the scale of the motion either where the influence of surface irregularity ceases or where buoyancy forces begin to dominate the motion. In these cases, the scale z_1 should be identified either with a scale of the roughness or, if the shear layer is one of constant Reynolds stress and heat flux, with the Monin–Obukhov length.

CHAPTER 9

TURBULENT FLOW WITH CURVATURE OF THE MEAN VELOCITY STREAMLINES

9.1 Mean value equations for curved flow: the analogy between the effects of flow curvature and density stratification

If streamlines of the mean flow are appreciably curved, energy may be transferred between the mean flow and the turbulent motion in a way that resembles the transfer of energy by buoyancy forces in stratified flows. Consider a two-dimensional mean flow, homogeneous in the Oy direction and satisfying a boundary layer approximation that gradients of mean values along streamlines are small compared with gradients in the direction of mean velocity variation at right angles. An example is a boundary layer on a curved surface. Then cylindrical polar co-ordinates can be chosen so that the Oy axis is a representative axis of curvature for the mean flow, ϕ is the angle variable, and r is distance from the Oy axis (fig. 9.1). To the approximation, the equations of mean motion are

$$
\left.
\begin{aligned}
\left(\frac{U}{r}\frac{\partial}{\partial\phi}+W\frac{\partial}{\partial r}\right)U+\frac{1}{r}\frac{\partial(\overline{uw}r)}{\partial r} &= -\frac{1}{r}\frac{\partial P}{\partial\phi}+\frac{\nu}{r}\frac{\partial}{\partial r}\left(r^2\frac{\partial(U/r)}{\partial r}\right) \\
\text{and} \qquad\qquad\qquad & \\
\left(\frac{U}{r}\frac{\partial}{\partial\phi}+W\frac{\partial}{\partial r}\right)W-\frac{\overline{u^2}}{r} &= -\frac{\partial P}{\partial r}+\frac{U^2}{r}
\end{aligned}
\right\} \quad (9.1.1)
$$

and the equations for the kinetic energies of the three components of the velocity fluctuation are

$$
\left.
\begin{aligned}
\left(\frac{U}{r}\frac{\partial}{\partial\phi}+W\frac{\partial}{\partial r}\right)\tfrac12\overline{u^2}+\frac{\overline{uw}}{r}\frac{\partial(Ur)}{\partial r}+\frac{1}{r}\frac{\partial}{\partial r}(\tfrac12\overline{u^2w}r) &= -\frac{1}{r}\overline{u\frac{\partial p}{\partial\phi}}-\varepsilon_u, \\
\left(\frac{U}{r}\frac{\partial}{\partial\phi}+W\frac{\partial}{\partial r}\right)\tfrac12\overline{v^2}+\frac{1}{r}\frac{\partial}{\partial r}(\tfrac12\overline{v^2w}r) &= -\overline{v\frac{\partial p}{\partial y}}-\varepsilon_v, \\
\left(\frac{U}{r}\frac{\partial}{\partial\phi}+W\frac{\partial}{\partial r}\right)\tfrac12\overline{w^2}-\frac{2\overline{uw}}{r}U+\frac{1}{r}\frac{\partial}{\partial r}(\tfrac12\overline{w^3}r) &= -\overline{w\frac{\partial p}{\partial r}}-\varepsilon_w.
\end{aligned}
\right\} \quad (9.1.2)
$$

If the Reynolds number of the turbulent motion is large, the rates of viscous dissipation for the three components, ε_u, ε_v and ε_w, are equal as a consequence of the isotropy of the smallest eddies.

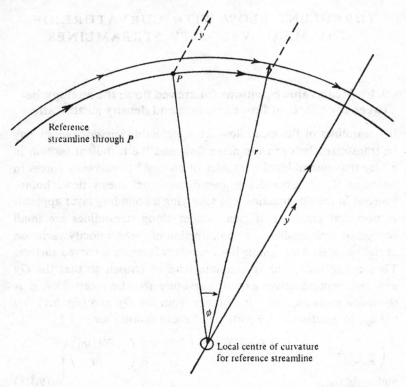

Fig. 9.1. Polar co-ordinates for nearly axisymmetric flow with curved streamlines.

The terms in the energy equations have the usual meanings but, unlike the equations for rectilinear flow, energy transfer terms appear both in the equation for the longitudinal component and in the equation for the transverse component. A measure of the effects of the flow curvature is provided by the ratio of the rate of energy transfer from the transverse fluctuations, $-2\overline{uw}U/r$, to the total rate of transfer from the mean flow, $-\overline{uw}r\partial(U/r)/\partial r$,

$$R_c = 2S/(1 - S), \tag{9.1.3}$$

where

$$S = (U/r)/(\partial U/\partial r) \tag{9.1.4}$$

is the curvature parameter of the mean flow. Bradshaw (1969) has suggested that the transfer from the transverse motion is analogous with the transfer of energy by buoyancy forces in stratified flow, and that the parameter R_c describes the modification of the flow by curvature in the same way as the flux Richardson number describes the effects of density stratification (§ 8.14).

The physical processes involved in the transfer of energy from the transverse motion may be described in terms of the angular momentum of the flow about the axis of flow curvature. Consider the equation for the component parallel to that axis,

$$K_2 + k_2 = [\mathbf{x} \times (\mathbf{U} + \mathbf{u})]_2,$$

$$\frac{D}{Dt}(K_2 + k_2) = \left[\mathbf{x} \times \frac{D}{Dt}(\mathbf{U} + \mathbf{u})\right]_2$$

$$= \frac{\partial}{\partial \phi}(P + p) + r\nu\nabla^2(U + u). \tag{9.1.5}$$

If pressure gradients in the direction of flow may be neglected, this component of the angular momentum is convected and diffused by the fluid in a similar way to temperature, and radial transfer of fluid with conservation of angular momentum may absorb or release energy of the mean flow. If two unit masses, originally moving with the mean velocities for their positions in the flow are interchanged between positions at radii r and $r + \delta r$ without changing their angular momenta, the energy change of the flow is

$$\delta E = \frac{1}{2}\left[\frac{(K_2 + \delta K_2)^2}{r^2} + \frac{K_2{}^2}{(r + \delta r)^2} - \frac{K_2{}^2}{r^2} - \frac{(K_2 + \delta K_2)^2}{(r + \delta r)^2}\right]$$

$$= \frac{1}{r^3}\frac{dK_2{}^2}{dr}(\delta r)^2 \tag{9.1.6}$$

if $\delta r/r$ is small. With the square of the angular momentum, $K_2{}^2 = U^2 r^2$, increasing outwards, it is necessary to supply energy for the interchange but, if it decreases, energy is released and may appear as energy of turbulent motion. The expression for energy release may be compared with that for similar interchange in stratified fluid,

$$\delta E = \frac{g}{T_a}\frac{dT}{dz}(\delta z)^2. \tag{9.1.7}$$

The interpretation of equation (9.1.6) is that flow curvature allows transfer of energy between the mean flow and the turbulence by simple mixing movements, similar to those that transfer energy from the potential form in stratified fluids, and the analogy implies that the ordinary Richardson number for stratified flow has as equivalent in curved flow the parameter,

$$R_{ic} = r^{-3} \frac{dK_2{}^2}{dr} \bigg/ \left(\frac{\partial U}{\partial r}\right)^2$$

$$= 2S(1+S). \tag{9.1.8}$$

In effect the number compares the limiting frequency of inertial waves in the local mean flow (or the divergence rate if the radial distribution of angular momentum is unstable) with the rate of shear. Unlike shear flows with density stratification, the two 'Richardson numbers' for curved flow are related explicitly (by $R_{ic} = R_c(1 + R_c)/(1 + \frac{1}{2}R_c)^2$) and a comparison of the properties of stratified and curved flows on the basis of Richardson number is unlikely to be valid unless the numbers are small in magnitude.

In spite of the similarities between the expressions for energy release, there are important differences between flows with density stratification and flows with curved streamlines. Consider the equations for the kinetic energies of the longitudinal motion, $\frac{1}{2}\overline{u^2}$, and of the transverse motion, $\frac{1}{2}(\overline{v^2} + \overline{w^2})$,

$$\left(\frac{U}{r}\frac{\partial}{\partial\phi} + W\frac{\partial}{\partial r}\right)\frac{1}{2}\overline{u^2} + \frac{\overline{uw}}{r}\frac{\partial(Ur)}{\partial r} + \frac{1}{r}\frac{\partial}{\partial r}(\frac{1}{2}\overline{u^2}wr) = -\frac{1}{r}\overline{u\frac{\partial p}{\partial\phi}} - \varepsilon_u,$$

$$\left.\begin{aligned}\left(\frac{U}{r}\frac{\partial}{\partial\phi} + W\frac{\partial}{\partial r}\right)\frac{1}{2}(\overline{v^2}+\overline{w^2}) - \frac{2\overline{uw}U}{r} + \\ + \frac{1}{r}\frac{\partial}{\partial r}[(\overline{pw}+\frac{1}{2}\overline{v^2w}+\frac{1}{2}\overline{w^3})r] = \frac{1}{r}\overline{u\frac{\partial p}{\partial\phi}} - \varepsilon_v - \varepsilon_w.\end{aligned}\right\} \tag{9.1.9}$$

If the longitudinal pressure gradient term, $r^{-1}\overline{u\,\partial p/\partial\phi}$, can be neglected, the two equations resemble closely those for fluctuation entropy and turbulent kinetic energy in stratified flow ((8.19.5) and (8.19.4)) with angular momentum the analogue of temperature and the quantity $2U/r^2$ the analogue of the buoyancy coefficient g/T_a.[†]

[†] It is supposed that the radius variation is comparatively small so that, if the equation for $\frac{1}{2}\overline{u^2}$ is multiplied by the approximately constant quantity r^2, it becomes an equation for the mean square fluctuation of angular momentum.

A difficulty is that, while the transverse motion may appear to be independent of the longitudinal motion, the partial dissipation rates are related by $\varepsilon_u = \varepsilon_v = \varepsilon_w = \frac{1}{3}\varepsilon$ and energy transfer through the pressure–velocity correlation term is always appreciable. The close resemblance between the two kinds of flow depends probably on the energy transfer between the longitudinal and transverse motions taking place at scales an order of magnitude less than the scales of the motions transferring angular momentum.

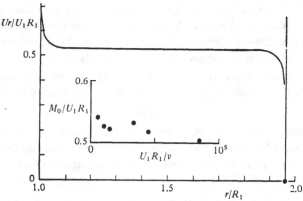

Fig. 9.2. Distribution of angular momentum between rotating cylinders: outer cylinder stationary (after Taylor 1935). Inset shows variation of central value with Reynolds number.

In the analogy, the two-dimensional turbulence described by the transverse velocity components is the counterpart of the three-dimensional motion in stratified flow, while the longitudinal velocity fluctuation, determining the instantaneous angular momentum, corresponds with the temperature fluctuation. Thus, radial transport of angular momentum, $\overline{uwr}$, releases energy to the transverse turbulence at a rate $(\overline{uwr})2U/r^2$ and generates fluctuations of angular momentum at a rate $-\overline{uwr}\ \partial(Ur)/\partial r$. Choosing the sense of the angle co-ordinate so that $-\overline{uw}$ is positive, i.e. $\partial(U/r)/\partial r$ is positive, the energy released to the transverse turbulence is positive if U is negative and conversely. If U is negative, elongated two-dimensional eddies with small longitudinal pressure gradients are efficient agents for releasing flow energy to the transverse motion and they can be expected to produce a nearly uniform distribution of angular momentum. If

U is positive, energy is removed from the transverse motion and the transfer of angular momentum is inhibited.

By assuming that the changes in the turbulent energy balance caused by flow curvature are the same as those caused by density stratification if the Richardson numbers are the same, Bradshaw (1969) has calculated the effects of flow curvature on flows, using measurements of atmospheric boundary layers. The comparative success of the procedure probably depends on the naturally elongated form of the eddies in rectilinear shear flow which allows the equations (9.1.9) to be a good guide to the changes in the effective rates of energy production. As an example, consider pressure flow between equidistant curved walls with separation $2D$. If the curvature is small, observations show that dU/dr is about $6\tau_0^{1/2}/D$ in the central region. Since $\tau_0^{1/2}/U$ is near 0.04 at ordinary Reynolds numbers, the curvature parameter for small curvatures is given by

$$S = 4D/r.$$

In stratified boundary layers, considerable changes of the velocity profiles are observed if the Richardson number exceeds 0.05, and effects of curvature may be expected if the radius of curvature is less than about $160D$.

9.2 Couette flow between rotating cylinders

The simplest kind of flow with curved streamlines is Couette flow between concentric, rotating cylinders, of radii R_1 and R_2 ($R_1 < R_2$) with peripheral velocities U_1 and U_2 respectively. If the cylinders are very long, the flow should be statistically homogeneous in the y and ϕ directions and all mean values depend only on the radial co-ordinate. In the absence of axial flow, the only non-zero component of mean velocity is in the ϕ direction, and the equations of mean motion reduce to two,

$$\left.\begin{aligned}
\frac{\partial}{\partial r}(\overline{uw}r^2) &= \nu \frac{\partial}{\partial r}\left(r^3 \frac{\partial}{\partial r}(U/r)\right), \\[2mm]
\text{and} \quad \frac{1}{r}\frac{\partial}{\partial r}(\overline{w^2}r) - \frac{\overline{u^2}}{r} &= -\frac{\partial P}{\partial r} + \frac{U^2}{r}.
\end{aligned}\right\} \tag{9.2.1}$$

The first equation may be integrated to give

$$-\overline{uw}r^2 + vr^3 \frac{\partial}{\partial r}(U/r) = G, \qquad (9.2.2)$$

where $2\pi G$ is the (kinematic) couple per unit length exerted on the inner cylinder by the fluid. Its meaning is that the sum of the couples transmitted by viscous and turbulent stresses across surfaces of constant radius is independent of the radius. The second equation describes the pressure distribution necessary to maintain the centri-petal acceleration of the mean flow, and the terms, $-\overline{u^2}/r$ and $(1/r)(\partial/\partial r)(\overline{w^2}r)$, are contributions from the normal stresses.

The boundary conditions are specified by the radii of the cylinders and by their peripheral velocities, and the non-dimensional parameters of the flow are a Reynolds number, the radius ratio and the velocity ratio. Then the transmitted torque is given by an expression of the form,

$$G = v^2 Fn(U_1 R_1/v, R_2/R_1, U_2/U_1) \qquad (9.2.3)$$

and it is possible and convenient to use a Reynolds number based on the torque,

$$R_g = G^{1/2}/v \qquad (9.2.4)$$

having defined the direction of the angle co-ordinate so that the torque is positive.

The laminar, time-independent solution of the equations of motion,

$$\frac{U}{r} - \frac{U_1}{R_1} = \left(\frac{U_2}{R_2} - \frac{U_1}{R_1}\right)\left(\frac{1}{R_1{}^2} - \frac{1}{r^2}\right)\bigg/\left(\frac{1}{R_1{}^2} - \frac{1}{R_2{}^2}\right) \qquad (9.2.5)$$

is stable to disturbance if the Reynolds number is not too large, but the flow becomes three-dimensional and eventually turbulent at large Reynolds numbers. In a turbulent flow, Reynolds stresses are large compared with mean viscous stresses over most of the flow volume but necessarily approach zero at a solid boundary. It follows from (9.2.2) that the gradients of angular velocity are much larger close to the cylinders than near the centre of the flow and that the curvature parameter becomes small in the neighbourhood of either cylinder. If the effects of rotation are small in part of the fully turbulent flow near a cylinder, the mean velocity distribution will have the logarithmic form for an equilibrium layer, i.e.

$$U = U_1 + \frac{G^{1/2}}{kR_1}\left[\log\left(\frac{G^{1/2}}{v}\frac{r-R_1}{R_1}\right)+A\right] \qquad (9.2.6)$$

near the inner cylinder, and

$$U = U_2 - \frac{G^{1/2}}{kR_2}\left[\log\left(\frac{G^{1/2}}{v}\frac{R_2-r}{R_2}\right)+A\right] \qquad (9.2.7)$$

near the outer cylinder. The values of the curvature parameter calculated from the distributions are

$$S = \frac{r-R_1}{R_1}\left[\frac{kU_1R_1}{G^{1/2}}+\log\left(\frac{G^{1/2}}{v}\frac{r-R_1}{R_1}\right)+A\right] \qquad (9.2.8)$$

and

$$S = \frac{R_2-r}{R_2}\left[\frac{kU_2R_2}{G^{1/2}}-\log\left(\frac{G^{1/2}}{v}\frac{R_2-r}{R_2}\right)-A\right] \qquad (9.2.9)$$

and they will be small in the turbulent part of the equilibrium layer, say where $(G^{1/2}/v)(r - R_1)/R_1 = 30$, for large Reynolds numbers.

If the positive direction of mean flow is chosen for positive torque, energy can be supplied to the flow only if the angular velocity of the outer cylinder exceeds that of the inner one. Three kinds of flow may be distinguished, depending on the sign of the energy release to the transverse motion or, equivalently, on the stability of the velocity distribution to convection with conservation of angular momentum:

(1) Mean velocity everywhere positive. Since

$$(U_2R_2)/(U_1R_1) \equiv (R_2/R_1)^2(U_2/R_2)/(U_1/R_1)$$

angular momentum increases outwards, energy is absorbed from the transverse motion, and the flow is convectively stable.

(2) Mean velocity everywhere negative. Angular momentum decreases outwards, energy is released to the transverse motion, and the flow is convectively unstable.

(3) Mean velocity changes sign, i.e. the directions of rotation are opposed. The inner part of the flow with negative peripheral velocity is unstable and the outer part stable.

9.3   Flow with the outer cylinder stationary

If only the inner cylinder rotates, the time-independent solution of the equations of motion,

$$U/r = U_1/R_1 - U_1/R_1\left(\frac{1}{R_1{}^2} - \frac{1}{r^2}\right) \Big/ \left(\frac{1}{R_1{}^2} - \frac{1}{R_2{}^2}\right) \qquad (9.3.1)$$

is stable and the torque is related to rotation speed by

$$G = -2\nu\frac{U_1}{R_1} \Big/ \left(\frac{1}{R_1{}^2} - \frac{1}{R_2{}^2}\right) \qquad (9.3.2)$$

unless the flow Reynolds number,

$$R_g = G^{1/2}/\nu = \left[-2\frac{U_1 R_1}{\nu} \Big/ (1 - R_1{}^2/R_2{}^2) \right]^{1/2}, \qquad (9.3.3)$$

exceeds a critical value, dependent on the radius ratio. Above that Reynolds number, the laminar flow is unstable to the development of regular toroidal vortices, uniformly spaced along the length of the cylinders (Taylor 1923). The flow remains coherently periodic to rotation speeds more than ten times the critical speed (Coles 1965), and organised, spatially periodic flow has been observed at much greater speeds (Pai 1943). The difficulty of attaining turbulent flow of the expected irregularity may be compared with the persistence of periodicity in Bénard convection.

Close to each cylinder, the effects of flow curvature are likely to be small and the velocity distributions should be nearly those for equilibrium layers with the appropriate surface stress, but the effects of curvature may become large near the centre of the flow. In these flows, the effect of curvature is comparable with the effect of an unstable density stratification on shear flow at a Richardson number of approximately $2S$, and, since the angular momentum is analogous with temperature, its distribution may be of the form,

$$\frac{\mathrm{d}(Ur)}{\mathrm{d}r} = \frac{G^{1/2}}{kz}F_c(-2S), \qquad (9.3.4)$$

where the function F_c is similar in behaviour to the function of equation (8.15.1),† and z is distance from the relevant cylinder. Observations in stratified boundary layers show that the function has become much less than one for Richardson numbers of order 0.1 and the temperature is nearly independent of height above the

† For small values of s, either of the curvature forms for Richardson number is approximated by $2S$, and $-z/L$ is nearly equal to the Richardson number in the stratified flow if it is small.

level where that value is attained. Approximately, then, the asymptotic value of angular momentum is given by the equilibrium distributions for values of z such $-2S = \alpha_0$, where α_0 is a number near 0.1. These values are, from equations (9.2.6–9), $z_c{}'$ and $z_c{}''$ defined by

$$\alpha_0 = -\frac{2z_c{}'}{R_1}\left[\frac{kU_1R_1}{G^{1/2}} + \log\left(\frac{G^{1/2}}{\nu}\frac{z_c{}'}{R_1}\right) + A\right]$$

$$= \frac{2z_c{}''}{R_2}\left[\log\left(\frac{G^{1/2}}{\nu}\frac{z_c{}''}{R_2}\right) + A\right]. \qquad (9.3.5)$$

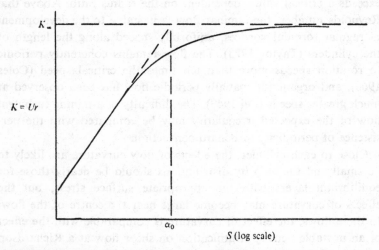

Fig. 9.3. Diagram of variation of angular momentum with distance from cylinder surface, showing inner equilibrium layer and transition layer to central layer of constant angular momentum.

If the gap between the cylinders is sufficient for the establishment of a substantial region of constant angular momentum, where $Ur = K_c$, the equations show that

$$\log\left(\frac{G^{1/2}}{\nu}\right) = \log\left(\frac{-kK_c}{G^{1/2}}\right) + \frac{k(K_c - U_1R_1)}{G^{1/2}} + \log\frac{2}{\alpha_0} - A$$

$$= \log\left(\frac{-kK_c}{G^{1/2}}\right) - \frac{kK_c}{G^{1/2}} + \log\frac{2}{\alpha_0} - A \qquad (9.3.6)$$

using the distributions at each surface. For consistency,

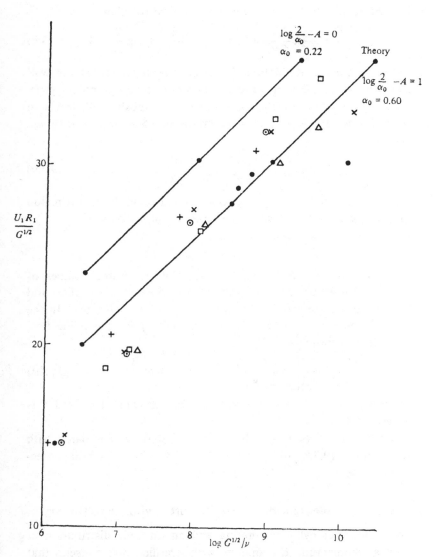

Fig. 9.4. Comparison of observed values of torque coefficient with Reynolds number dependence (9.3.10) (values from Taylor 1936a). R_2/R_1: ●, 1.042; +, 1.266; ⊙, 1.174; ×, 1.130; □, 1.102; △, 1.084.

14

$$K_c = \tfrac{1}{2} U_1 R_1 \qquad (9.3.7)$$

and the relation between torque and speed of rotation is

$$\log \frac{G^{1/2}}{v} = -\tfrac{1}{2} k \frac{U_1 R_1}{G^{1/2}} + \log\left(-\tfrac{1}{2} k \frac{U_1 R_1}{G^{1/2}}\right) + \log \frac{2}{\alpha_0} - A. \qquad (9.3.8)$$

The torque–speed relation (9.3.8) can be valid only if the wall layers in which Ur varies occupy part only of the annular space, that is, if the sum of the distances $z_c{}'$ and $z_c{}''$ is substantially less than the difference of the radii. From equations (9.3.5–8), it may be shown that

$$\frac{z_c{}' + z_c{}''}{R_2 - R_1} = \frac{\alpha_0}{2k} \frac{R_1 + R_2}{R_2 - R_1} \frac{G^{1/2}}{U_1 R_1}. \qquad (9.3.9)$$

Typical measured values of $U_1 R/G^{1/2}$ are around 30, and a region of nearly constant angular momentum should be found if

$$\frac{R_2 - R_1}{R_1 + R_2} > 0.04 \quad \text{or} \quad \frac{R_2}{R_1} > 1.08.$$

If the radius ratio is less than about 1.08, asymptotic values of angular momentum are not attained, and the velocity profiles and the surface stresses are nearly those for plane Couette flow. In the present context, the friction equation (5.6.16) for the plane flow appears as a torque–speed relation,

$$\frac{U_1 R_1}{G^{1/2}} = \frac{2}{k}\left[\log\left(\frac{G^{1/2}}{v} \frac{R_2 - R_1}{R}\right) + A - C_0\right], \qquad (9.3.10)$$

where C_0 (≈ 1.2) is the flow constant for the plane flow, and R is a mean radius.

Measurements of torque by Taylor (1936a) are compared with the relations (9.3.8) and (9.3.10) in fig. 9.4, with reasonable agreement at the higher speeds of rotation.

9.4 Turbulent motion with the outer cylinder stationary

With the inner cylinder rotating and an unstable distribution of angular momentum, the analogy with stratified flow suggests that the motion in a central region of constant angular momentum will resemble that in Bénard convection, with angular momentum fluctuations analogous to temperature fluctuations. Then the intensity of the longitudinal fluctuations should be smaller near the centre of the flow and larger near the walls, while the intensities of the transverse fluctuations, $\overline{v^2}$ and $\overline{w^2}$, should be respectively almost uniformly distributed and larger near the centre than near the walls. Remarkably little information about the intensity distributions has been published, but the measurements of MacPhail (1941) do offer some support (fig. 9.5).

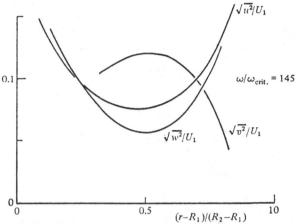

Fig. 9.5. Distributions of turbulent intensities with outer cylinder stationary (after MacPhail 1941).

The mean flow environment of turbulence near the centre of the annulus is not the usual one for shear flow. Since the mean angular momentum is nearly constant, the mean flow is nearly irrotational, and a parcel of turbulent fluid undergoes irrotational distortion with principal axes of strain rotating with angular velocity U/r. Relative to stationary rectangular axes, instantaneously coincident with the directions of the polar co-ordinates, the relevant components of the rate-of-strain tensor are

$$\frac{\partial U_i}{\partial x_j} = \frac{U}{r}\begin{pmatrix} 0 & -1 \\ -1 & 0 \end{pmatrix} \quad (i, j \text{ have the values 1, 3 only).} \quad (9.4.1)$$

The tensor for a particular parcel is

$$-U/r\begin{pmatrix} \sin 2\omega t & \cos 2\omega t \\ \cos 2\omega t & -\sin 2\omega t \end{pmatrix}, \quad (9.4.2)$$

where $\omega = U/r$ and t is elapsed time. Although considerable interaction between all parts of the flow is likely in conditions of convective instability, it is of interest to calculate the effect of the

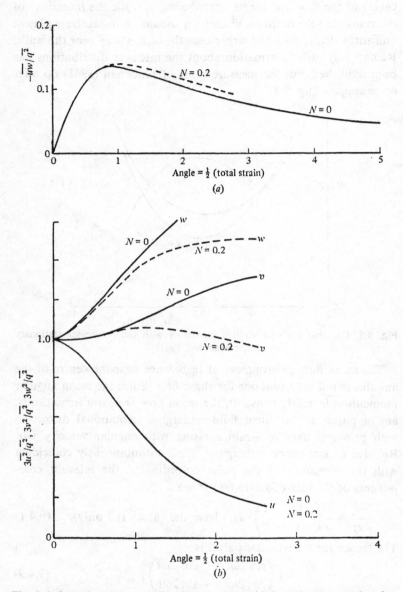

Fig. 9.6. Intensity ratios calculated from the rapid-distortion assumption for constant angular momentum of the mean flow. (a) Stress-intensity ratio. (b) Relative intensities of the velocity components.

rotating strain pattern on initially isotropic turbulence, neglecting the non-linear terms in the equations of motion.

With irrotational mean flow, the vorticity at a particular fluid particle remains the same multiple of the vector length of a material line element in the same direction, and, to the rapid-distortion approximation, the change of vorticity at a point moving with the mean flow is given by

$$\left.\begin{array}{l} \xi = \xi_0 - 2\omega t \zeta_0, \\ \eta = \eta_0, \\ \zeta = \zeta_0, \end{array}\right\} \qquad (9.4.3)$$

where ξ, η, ζ are the components in axes rotating with the angular velocity of the mean flow, and ξ_0, η_0, ζ_0 are their values at zero time. The effect of continuing distortion is to align vorticity nearly in the local direction of mean flow, leading to weak longitudinal fluctuations of velocity and strong transverse ones. Fig. 9.6 displays values of intensities and stresses calculated, using the rapid-distortion equations of § 3.10 and supposing the initial turbulent fluctuations to be isotropic. In contrast to the results for simple shearing, the energy extracted from the mean flow appears as energy of the transverse motion, not the longitudinal motion which loses energy. The variation of the stress–intensity ratio, $|-\overline{uw}|/\overline{q^2}$, with total strain is similar in form but the maximum value of about 0.11 is considerably less.

Although penetrative movements found in convective flows make questionable the basic assumption of homogeneity that is the basis of the rapid-distortion approximation, the predictions are not in disagreement with the few observations, and they supplement the inferences from the energy equations concerning the differences between flows with simple shearing and flows with irrotational shearing.

9.5   Flow with the outer cylinder rotating

If only the outer cylinder rotates, the laminar flow remains stable to small disturbances, and the observed critical speed for the onset of unsteady flow is much larger than if only the inner cylinder moves. The unsteady flow probably arises from the growth of finite amplitude disturbances and it has the usual characteristics of turbulent motion, even for rotation speeds not very large compared with the critical speed. Near the cylinders, the effects of flow curvature are small and the distributions of mean velocity and curvature parameter are given by equations (9.2.6–9). If the curvature parameter becomes significant, the effect is now to remove energy from the transverse motion and the flow is convectively stable. In the analogous case of a stably stratified boundary layer, nearly all the energy obtained by working of the mean flow against Reynolds stresses is absorbed by working against buoyancy forces if the local Richardson number is large but, in curved flow, the analogues of temperature and velocity are not distinct and turbulent dissipation of energy can never become negligible. However, it is conceivable and even likely that an asymptotic state can be reached with a constant ratio of transfer from the transverse motion to total energy transfer from the mean flow. That is, the ratio

$$\frac{-2\overline{uw}U/r}{-\overline{uw}r\,\partial(U/r)/\partial r} \equiv \frac{2S}{1-S} \qquad (9.5.1)$$

approaches a constant value β_0 at distances from the cylinder such that the values of S calculated from equations (9.2.8, 9.2.9) exceed $\beta_0/(2 + \beta_0)$. Then the distribution of mean velocity is

$$U = Dr^{1+2/\beta_0}, \qquad (9.5.2)$$

where D is a constant.

Experimental evidence for the validity of the assumption of constant curvature parameter in the central flow is hardly plentiful. Velocity measurements by Taylor (1936b) can be described by the power law relation (9.5.2) with $\beta_0 = 0.33$ (fig. 9.7), but the range of rotation speeds is small and close to the critical speed. Measurements by Wang & Gelhar (1970) with the inner cylinder rotating in the opposite direction to the outer one could be described using the value $\beta_0 = 0.22$ for $U_2R_1/(U_1R_2) = -2.0$, but the curvature parameter is small over most of the central flow and the region of apparent constancy is rather small.

If the annular space is wide enough to allow a considerable depth of strongly 'stable' flow with the power law velocity distribution, a relation between torque and rotation speed can be derived. To a fair approximation, the constant value of Ur^{-1-2/β_0}, say J_∞, is that

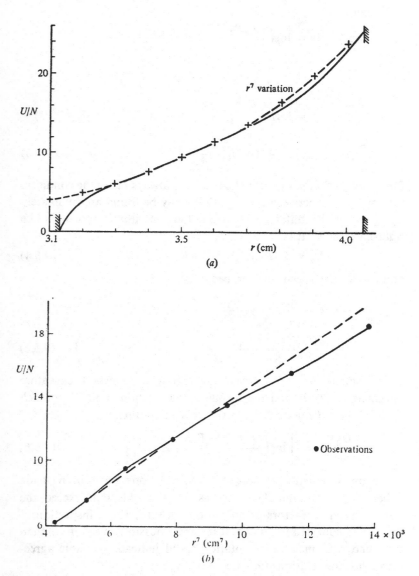

Fig. 9.7. Comparison between measured mean velocity distribution with inner cylinder stationary, outer cylinder rotating, and power-law variation (measurements from Taylor 1936b). R_2, 4.05 cm; R_1, 3.13 cm; N, 33.0 r.p.s.

given by the logarithmic profiles for values of r such that the calculated values of S equal $\beta_0/(2 + \beta_0)$, a condition that leads to the relations,

$$\log G^{1/2}/\nu = \log\left(\frac{kJ_\infty R_1^{2+2/\beta_0}}{G^{1/2}}\right) + \frac{kJ_\infty R_1^{2+2/\beta_0}}{G^{1/2}} -$$
$$- A + \log\left(\frac{2+\beta_0}{\beta_0}\right)$$
$$= \log\left(\frac{kJ_\infty R_2^{2+2/\beta_0}}{G^{1/2}}\right) + \frac{k}{G^{1/2}}(U_2 R_2 - J_\infty R_2^{2+2/\beta_0}) -$$
$$- A + \log\left(\frac{2+\beta_0}{\beta_0}\right). \tag{9.5.3}$$

Since the friction parameter $U^2 R^2/G^{1/2}$ is always large, the condition for consistency between the equations may be found approximately by ignoring the difference between the logarithmic terms on the right-hand sides. It is

$$J_\infty = U_2 R_2/(R_1^{2+2/\beta_0} + R_2^{2+2/\beta_0}) \tag{9.5.4}$$

and the torque–speed relation becomes

$$\log \frac{G^{1/2}}{\nu} = \log\left[\frac{kU_2 R_2}{G^{1/2}} \frac{R_2^{2+2/\beta_0}}{R_1^{2+2/\beta_0} + R_2^{2+2/\beta_0}}\right] +$$
$$+ \frac{kU_2 R_2}{G^{1/2}} \frac{R_1^{2+2/\beta_0}}{R_1^{2+2/\beta_0} + R_2^{2+2/\beta_0}} - A + \log\left(\frac{2+\beta_0}{\beta_0}\right). \tag{9.5.5}$$

For small annular spacing, no region of constant curvature parameter is likely and the torque speed relation should approach that for plane Couette flow, in the present context,

$$\frac{U_2 R_2}{G^{1/2}} = \frac{2}{k}\left[\log\left(\frac{G^{1/2}}{\nu} \frac{R_2 - R_1}{R}\right) + A - C_0\right]. \tag{9.5.6}$$

Most measurements of torque–speed relations have been made either with small annular spacings or at speeds that exceed the critical speed by factors of no more than four. From inspection of the results collected in fig. 9.8, it is hardly possible to say more than the trend with increase of rotation speed indicates possible agreement with the relations (9.5.5, 9.5.6).

Wang & Gelhar (1970) have measured mean velocities and turbulent intensities between cylinders rotating in opposite directions.

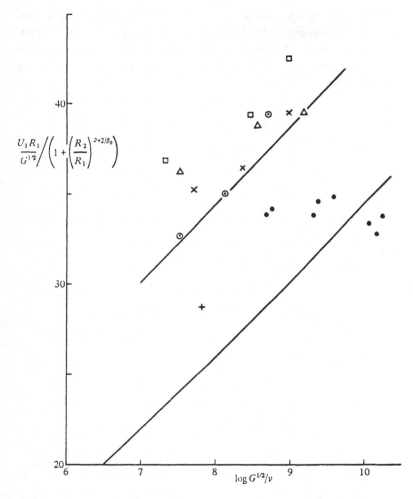

Fig. 9.8. Comparison of observed values of torque with Reynolds number dependence given by (9.5.5) with outer cylinder rotating (values from Taylor 1936a). R_2/R_1: ●, 1.042; +, 1.266; ⊙, 1.174; ×, 1.130; □, 1.102; △, 1.084.

For this configuration, the mean velocity over most of the annular space is positive, that is to say, in the direction of motion of the outer cylinder, and the curvature parameter is everywhere either small or positive. The measurements are so representative of convectively stable flow in which energy is withdrawn from the transverse motion. It is not surprising to find that the measured intensity of the radial component of fluctuation, $\overline{w^2}$, is approximately one-quarter the intensity of the longitudinal fluctuations, $\overline{u^2}$. In contrast, the measurements of MacPhail (1941) in convectively unstable flow show that $\overline{w^2}$ exceeds $\overline{u^2}$ by a factor of four near the centre of the annular space.

REFERENCES

A. Books

Batchelor, G. K. (1953). *The theory of homogeneous turbulence.* Cambridge University Press.

Batchelor, G. K. & Davies, R. M. (Eds.) (1956). *Surveys in Mechanics.* Cambridge University Press.

Batchelor, G. K. & Moffat, H. K. (Eds.) (1971). *Proceedings of the Boeing Symposium on Turbulence* (reprinted from *J. Fluid Mech.*, **41**, 1–480). Cambridge University Press.

Bradshaw, P. (1971). *An introduction to turbulence and its measurement.* Pergamon Press: London.

Favre, A. (Ed.) (1962). *Mécanique de la turbulence.* Éditions du centre national de la récherche scientifique: Paris.

Goldstein, S. (Ed.) (1938). *Modern development in fluid dynamics.* Oxford University Press.

Hinze, J. O. (1959). *Turbulence.* McGraw-Hill: New York.

Lin, C. C. (Ed.) (1959). *Turbulent flows and heat transfer.* Oxford University Press.

Lumley, J. L. & Panofsky, H. A. (1964). *The structure of atmospheric turbulence.* Interscience Publishers: New York.

Monin, A. S. & Yaglom, A. M. (1971). *Statistical fluid mechanics.* M.I.T. Press: Cambridge, Mass.

Plate, E. J. (1971). *Aerodynamic characteristics of atmospheric boundary layers.* U.S. Atomic Energy Commission: Oak Ridge, Tennessee.

Reynolds, A. J. (1974). *Turbulent flows in Engineering.* Interscience Publishers: New York.

Schlichting, H. (1955). *Boundary layer theory.* Pergamon Press: London.

Taylor, G. I. (1960). *The scientific papers of Sir Geoffrey Ingram Taylor* (Ed. G. K. Batchelor). Cambridge University Press.

Tatarski, V. I. (1961). *Wave propagation in a turbulent medium.* McGraw-Hill: New York.

Tennekes, H. & Lumley, J. L. (1972). *A first course in turbulence.* M.I.T. Press: Cambridge, Mass.

Townsend, A. A. (1956). *The structure of turbulent shear flow.* 1st edn. Cambridge University Press.

B. Articles

Antonia, R. A. & Luxton, R. E. (1971). The response of a turbulent boundary layer to a step change in surface roughness: Part I. Smooth to rough. *J. Fluid Mech.* **48**, 721–62.

Antonia, R. A. & Luxton, R. E. (1972). The response of a turbulent boundary layer to a step change in surface roughness: Part II. Rough to smooth. *J. Fluid Mech.* **53**, 737–58.

Arya, S. P. S. (1972). The critical condition for the maintenance of turbulence in stratified flow. *Quart. J. Roy. Met. Soc.* **98**, 264–73.

Van Atta, C. W. & Chen, W. Y. (1969). Measurements of spectral energy transfer in grid turbulence. *J. Fluid Mech.* **38**, 743–64.

van Atta, C. W. & Chen, W. Y. (1970). Structure functions of turbulence in the atmospheric boundary layer over the ocean. *J. Fluid Mech.* **44**, 145–60.

Bakewell, H. P. (1966). An experimental investigation of the viscous sub-layer in turbulent pipe flow. *Rep. Dept. Aerosp. Eng. Penn. State Univ.*

Bakewell, H. P. & Lumley, J. L. (1967). Viscous sublayer and adjacent region in turbulent pipe flow. *Phys. Fluids*, **10**, 1880–89.

Baldwin, L. V. & Sandborn, V. A. (1968). Intermittency of far wake turbulence. *A.I.A.A. J.* **6**, 1163–4.

Batchelor, G. K. (1952). The effect of homogeneous turbulence on material lines and surfaces. *Proc. Roy. Soc.* **A213**, 349–66.

Batchelor, G. K. (1959). Small-scale variation of convected quantities like temperature in turbulent fluid. Part 1. General discussion and the case of small conductivity. *J. Fluid Mech.* **5**, 113–33.

Batchelor, G. K., Binnie, A. M. & Phillips, O. M. (1955). The mean velocity of discrete particles in turbulent flow in a pipe. *Proc. Phys. Soc. Lond.* **68**, 1095–1104.

Batchelor, G. K., Howells, I. D. & Townsend, A. A. (1959). Small-scale variation of convected quantities like temperature in turbulent fluid. Part 2. The case of large conductivity. *J. Fluid Mech.* **5**, 134–9.

Batchelor, G. K. & Proudman, I. (1954). The effect of rapid distortion on a fluid in turbulent motion. *Quart. J. Mech. Appl. Math.* **7**, 83–103.

Batchelor, G. K. & Townsend, A. A. (1947). Decay of vorticity in isotropic turbulence. *Proc. Roy. Soc.* **A190**, 534–50.

Batchelor, G. K. & Townsend, A. A. (1949). The nature of turbulent motion at large wave-numbers. *Proc. Roy. Soc.* **A199**, 238–55.

Bearman, P. W. (1972). Some measurements of the distortion of turbulence approaching a two-dimensional bluff body. *J. Fluid Mech.* **53**, 461–68.

Bellhouse, B. J. & Schultz, D. L. (1968). The measurement of fluctuating skin friction with heated thin-film gauges. *J. Fluid Mech.* **32**, 675–80.

Black, T. J. & Sarnecki, A. J. (1958). The turbulent boundary layer with suction or injection. *Aero. Res. Counc., Lond., Rep.* no. 20,501.

Blackadar, A. K. (1962). The vertical distribution of wind and turbulent exchange in a neutral atmosphere. *J. Geophys. Res.* **67**, 3095–3102.

Blackadar, A. K., Panofsky, H. A., Glass, P. E. & Boogaard, J. F. (1967). Determination of the effect of roughness change on the wind profile. *Phys. Fluids*, **10**, S209–11.

Blake, W. K. (1970). Turbulent boundary-layer wall-pressure fluctuations on smooth and rough walls. *J. Fluid Mech.* **44**, 637–60.

Bowden, K. F. (1962). Measurements of turbulence near the sea bed in a tidal current. *J. Geophys. Res.* **67**, 3181–6.

Bowden, K. F. & Howe, M. R. (1963). Observations of turbulence in a tidal current. *J. Fluid Mech.* **17**, 271–84.

Bradbury, L. J. S. (1965). The structure of a self-preserving turbulent plane jet. *J. Fluid Mech.* **23**, 31–64.

Bradbury, L. J. S. & Riley, J. (1967). The spread of a turbulent jet issuing into a parallel moving air-stream. *J. Fluid Mech.* **27**, 381–94.

Bradley, E. F. (1968). A micrometeorological study of velocity profiles and surface drag in the region modified by a change in surface roughness. *Quart. J. Roy. Met. Soc.* **94**, 361–79.

Bradshaw, P. (1965). The effect of wind-tunnel screens on nominally two-dimensional boundary layers. *J. Fluid Mech.* **22**, 679–87.

Bradshaw, P. (1966). The effect of initial conditions on the development of a free shear layer. *J. Fluid Mech.* **26**, 225–36.

Bradshaw, P. (1967a). Irrotational fluctuations near a turbulent boundary layer. *J. Fluid Mech.* **27**, 209–30.

Bradshaw, P. (1967b). The turbulence structure of equilibrium boundary layers. *J. Fluid Mech.* **29**, 625–46.

Bradshaw, P. (1967c). 'Inactive' motion and pressure fluctuations in turbulent boundary layers. *J. Fluid Mech.* **30**, 241–58.

Bradshaw, P. (1967d). Conditions for the existence of an inertial subrange in turbulent flow. *N. P. L., Aero. Rep.* 1220.

Bradshaw, P. (1967e). Negative entrainment in turbulent shear flow. *N. P. L., Aero. Rep.* 1249.

Bradshaw, P. (1969). The analogy between streamline curvature and buoyancy in turbulent shear flow. *J. Fluid Mech.* **36**, 177–92.

Bradshaw, P. (1971). Calculation of three-dimensional turbulent boundary layers. *J. Fluid Mech.* **46**, 417–45.

Bradshaw, P. & Ferriss, D. H. (1966). Calculation of boundary layer development using the turbulent energy equation: II – Compressible flow on adiabatic walls. *N. P. L., Aero. Rep.* 1217.

Bradshaw, P., Ferriss, D. H. & Atwell, N. P. (1967). Calculation of boundary layer development using the turbulent energy equation. *J. Fluid Mech.* **28**, 593–616.

Bradshaw, P., Ferriss, D. H. & Johnson, R. F. (1964). Turbulence in the noise-producing region of a circular jet. *J. Fluid Mech.* **19**, 591–624.

Bradshaw, P. & Galea, P. V. (1967). Step-induced separation of a turbulent boundary layer in incompressible flow. *J. Fluid Mech.* **27**, 111–30.

Bradshaw, P. & Wong, F. Y. F. (1972). The reattachment and relaxation of a turbulent shear layer. *J. Fluid Mech.* **52**, 113–36.

Brundrett, E. & Baines, W. D. (1964). The production and diffusion of vorticity in duct flow. *J. Fluid Mech.* **19**, 375–94.

Bull, M. K. (1967). Wall-pressure fluctuations associated with subsonic turbulent boundary layer flow. *J. Fluid Mech.* **28**, 719–54.

Busch, N. E. & Panofsky, H. A. (1968). Recent spectra of atmospheric turbulence. *Quart. J. Roy. Met. Soc.* **94**, 132–48.

Champagne, F. H., Harris, V. G. & Corrsin, S. (1970). Experiments on nearly homogeneous shear flow. *J. Fluid Mech.* **41**, 81–140.

Clauser, F. H. (1954). Turbulent boundary layers in adverse pressure gradients. *J. Aero. Sci.* **21**, 91–108.

Clauser, F. H. (1956). The turbulent boundary layer. *Advances in Mechanics*, **4**, 1–51.

Coantic, M. (1965). Rémarques sur la structure de la turbulence à proximité d'une paroi. *C.R. Acad. Sci. Paris*, **260**, 2981–4.

Coles, D. (1955). The law of the wall in turbulent shear flow. In *50 Jahre Grenzschicht-forschung*, ed. H. Gortler & W. Tollmien, pp. 153–63. Vieweg: Braunschweig.

Coles, D. (1956). The law of the wake in the turbulent boundary layer. *J. Fluid Mech.* **1**, 191–226.

Coles, D. (1957). Remarks on the equilibrium turbulent boundary layer. *J. Aero. Sci.* **24**, 495–506.

Coles, D. (1965). Transition in circular Couette flow. *J. Fluid Mech.* **21**, 385–425.

Comte-Bellot, G. (1959). Entrainement turbulent de l'air dans un tunnel bidimensionalle à parois parallèles. *C.R. Acad. Sci. Paris*, **248**, 2710–12.

Comte-Bellot, G. (1961). Sur la representation adimensionelle des spectres des fluctuations de vitesse dans un tunnel bidimensionelle à parois parallèles. *C.R. Acad. Sci. Paris*, **253**, 2457–9, 2846–8.

Comte-Bellot, G. & Corrsin, S. (1966). The use of a contraction to improve the isotropy of grid turbulence. *J. Fluid Mech.* **25**, 657–82.

Corcos, G. M. (1964). The structure of the turbulent pressure field in boundary layer flows. *J. Fluid Mech.* **18**, 353–78.

Corrsin, S. (1943). Investigation of flow in an axially symmetrical heated jet of air. *Nat. Adv. Ctee Aero., Wash., Wartime Rep.* W-94.

Corrsin, S. & Kistler, A. L. (1955). Free-stream boundaries of turbulent flows. *Nat. Adv. Ctee Aero., Wash., Rep.* no. 1244.

Corrsin, S. & Uberoi, M. S. (1950). Further experiments on the flow and heat transfer in a heated turbulent air jet. *Nat. Adv. Ctee Aero., Wash., Rep.* no. 998.

Craya, A. & Milliat, J.-P. (1955). Etude de l'écoulement turbulent dans un divergent. *C.R. Acad. Sci., Paris*, **241**, 542–4 and 587–92.

Crow, S. (1968). Viscoelastic properties of fine-grained incompressible turbulence. *J. Fluid Mech.* **33**, 1–20.

Csanady, G. T. (1963). The energy balance in a mixing layer. *J. Fluid Mech.* **15**, 545–59.

Davies, E. B. & Young, A. D. (1964). Streamwise edge effects in the turbulent boundary layer on a flat plate of finite aspect ratio. *Aero. Res. Ctee, Lond. R. & M.* 3367.

Davies, P. O. A. L., Fisher, M. J. & Barrett, M. J. (1963). Turbulence in the mixing region of a round jet. *J. Fluid Mech.* **15**, 337–67.

Deardorff, J. W. (1970). A numerical study of three-dimensional channel flow at large Reynolds numbers. *J. Fluid Mech.* **41**, 453–80.

Deardorff, J. W. & Willis, G. E. (1967a). Investigation of turbulent thermal convection between horizontal plates. *J. Fluid Mech.* **28**, 675–704.

Deardorff, J. W. & Willis, G. E. (1967b). The free-convection temperature profile. *Quart. J. Roy. Met. Soc.* **93**, 166–175.

Deissler, R. G. (1958). On the decay of homogeneous turbulence before the final period. *Phys. Fluids*, **1**, 111–21.

Deissler, R. G. (1965). Problem of steady-state shear-flow turbulence. *Phys. Fluids*, **8**, 391–398.

Dorrance, W. H. & Dore, F. J. (1954). The effect of mass transfer on the compressible turbulent boundary-layer skin friction and heat transfer. *J. Aero. Sci.* **21**, 404–10.

Elder, J. W. (1960). The flow past a flat plate of finite width. *J. Fluid Mech.* **9**, 133–53.

Elliott, W. P. (1958). The growth of the atmospheric internal boundary layer. *Trans. Amer. Geophys. Union*, **39**, 1048–54.

Ellison, T. H. (1957). Turbulent transport of heat and momentum from an infinite rough plane. *J. Fluid Mech.* **2**, 456–66.

Ellison, T. H. (1960). A note on the velocity profile and longitudinal mixing in a broad open channel. *J. Fluid Mech.* **8**, 33–40.

Elswick, R. C. (1967). Investigation of a theory for the structure of the viscous layer in wall turbulence. *Rep. Dept. Aerosp. Eng., Penn. State Univ.*

Eskinazi, S. & Yeh, H. (1956). An investigation on fully developed turbulent flows in a curved channel. *J. Aero. Sci.* **23**, 23–34.

Favre, A. J., Gaviglio, J.-J. & Dumas, R. J. (1957). Space–time double correlations and spectra in a turbulent boundary layer. *J. Fluid Mech.* **2**, 313–42.

Favre, A. J., Gaviglio, J.-J. & Dumas, R. J. (1958). Further space–time correlations of velocity in a turbulent boundary layer. *J. Fluid Mech.* **3**, 344–56.

Favre, A. J., Gaviglio, J.-J. & Dumas, R. J. (1962). Corrélations spatiotemporelles en écoulements turbulents. In *Mecanique de la turbulence*, ed. A. J. Favre, pp. 419–45.

Fekete, G. I. (1970). Two-dimensional, self-preserving jets in streaming flow. *Rep. Dept. Mech. Eng., McGill Univ., Montreal*, no. 70-11.

Fernholz, H. (1964). Three-dimensional disturbances in a two-dimensional incompressible turbulent boundary layer. *Aero. Res. Ctee, Lond. R. & M.* no. 3368.

Fiedler, H. & Head, M. R. (1966). Intermittency measurements in the turbulent boundary layer. *J. Fluid Mech.* **25**, 719–36.

Finley, P. J., Phoe, K. C. & Poh, C. J. (1966). Velocity measurements in a thin horizontal water layer. *Houille Blanche*, **21**, 713–21.

Fisher, M. J. & Davies, P. O. A. L. (1964). Correlation patterns in a non-frozen pattern of turbulence. *J. Fluid Mech.* **18**, 97–116.

Frenkiel, P. N. & Klebanoff, P. S. (1965). Two-dimensional probability distribution in a turbulent field. *Phys. Fluids*, **8**, 2291–3.

Frenkiel, P. N. & Klebanoff, P. S. (1967a). Higher order correlations in a turbulent field. *Phys. Fluids*, **10**, 507–20.

Frenkiel, P. N. & Klebanoff, P. S. (1967b). Correlation measurements in a turbulent flow using high-speed computing methods. *Phys. Fluids*, **10**, 1737–47.

Frenzen, P. (1965). Determination of turbulence dissipation by Eulerian variance analysis. *Quart. J. Roy. Met. Soc.* **91**, 28–34.

Gartshore, I. S. (1966). An experimental examination of the large-eddy equilibrium hypothesis. *J. Fluid Mech.* **24**, 89–98.

Gartshore, I. S. (1967). Two-dimensional turbulent wakes. *J. Fluid Mech.* **30**, 547–60.

Gibson, M. M. (1963). Spectra of turbulence in a round jet. *J. Fluid Mech.* **15**, 161–73.

Gibson, C. H. (1968). Fine structure of scalar fields mixed by turbulence: I. Zero-gradient points and minimal gradient surfaces. II. Spectral theory. *Phys. Fluids*, **11**, 2305–27.

Gibson, C. H. & Schwarz, W. H. (1963). The universal equilibrium spectra of turbulent velocity and scalar fields. *J. Fluid Mech.* **16**, 365–84.

Gibson, C. H., Stegen, G. R. & Williams, R. B. (1970). Statistics of the fine structure of turbulent velocity and temperature fields measured at high Reynolds number. *J. Fluid Mech.* **41**, 153–68.

Grant, H. L. (1958). The large eddies of turbulent motion. *J. Fluid Mech.* **4**, 149–90.

Grant, H. L., Hughes, B. A., Vogel, W. M. & Moilliet, A. (1968). The spectrum of temperature fluctuations in turbulent flow. *J. Fluid Mech.* **34**, 423–42.

Grant, H. L. & Nisbet, I. C. T. (1957). The inhomogeneity of grid turbulence. *J. Fluid Mech.* **2**, 263–72.

Grant, H. L., Stewart, R. W. & Moilliet, A. (1962a). Turbulence spectra from a tidal channel. *J. Fluid Mech.* **12**, 241–68.

Grant, H. L., Stewart, R. W. & Moilliet, A. (1962b). The spectrum of a cross-stream component of turbulence in a tidal stream. *J. Fluid Mech.* **13**, 237–40·

Grass, A. J. (1971). Structural features of turbulent flow over smooth and rough boundaries. *J. Fluid Mech.* **50**, 223–56.

Gupta, A. K., Laufer, J. & Kaplan, R. E. (1971). Spatial structure in the viscous sublayer. *J. Fluid Mech.* **50**, 493–512.

Hackett, J. E. & Cox, D. K. (1970). The three-dimensional mixing-layer between two grazing perpendicular streams. *J. Fluid Mech.* **43**, 77–96.

Halleen, R. M. (1964). A literature review on subsonic turbulent shear flow· *Dept. Mech. Eng., Stanford Univ. Rep.* MD-11.

Hama, F. R. (1954). Boundary layer characteristics for smooth and rough surfaces. *Soc. Naval Archit. Marine Eng.*, New York, Paper no. 6.

Hanjalic, K. & Launder, B. E. (1972). Fully developed asymmetric flow in a plane channel. *J. Fluid Mech.* **51**, 301–36.

Hanratty, T. J. (1967). Study of turbulence close to a solid wall. In *Boundary layers and turbulence: Phys. Fluids (supplement)*, **10**, S126–33.

Head, M. R. & Bradshaw, P. (1971). Zero and negative entrainment in turbulent shear flow. *J. Fluid Mech.* **46**, 385–94.

van der Hegge Zijnen, B. G. (1900). Measurements of the velocity distribution in a plane turbulent jet of air. *Appl. Sci. Res.* A7, 256–75.

Heisenberg, W. (1948). On the theory of statistical and isotropic turbulence. *Proc. Roy. Soc.* A195, 402–6.

Hornung, H. G. & Joubert, P. N. (1963). The mean velocity profile in three-dimensional turbulent boundary layers. *J. Fluid Mech.* **15**, 368–84.

Hwang, N. H. C. & Baldwin, L. V. (1966). The decay of turbulence in axi-symmetric wakes. *J. Basic Eng.* **88**.

Imaki, K. (1968). Structure of superlayer in the turbulent boundary layer. *Bull. Inst. Space Aero. Sci., Univ. Tokyo*, **4**, 348–67.

Johnson, D. S. (1959). Velocity and temperature fluctuation measurements in a turbulent boundary layer downstream of a stepwise discontinuity in wall temperature. *J. Appl. Mech.* **26**, 325–36.

Johnston, J. P. (1960). On the three-dimensional turbulent boundary layer generated by secondary flow. *Trans. A.S.M.E.* D82, 233–48.

Johnston, J. P. (1970). Measurements in a three-dimensional boundary layer induced by a swept, forward-facing step. *J. Fluid Mech.* **42**, 823–44.

Karabelas, A. J. & Hanratty, T. J. (1968). Determination of the direction of surface velocity gradients in three-dimensional boundary layers. *J. Fluid Mech.* **34**, 159–62.

Keffer, J. F. (1965). The uniform distortion of a turbulent wake. *J. Fluid Mech.* **22**, 135–60.

Keffer, J. F. (1967). A note on the expansion of turbulent wakes. *J. Fluid Mech.* **28**, 183–94.

Kim, H. T., Kline, S. J. & Reynolds, W. C. (1971). The production of turbulence near a smooth wall in a turbulent boundary layer. *J. Fluid Mech.* **50**, 133–60.

Kistler, A. L. & Vrebalovich, T. (1966). Grid turbulence at large Reynolds numbers. *J. Fluid Mech.* **26**, 37–48.

Klebanoff, P. S. (1955). Characteristics of turbulence in a boundary layer with zero pressure gradient. *Nat. Adv. Ctee Aero., Wash., Rep.* no. 1247.

Kline, S. J., Reynolds, W. C., Schraub, F. A. & Runstadler, P. W. (1967). The structure of turbulent boundary layers. *J. Fluid Mech.* **30**, 741–76.

Kolmogorov, A. N. (1962). A refinement of previous hypotheses concerning the local structure in a viscous incompressible fluid at high Reynolds numbers. *J. Fluid Mech.* **13**, 82–5.

Kovasznay, L. S. G. (1948). Spectrum of locally isotropic turbulence. *J. Aero. Sci.* **15**, 745–53.

Kovasznay, L. S. G. (1967). Structure of the turbulent boundary layer. *Phys. Fluids (supplement)*, **10**, S25–S30.

Kovasznay, L. S. G., Kibens, V. & Blackwelder, R. F. (1970). Large-scale motion in the intermittent region of a turbulent boundary layer. *J. Fluid Mech.* **41**, 283–326.

Kuo, A. Y.-S. & Corrsin, S. (1971). Experiments on internal intermittency and fine-structure distribution functions in fully turbulent fluid. *J. Fluid Mech.* **50**, 285–320.

Laufer, J. (1950). Some recent measurements of turbulent flow in a two-dimensional channel. *J. Aero. Sci.* **17**, 277–87.

Laufer, J. (1951). Investigation of turbulent flow in a two-dimensional channel. *Nat. Adv. Ctee Aero., Wash., Rep.* no. 1053.

Laufer, J. (1955). The structure of turbulence in fully developed pipe flow. *Nat. Adv. Ctee Aero., Wash., Rep.* no. 1174.

Launder, B. E. & Ying, W. M. (1972). Secondary flows in ducts of square cross-section. *J. Fluid Mech.* **54**, 289–96.

Laurence, J. C. (1956). Intensity, scale and spectra of turbulence in mixing region of free subsonic jet. *Nat. Adv. Ctee Aero., Wash., Rep.* no. 1292.

Liepmann, H. W. & Laufer, J. (1947). Investigation of free turbulent mixing. *Nat. Adv. Ctee Aero., Wash., Tech. Note* no. 1257.

Lighthill, M. J. (1952a). On sound generated aerodynamically: I. General theory *Proc. Roy. Soc.* **A211**, 564–87.

Lighthill, M. J. (1952b). On sound generated aerodynamically: II. Turbulence as a source of sound. *Proc. Roy. Soc.* **A222**, 1–32.

Lighthill, M. J. (1962). On sound generated aerodynamically. *Proc. Roy. Soc.* **A267**, 147–82.

Liu, C. K., Kline, S. J. & Johnston, J. P. (1966). An experimental study of turbulent boundary layer on rough walls. *Dept. Mech. Eng., Stanford Univ. Rep.* no. MD-15.

Lumley, J. L. (1965). The structure of inhomogeneous turbulent flows. In *Atmospheric Turbulence and Radio Wave Propagation: Proc. Intern. Colloq., Moscow,* June 15–22, pp. 166–76, Publishing House 'NAUK': Moscow, 1967.

MacPhail, D. C. (1941). Turbulence in a distorted passage and between rotating cylinders. Ph.D. dissertation, University of Cambridge.

Maczynski, J. F. J. (1962). A round jet in an ambient coaxial stream. *J. Fluid Mech.* **13**, 597–608.

Malkus, W. V. R. (1956). Outline of a theory of turbulent shear flow. *J. Fluid Mech.* **1**, 521–39.

Maréchal, J. (1967). Anisotropie d'une turbulence de grille déformée par un champ de vitesse moyenne homogène. *C.R. Acad. Sci. Paris*, **265A**, 478–81.

Margolis, D. P. & Lumley, J. L. (1965). Curved turbulent mixing layer. *Phys. Fluids*, **8**, 1775–84.

Mellor, G. L. & Gibson, D. M. (1966). Equilibrium turbulent boundary layers. *J. Fluid Mech.* **24**, 225–54.

Milliat, J. P. (1957). Study of the turbulent flow in a weakly divergent channel. *9me. Congrès Intern. Mécan, Appl.*, *Univ. Bruxelles*, **3**, 419–34.

Mills, R. R., Kistler, A. L., O'Brien, V. & Corrsin, S. (1958). Turbulence and temperature fluctuations behind a heated grid. *Nat. Adv. Ctee Aero.*, *Wash.*, *Tech. Note* no. 4288.

Mitchell, J. E. & Hanratty, T. J. (1966). A study of turbulence at a wall using an electrochemical wall shear-stress meter. *J. Fluid Mech.* **26**, 199–221.

Mobbs, F. R. (1968). Spreading and contraction at the boundaries of free turbulent flows. *J. Fluid Mech.* **33**, 227–40.

Morton, B. R., Taylor, G. I. & Turner, J. S. (1956). Turbulent gravitational convection from maintained and instantaneous sources. *Proc. Roy. Soc.* **A234**, 1–23.

Narahari Rao, K., Narasimha, R. & Badri, M. A. (1971). The 'bursting' phenomenon in a turbulent boundary layer. *J. Fluid Mech.* **48**, 339–52.

Nee, V. W. & Kovasznay, L. S. G. (1969). Simple phenomenological theory of turbulent shear flows. *Phys. Fluids*, **12**, 473–84.

Newman, B. G. (1967). Turbulent jets and wakes in a pressure gradient. In *Fluid dynamics of internal flow*, ed. G. Sovran, pp. 170–209. Elsevier: Amsterdam.

Nicholl, C. I. H. (1970). Some dynamical effects of heat on a turbulent boundary layer. *J. Fluid Mech.* **40**, 361–84.

Nye, J. O. & Brodkey, R. S. (1967). The scalar spectrum in the viscous convective range. *J. Fluid Mech.* **29**, 151–64.

Obukhov, A. M. (1941). On the distribution of energy in the spectrum of turbulent flow. *Doklady Akad. Nauk SSSR*, **32**, 19–21.

Obukhov, A. M. (1962). Some specific features of atmospheric turbulence. *J. Fluid Mech.* **13**, 77–81.

Owen, P. R. & Thomson, W. R. (1963). Heat transfer across rough surfaces. *J. Fluid Mech.* **15**, 321–34.

Pai, S. I. (1943). Turbulent flow between rotating cylinders. *Nat. Adv. Ctee Aero.*, *Wash.*, *Tech. Note* no. 892.

Panofsky, H. A. (1963). Determination of stress from wind and temperature measurements. *Quart. J. Roy. Met. Soc.* **89**, 85–94.

Panofsky, H. A. & Pasquill, F. (1963). The constant of the Kolmogorov law. *Quart. J. Roy. Met. Soc.* **89**, 550–1.

Panofsky, H. A. & Townsend, A. A. (1964). Change of terrain roughness and the wind profile. *Quart. J. Roy. Met. Soc.* **90**, 147–55.

Patel, R. P. (1964). The effects of wind tunnel screens and honeycombs on the spanwise variation of skin friction in 'two-dimensional' turbulent boundary layers. *Mech. Eng. Res. Lab.*, *McGill Univ.*, *Tech. Note* 64–7.

Patel, R. P. & Newman, B. G. (1961). Self-preserving two-dimensional jets and wall-jets in a moving stream. *Mech. Eng. Res. Lab.*, *McGill Univ.*, *Report* no. Ae 5.

Payne, F. R. (1966). Large eddy structure of a turbulent wake. *Rep. Dept. Aerosp. Eng. Penn. State Univ.*

Payne, F. R. & Lumley, J. L. (1967). Large eddy structure of the turbulent wake behind a circular cylinder. *Phys. Fluids (supplement)*, **10**, S194–S196.

Pearson, J. R. A. (1959). The effect of uniform distortion on weak homogeneous turbulence. *J. Fluid Mech.* **5**, 274–88.

Perry, A. E. (1966). Turbulent boundary layers in decreasing adverse pressure gradients. *J. Fluid Mech.* **26**, 481–506.

Perry, A. E., Bell, J. B. & Joubert, P. N. (1966). Velocity and temperature profiles in adverse pressure gradient turbulent boundary layers. *J. Fluid Mech.* **25**, 299–320.

Perry, A. E. & Joubert, P. N. (1965). A three-dimensional turbulent boundary layer. *J. Fluid Mech.* **22**, 285–304.

Perry, A. E., Schofield, W. H. & Joubert, P. N. (1969). Rough wall turbulent boundary layers. *J. Fluid Mech.* **37**, 383–413.

Phillips, O. M. (1955). The irrotational motion outside a free turbulent boundary. *Proc. Camb. Phil. Soc.* **51**, 220–9.

Phillips, O. M. (1967). The maintenance of Reynolds stress in turbulent shear flow. *J. Fluid Mech.* **27**, 131–44.

Phillips, O. M. (1972). The entrainment interface. *J. Fluid Mech.* **51**, 97–118.

Protheroe, W. M. (1964). The motion and structure of stellar shadow-band patterns. *Quart. J. Roy. Met. Soc.* **90**, 27–42.

Record, F. A. & Cramer, H. E. (1966). Turbulent energy dissipation rates and exchange processes above a non-homogeneous surface. *Quart. J. Roy. Met. Soc.* **92**, 519–32.

Reichardt, H. (1941). Gesetzmässigkeit der freien Turbulenz. *Z. angew. Math. Mech.* **21**, 257–64.

Reichardt, H. (1956). Geschwindigkeitverteilung in einer geradlinigen Couetteströmung. *Z. angew. Math. Mech.* **36**, S26–9.

Reynolds, A. J. (1963). Analysis of turbulent bearing films. *J. Mech. Eng. Sci.* **5**, 258–72.

Reynolds, A. J. (1968). The distortion of initially anisotropic turbulence. (Unpublished.)

Richards, J. M. (1961). Experiments on the penetration of an interface by buoyant thermals. *J. Fluid Mech.* **11**, 369–84.

Richardson, L. F. (1920). The supply of energy from and to atmospheric eddies. *Proc. Roy. Soc.* **A97**, 354–73.

Robertson, J. M. (1959). On turbulent plane-Couette flow. In *Proc. 6th ANN. Midwest Conf. Fluid Mech.*, *Univ. Texas*, pp. 169–82.

Rose, W. G. (1966). Results of an attempt to generate a homogeneous turbulent shear flow. *J. Fluid Mech.* **25**, 97–120.

Rose, W. G. (1970). Interaction of grid turbulence with a uniform mean shear. *J. Fluid Mech.* **44**, 767–80.

Ruetenik, J. R. (1954). Investigation of equilibrium flow in a slightly divergent channel. *Johns Hopkins Univ.*, *Dept. Mech. Eng. Rep.* I-19.

Ruetenik, J. R. & Corrsin, S. (1955). In *50 Jahre Grenzschichtforschung*, ed. H. Görtler & W. Tollmien, pp. 446ff. Vieweg, Braunschweig.

Rust, J. H. & Sesonske, A. (1966). Turbulent temperature fluctuations in mercury and ethylene glycol in pipe flow. *J. Heat Mass Transfer*, **9**, 215–27.

Sabin, C. M. (1963). An analytical and experimental study of the plane, incompressible, turbulent shear layer with arbitrary velocity ratio and pressure gradient. *Dept. Mech. Eng., Stanford Univ. Rep.* MD-9.

Sandborn, V. A. (1959). Measurements of intermittency of turbulent motion in a boundary layer. *J. Fluid Mech.* 6, 221–40.

Schoenherr, K. E. (1932). Resistance of flat surfaces moving through a fluid. *Trans. Soc. Nav. Archit., New York*, 40, 279–313.

Schubauer, G. B. (1954). Turbulent processes as observed in boundary layer and pipe. *J. Appl. Phys.* 25, 188–96.

Schubauer, G. B. & Klebanoff, P. S. (1951). Investigation of separation of the turbulent boundary layer. *Nat. Adv. Ctee Aero., Wash., Rep.* no. 1030.

Schubert, G. & Corcos, G. M. (1967). The dynamics of turbulence near a wall according to a linear model. *J. Fluid Mech.* 29, 113–36.

Scorer, R. S. (1957). Experiments on convection of isolated masses of buoyant fluid. *J. Fluid Mech.* 2, 583–94.

Sirkar, K. K. & Hanratty, T. J. (1970a). Relation of turbulent mass transfer to a wall at high Schmidt number to the velocity field. *J. Fluid Mech.* 44, 589–604.

Sirkar, K. K. & Hanratty, T. J. (1970b). The limiting behaviour of the turbulent transverse velocity component close to a wall. *J. Fluid Mech.* 44, 605–14.

Somerscales, E. F. C. & Dropkin, D. (1966). Experimental investigation of the temperature distribution in a horizontal layer of fluid heated from below. *Int. J. Heat Mass Transfer*, 9, 1189–204.

Somerscales, E. F. C. & Gazda, I. W. (1969). Thermal convection in high Prandtl number liquids at high Rayleigh numbers. *Int. J. Heat Mass Transfer*, 12, 1491–511.

Stevenson, M. (1958). Experiment on turbulent shear flows in smooth two-dimensional tunnels. *Univ. Maryland, Tech. Note* BN-147.

Stewart, R. W. (1951). Triple velocity correlations in isotropic turbulence. *Proc. Camb. Phil. Soc.* 47, 146–57.

Stewart, R. W. & Townsend, A. A. (1951). Similarity and self-preservation in isotropic turbulence. *Phil. Trans. Roy. Soc. Lond.* 243, 359–86.

Stewart, R. W., Wilson, J. R. & Burling, R. W. (1970). Some statistical properties of small-scale turbulence in an atmospheric boundary layer. *J. Fluid Mech.* 41, 141–52.

Stratford, B. S. (1959a). The prediction of separation of the turbulent boundary layer. *J. Fluid Mech.* 5, 1–16.

Stratford, B. S. (1959b). An experimental flow with zero skin friction throughout its region of pressure rise. *J. Fluid Mech.* 5, 17–35.

Szablewski, W. (1960). Analyse von Messungen turbulenter Grenzschichten mittels der Wandgesetze. *Ing.-Archiv*, 29, 291–300.

Taylor, G. I. (1921). Diffusion by continuous movements. *Proc. Lond. Math. Soc.* (2)20, 196–212.

Taylor, G. I. (1923). Stability of a viscous liquid contained between two rotating cylinders. *Phil. Trans. Roy. Soc. Lond.* A223, 289–343.

Taylor, G. I. (1935). Distribution of velocity and temperature between concentric rotating cylinders. *Proc. Roy. Soc. Lond.* A151, 494–512.

Taylor, G. I. (1936a). Fluid friction between rotating cylinders: Part I. Torque measurements. *Proc. Roy. Soc. Lond.* A157, 546–64.

Taylor, G. I. (1936b). Fluid friction between rotating cylinders: Part II. Distribution of velocity between concentric cylinders when outer one is rotating and inner one is at rest. *Proc. Roy. Soc. Lond.* A157, 565–78.

Taylor, G. I. (1938). The spectrum of turbulence. *Proc. Roy. Soc. Lond.* A164, 476–90.

Taylor, G. I. (1954). The dispersion of matter in turbulent flow through a pipe. *Proc. Roy. Soc. Lond.* A223, 446–68.

Taylor, R. J. (1961). A new approach to the measurement of turbulent fluxes in the lower atmosphere. *J. Fluid Mech.* 10, 449–58.

Thomas, R. M. (1973). Conditional sampling and other measurements in a plane turbulent wake. *J. Fluid Mech.* 57, 549–82.

Townsend, A. A. (1947). Measurements in the turbulent wake of a cylinder. *Proc. Roy. Soc. Lond.* A190, 551–61.

Townsend, A. A. (1949a). Momentum and energy diffusion in the turbulent wake of a cylinder. *Proc. Roy. Soc. Lond.* A190, 551–61.

Townsend, A. A. (1949b). The fully developed turbulent wake of a circular cylinder. *Aust. J. Sci. Res.* 2, 451–68.

Townsend, A. A. (1951a). The passage of turbulence through wire gauzes. *Quart. J. Mech. Appl. Math.* 4, 308–20.

Townsend, A. A. (1951b). On the fine-scale structure of turbulence. *Proc. Roy. Soc. Lond.* A208, 534–42.

Townsend, A. A. (1954). The uniform distortion of homogeneous turbulence. *Quart. J. Mech. Appl. Math.* 7, 704–27.

Townsend, A. A. (1956). The properties of equilibrium boundary layers. *J. Fluid Mech.* 1, 561–73.

Townsend, A. A. (1959). Temperature fluctuations over a heated horizontal surface. *J. Fluid Mech.* 5, 209–41.

Townsend, A. A. (1960). The development of boundary layers with negligible wall stress. *J. Fluid Mech.* 8, 143–55.

Townsend, A. A. (1961a). Equilibrium layers and wall turbulence. *J. Fluid Mech.* 11, 97–120.

Townsend, A. A. (1961b). The behaviour of a turbulent boundary layer near separation. *J. Fluid Mech.* 12, 536–54.

Townsend, A. A. (1965a). Self-preserving flow inside a turbulent boundary layer. *J. Fluid Mech.* 22, 773–98.

Townsend, A. A. (1965b). The response of a turbulent boundary layer to abrupt changes in surface conditions. *J. Fluid Mech.* 22, 799–822.

Townsend, A. A. (1965c). Self-preserving development within turbulent boundary layers in strong adverse pressure gradients. *J. Fluid Mech.* 23, 767–78.

Townsend, A. A. (1965d). The interpretation of stellar shadow bands as a consequence of turbulent mixing. *Quart. J. Roy. Met. Soc.* 91, 1–9.

Townsend, A. A. (1966a). The flow in a turbulent boundary layer after a change in surface roughness. *J. Fluid Mech.* 26, 255–66.

Townsend, A. A. (1966b). The mechanism of entrainment in free turbulent flows. *J. Fluid Mech.* 26, 689–715.

Townsend, A. A. (1968). Excitation of internal waves in a stably-stratified atmosphere with considerable wind-shear. *J. Fluid Mech.* 32, 145–72.

Townsend, A. A. (1970). Entrainment and the structure of turbulent flow. *J. Fluid Mech.* 41, 13–46.

Townsend, A. A. (1972). Mixed convection over a heated horizontal plane. *J. Fluid Mech.* 55, 209–27.

Tritton, D. J. (1967). Some new correlation measurements in a turbulent boundary layer. *J. Fluid Mech.* **28,** 439–62.

Tucker, H. J. & Reynolds, A. J. (1968). The distortion of turbulence by irrotational plane strain. *J. Fluid Mech.* **32,** 657–73.

Turner, J. S. (1957). Buoyant vortex rings. *Proc. Roy. Soc. Lond.* **A239,** 61–75.

Uberoi, M. S. (1956). Effect of wind-tunnel contraction on free-stream turbulence. *J. Aero. Sci.* **23,** 754–64.

Wang, A. K. M. & Gelhar, L. W. (1970). Turbulent flow between concentric rotating cylinders. *Dept. Civil Eng. M. I. T., Rep.* no. 132.

Watt, W. E. (1967). The velocity-temperature mixing layer. *Dept. Mech. Eng., Univ. Toronto, Tech. Paper* 6705.

Willis, G. E. & Deardorff, J. W. (1965). Measurements on the development of thermal turbulence in air between horizontal plates. *Phys. Fluids,* **8,** 2225–9.

Willmarth, W. W. & Roos, F. W. (1965). Resolution and structure of the wall pressure field beneath a turbulent boundary layer. *J. Fluid Mech.* **22,** 81–94.

Willmarth, W. W. & Woolridge, C. E. (1962). Measurements of the fluctuating pressures beneath a thick turbulent boundary layer. *J. Fluid Mech.* **14,** 187–210.

Wills, J. A. B. (1964). On convection velocities in turbulent shear flows. *J. Fluid Mech.* **20,** 417–32.

Wygnanski, I. & Fiedler, H. E. (1968). Jets and wakes in tailored pressure gradients. *Phys. Fluids,* **11,** 2513–23.

Wygnanski, I. & Fiedler, H. E. (1969). Some measurements in the self-preserving jet. *J. Fluid Mech.* **38,** 577–612.

Wygnanski, I. & Fiedler, H. E. (1970). The two-dimensional mixing region. *J. Fluid Mech.* **41,** 327–62.

INDEX

attached eddies, *see* equilibrium wall layers
auto-correlation coefficient, eddy amplitude, 26, 63
auto-correlation function, Lagrangian velocity, 337
averages
ensemble, 4, 5
conditional, 235
space or time, 5
zone, 213

boundary-layer approximation, 188–93
boundary layers
general development: Bradshaw, Ferriss & Atwell scheme, 267, 295; convection velocity for energy, 296, 300; Mellor & Gibson scheme, 266, 277 internal, 299: change of roughness, 308–13; separation condition, 303, 307; stress profile, 310; strong adverse pressure gradient, 301–8 self-preserving, 272–6: adjustment fetches, 284; almost self-preserving development, 283–7; Clauser relation, 266, 276; conditions for self-preserving development, 262, 280; converging layers, 280; energy balances, 293; equations for mean velocity, 272; flow constants, 274, 282, 287; friction coefficient, 287–8; intermittency, 294; mean velocity distributions, 272, 279; momentum condition, 273; rough surface flow, 286; turbulent motion, 289–95; zero-stress layer, 277, 279
three-dimensional mean flow, 313–16: Ekman layer in rotating system, 319–23; finite width effects, 323–8; instability of lateral distribution of wall stress, 331; rapid change of free stream velocity, 316; triangle result, 319; stress-velocity relation, 315
Boussinesq approximation
equations of motion, 33, 36, 334
mean value equations, 339
buoyancy forces
effect on free turbulence, 372–5
effect on constant-stress layer, 375–8
transient effects, 379–80

cascade energy transfer between eddies, 55, 89
channel flow
axisymmetric, *see* pipe flow
two-dimensional: central flow, 143–50; Couette flow, 148, 150, 170; defect law, 135; energy relations, 136; flow constants, 146, 150, 168–9; free surface flow, 148; friction equations, 146; friction velocity, 135; mean value equations, momentum, 131; pressure fluctuations at wall, 165–8; Reynolds number, 131, (similarity), 133–5; rough walls, 139–43; stress distributions, 131; turbulent energy, 136; turbulent flow, 144, 150–62; velocity distributions, 143, 147, 170; viscous layer, 133, 162–4; wall stress (fluctuations), 155, 165, (mean value), 131
variable section, 172–6: self-preserving development, 173, (Clauser relation), 176, (maximum angle of divergence), 176, (wedge flow), 175
Clauser relation, 176, 266, 276
constant-stress layer
adiabatic: attached eddies, 152–4; conditions for existence, 138–9; flow similarity, 137, 156; Kármán

constant-stress layer—*contd.*
constant, 138, 156; turbulent flow, 144, 150–6, 158; viscous layer, 133, 144, 162–4
vertical heat transfer, 375–8: Monin–Obukhov length, 375; stable stratification (mean velocity and temperature), 376, (turbulent motion), 378; sudden change of heat flux, 379–80; unstable stratification (mean velocity and temperature), 378–9, (turbulent flow), 386–92
continuous movements, *see* diffusion
convection
forced, 335: internal thermal boundary layer, 361; longitudinal diffusion, 364–6; self-preserving development, free turbulent flows, 350–1; ratio of transfer coefficients for heat and momentum, 356, 358, 363; Reynolds analogy, 356; wall similarity, 352–6, (conductive layer, effect of Prandtl number), 353, (logarithmic distribution of temperature), 352
mixed: buoyancy effects, 372, 375; constant-stress layers, 375; critical Richardson number, 341, 373; rapid-distortion model, 342, 374
natural, 335, 366; Bénard convection between parallel planes, 381–4; mean value equations (entropy fluctuation), 382, (temperature), 382; plumes, self-preserving development, 367; scales of length, velocity, temperature, 384, 388, 392; similarity of velocity and temperature fields, 384, 385, 388–92; thermals (self-preserving development), 369, (angles of spread), 370; velocity and temperature fluctuations, 386, 390
of turbulent energy: bulk convection velocity, 125; diffusion coefficient, 180; nature of diffusion process, 178
convective turbulence
Bénard convection, 384
curved streamline flow, 394, *see also* rotating cylinder flow
continuity equation, 33, 37
co-ordinate systems, 5–6, 28
correlation functions
condition of incompressibility, 7, 16

construction from an assembly of simple eddies, 7
defining scalars (isotropic turbulence), 53, 59
double-velocity, 6, 53, 57, 59
longitudinal form, 108
temperature, 340
space–time, 6–7, 23, 62
convection velocity, 23–4
transverse form, 108
triple, 53, 57
Couette flow
plane, 148, 150, 170
rotating cylinder, 398

diffusion
of a scalar (temperature or concentration): eddy diffusion coefficient, 337; Eulerian description, 338, 365; Lagrangian description (continuous movements), 336, 365; material surfaces, increase of area, 337; Reynolds analogy, 356; turbulent transport rate, 356
of turbulent energy, 125, 128, 178, 383
of vorticity into free stream, 232
dissipation
energy by viscous stresses, 41
entropy by conduction, 43
length scale, 51, 61
temperature fluctuations, 43, 339
distortion of turbulent fluid, *see* homogeneous turbulence rapid distortion theory
distribution of fluctuations
velocity, 126, 128
velocity gradient or vorticity, 126

eddies
attached, 110, 123, 177
double-cone, 156
double-roller, 119
entrainment, 123
persistence, 64
size distribution, 11, 17
velocity patterns, 118
eddy diffusion coefficients
effective strain, 214
heat, 337
turbulent energy, 180, 209
eddy viscosity, 126, 147, 202
effective strain, *see* rapid distortion model
energy-containing eddies, 54

energy equations
  mean flow kinetic energy, 295
  total kinetic energy, 40
  turbulent kinetic energy, 39, 46
entrainment
  cycle, 241
  eddies, 123, 241
  effect of mean flow gradients, 234
  mechanism, 230, 232, 241
  parameters, 205
  ratios, 231, 247
  velocity, 231
equations
  of fluid motion: Boussinesq approximation, 36, 335; continuity (condition of incompressible flow), 33, 37; internal energy, 35; kinematic forms, 36; momentum, 34, 37, 335; potential temperature, 37, 335; total heat (enthalpy), 35, 37; viscous term, 35
  for mean values: continuity, 38; energy dissipation by viscosity, 41, 42; fluctuation entropy, 41, 339; kinetic energy (fluctuations), 39, 46, (mean flow), 295, (total), 40; momentum (mean velocity), 39, 339; potential temperature (mean), 40, 339, (mean square fluctuation), 41, 339
equilibrium wall layers, 130, 135
  attached eddies, 110, 123, 177–80
  blown or sucked, 184
  constant-stress, 135, 139, see also constant-stress layer
  diffusion of energy, 178, 180
  linear stress, 177, 180
  residence time of eddies, 122
  rough wall, 139
  similarity of motion, 137
  skewed, 186
  stress distributions, 155, 176
  velocity distributions, 180, 181, 185
  viscous layer, 137, 162, 354

flatness factor, 128
fluctuations, physical significance, 5
free turbulence
  general properties, 188
  boundary-layer approximation, 188: ratio of scales for turbulence and mean velocity, 189, 193
  integral conditions, 193
  jet in moving stream of constant velocity, 255

mean value equations, 190–2
  irrotational flow, 251
free turbulent shear flows, self-preserving
  conditions for development, 195
  effective strain, 214
  energy balances, 205
  flow constants, 205, 215, 220; Gartshore correlation, 213
  jets: axisymmetric, 200, 225; two-dimensional, 198, 220
  mean value equations, 197
  mean velocity distributions, 201–3: constant eddy viscosity forms, 202, 204
  mixing layers: skewed, 227; two-dimensional, 198, 227
  rates of approach to self-preserving development, 252
  small-defect family of wake-like flows, 199, 200, 224: limits, 225, 226
  structural similarity, 208, 220
  turbulent motion, 214–19
  Wakes: axisymmetric, 200, 226; two-dimensional, 199, 202, 206, 217, 224

grid turbulence
  anisotropy, 52, 66
  decay time, 50
  inhomogeneity, 50
  integral scale, 50, 59

Heisenberg term for eddy energy transfer, 101
homogeneous turbulence
  approximation by grid turbulence, 49, 51
  equations for turbulent kinetic energy, 46: pressure, 47; velocity fluctuation, 47
  Fourier representation, 46: equation for components amplitudes, 47; spectrum equation, 48, 61
  grid turbulence: approach to isotropy, 66; irrotational distortion, 68; uniform shear, 80
  persistence of eddies, 64

inactive motion, 123, 154, 161
integral length scale, 50, 59
intermittency
  conditional averages, 235
  convection velocities, 237, 239

intermittency—*contd.*
 dissipating eddies, 128, 233
 entrainment through surface, 232
 factor, 212
 folding of surface, 209: control
  mechanism, 243, 247; statistics of
  folding, 211, 235
 irrotational flow, 251
 surface, 188, 209
 velocity distribution near surface,
  235
 zone averages, 213
isotropic turbulence
 correlation function, 53, 57
 defining scalars: correlation, 53, 59;
  spectrum, 53
 dissipating eddies, 54
 energy decay, 56, 61
 Reynolds number similarity, 55
 self-preserving development, 59, 61
 space–time correlation, 62
 spectrum function, 54, 58
 Taylor length-scale, 85
 triple velocity correlation, 53, 57
 vorticity equation, 126

jets, *see* free turbulent shear flows

Kármán constant
 effect of inactive motion on, 156
 magnitude, 168
kinematic notation
 pressure, 38
 viscosity, 37
Kolmogorov theory of local similarity,
 see local similarity
Kovasznay term for eddy energy
 transfer, 101

Lagrangian
 auto-correlation, 337
 description of flow, 4
 description of diffusion, 336
large eddies, 105
 entrainment eddies, 123
 main motion, 106
local similarity
 velocity field: conditions for universal spectrum, 89; inertial range
  of eddies (spectrum function), 92,
  93, 98, (structure function), 93,
  96; mechanism of transfer of
  energy between eddies, 99; non-
  Newtonian fluids, 103; skewness
  factor for velocity differences, 98;

 viscous range of eddies, (scales of
  length, velocity), 92, (spectrum
  function), 100–3
 temperature field: convective–
  viscous range, 346; effect of
  Prandtl number, 344–7; inertial–
  convective range, 344–6; inertial–
  conductive range, 347

Mach number, 38
main turbulent motion
 correlation functions, 108
 energy-containing eddies, 54
 similarity in different flows, 106,
  110
 stress–intensity ratios, 107
mixing layers, *see* free turbulent shear
 flows

non-dimensional flow parameters
 Mach, 38
 Péclét, 336
 Prandtl, 335
 Rayleigh, 336
 Reynolds, 90, 133, 306, 335
 Richardson number, 335, 341
 Schmidt, 335
notation
 for shear flows: distinction between
  components, 6, 30; use of tensor
  notation, 5
 kinematic quantities, 38

Obukhov term for eddy transfer, 101
open channel flow, *see* channel flow

Péclét number, 336
pipe flow
 axisymmetric, 132
 central flow, 143
 energy balance, 150
 flow constants, 149
 friction equation, 149
 mean value equations, momentum,
  132
 Reynolds number, 133; similarity,
  133
 stress distributions, 132
 velocity distributions, 149
Prandtl number, 335
pressure fluctuations
 relation to velocity field, 43
 wall fluctuations, 165: convection
  velocity, 166; correlation function,
  166; correlation function, 166

rapid distortion model, 106
 effective strain: finite total strain, 120
  residence time of wall eddies, 122;
  elastic response to initial strain, 74
  irrotational distortion, 74: correlations, 79; intensity ratios, 76;
  scales, 75; spectra, 75
 simple shearing, 80: correlations,
  86, 88; eddy transfer coefficient,
  85, 121; intensity ratios, 84, 122;
  spectra, 83; stress–intensity ratio,
  84
 temperature field, 357
Rayleigh number, 336, 381
Reynolds number similarity, main
 motion, 53
Reynolds stress
 ratio to total intensity, 87
 relation to mean flow, 124
Richardson number, 335, 341
 analogue for flow with curved
  streamlines, 395
 flux form, 341, 395
rotating cylinder flow
 analogy with density stratification,
  396
 angular momentum, 395
 convective instability, 395
 mean value equations: angular
  momentum, 393; turbulent energy,
  393
 stable flow (inner cylinder at rest):
  distribution of angular momentum, 408; friction relation, 410
 turbulent flow, 404, 412
 unstable flow (outer cylinder at rest):
  distribution of angular momentum, 397; friction relation, 404;
  rapid distortion results, 405
 wall similarity, analogy with constant-stress layer, 401, 408
rough wall flow
 roughness length, 140
 roughness types, 140

scale height, 38
scattering of light, 348
secondary flow
 edge flow, 323
 lateral instability of wall flow, 328
 transient instability in viscous layer,
  331
skewness factor, 126
space–time correlation function, 23,
 62

spectrum function
 condition of incompressible flow, 12
 integrated, 13, 21, 93
 local significance, 12
 one-dimensional, 16, 20, 65, 93
 temperature, 342
 wave-number and frequency, 19, 27
structural similarity, 87, 106
structure function, 11, 93, 96

Taylor approximation of frozen flow,
 64
Taylor dissipation length, 85
temperature fluctuations
 correlation functions, 340
 dissipation by conduction, 43
 equation for intensity, 41, 339
 local similarity, see local similarity,
  temperature field
 thermometric conductivity, 37

viscosity
 effect on main motion, 54
viscous dissipation, 41
 length scale, 92
 spectrum of eddies, 101
vorticity
 diffusion across intermittency surface, 232
 equation for mean square vorticity,
  126
 rapid distortion changes, 74

wakes, see free turbulent shear flows
wall jets
 flow constants, 269
 friction parameters, 268
 integral conditions: momentum,
  268; total energy, 269
 possible exponents of velocity variation, 266, 271
 self-preserving development, 268
wall layers
 boundary layers, wall jets, 259
 mean value distributions: velocity,
  260, 264; stress, 261
 profile integrals, 264
 scales, 264
 self-preserving development, 262
 two-layer model, 259
wave-number space
 distortion (rapid-distortion model),
  47
 flow of turbulent energy, 48
 transfer function, 49

Printed in the United States
By Bookmasters